Geometry of Banach Spaces, Duality Mappings and Nonlinear Problems

Mathematics and Its Applications

Volume 62

Geometry of Banach Spaces, Duality Mappings and Nonlinear Problems

by

Ioana Cioranescu
Department of Mathematics,
University of Puerto Rico, Puerto Rico

KLUWER ACADEMIC PUBLISHERS
DORDRECHT / BOSTON / LONDON

Library of Congress Cataloging-in-Publication Data

Cioranescu, Ioana.
Geometry of banach spaces, duality mappings, and nonlinear problems / by Ioana Cioranescu.
p. cm. -- (Mathematics and it applications ; 62)
Includes bibliographical references and index.
ISBN 0-7923-0910-3 (alk. paper)
1. Banach spaces. 2. Duality theory (Mathematics) 3. Mappings (Mathematics) I. Title. II. Series: Mathematics and its applications (Kluwer Academic Publishers) ; v. 62.
QA322.2.C56 1990
515'.732--dc20 90-44194

ISBN 0-7923-0910-3

Published by Kluwer Academic Publishers,
P.O. Box 17, 3300 AA Dordrecht, The Netherlands.

Kluwer Academic Publishers incorporates
the publishing programmes of
D. Reidel, Martinus Nijhoff, Dr W. Junk and MTP Press.

Sold and distributed in the U.S.A. and Canada
by Kluwer Academic Publishers,
101 Philip Drive, Norwell, MA 02061, U.S.A.

In all other countries, sold and distributed
by Kluwer Academic Publishers Group,
P.O. Box 322, 3300 AH Dordrecht, The Netherlands.

Printed on acid-free paper

Printed in the Netherlands

To my father
the Romanian mathematician
Nicolae Cioranescu
In Memoriam

SERIES EDITOR'S PREFACE

'Et moi, ..., si j'avait su comment en revenir, je n'y serais point allé.'

Jules Verne

The series is divergent; therefore we may be able to do something with it.

O. Heaviside

One service mathematics has rendered the human race. It has put common sense back where it belongs, on the topmost shelf next to the dusty canister labelled 'discarded non-sense'.

Eric T. Bell

Mathematics is a tool for thought. A highly necessary tool in a world where both feedback and non-linearities abound. Similarly, all kinds of parts of mathematics serve as tools for other parts and for other sciences.

Applying a simple rewriting rule to the quote on the right above one finds such statements as: 'One service topology has rendered mathematical physics ...'; 'One service logic has rendered computer science ...'; 'One service category theory has rendered mathematics ...'. All arguably true. And all statements obtainable this way form part of the raison d'être of this series.

This series, *Mathematics and Its Applications*, started in 1977. Now that over one hundred volumes have appeared it seems opportune to reexamine its scope. At the time I wrote

> "Growing specialization and diversification have brought a host of monographs and textbooks on increasingly specialized topics. However, the 'tree' of knowledge of mathematics and related fields does not grow only by putting forth new branches. It also happens, quite often in fact, that branches which were thought to be completely disparate are suddenly seen to be related. Further, the kind and level of sophistication of mathematics applied in various sciences has changed drastically in recent years: measure theory is used (non-trivially) in regional and theoretical economics; algebraic geometry interacts with physics; the Minkowsky lemma, coding theory and the structure of water meet one another in packing and covering theory; quantum fields, crystal defects and mathematical programming profit from homotopy theory; Lie algebras are relevant to filtering; and prediction and electrical engineering can use Stein spaces. And in addition to this there are such new emerging subdisciplines as 'experimental mathematics', 'CFD', 'completely integrable systems', 'chaos, synergetics and large-scale order', which are almost impossible to fit into the existing classification schemes. They draw upon widely different sections of mathematics."

By and large, all this still applies today. It is still true that at first sight mathematics seems rather fragmented and that to find, see, and exploit the deeper underlying interrelations more effort is needed and so are books that can help mathematicians and scientists do so. Accordingly MIA will continue to try to make such books available.

If anything, the description I gave in 1977 is now an understatement. To the examples of interaction areas one should add string theory where Riemann surfaces, algebraic geometry, modular functions, knots, quantum field theory, Kac-Moody algebras, monstrous moonshine (and more) all come together. And to the examples of things which can be usefully applied let me add the topic 'finite geometry'; a combination of words which sounds like it might not even exist, let alone be applicable. And yet it is being applied: to statistics via designs, to radar/sonar detection arrays (via finite projective planes), and to bus connections of VLSI chips (via difference sets). There seems to be no part of (so-called pure) mathematics that is not in immediate danger of being applied. And, accordingly, the applied mathematician needs to be aware of much more. Besides analysis and numerics, the traditional workhorses, he may need all kinds of combinatorics, algebra, probability, and so on.

In addition, the applied scientist needs to cope increasingly with the nonlinear world and the extra mathematical sophistication that this requires. For that is where the rewards are. Linear models are honest and a bit sad and depressing: proportional efforts and results. It is in the nonlinear world that infinitesimal inputs may result in macroscopic outputs (or vice versa). To appreciate what I am hinting at: if electronics were linear we would have no fun with transistors and computers; we would have no TV; in fact you would not be reading these lines.

There is also no safety in ignoring such outlandish things as nonstandard analysis, superspace and anticommuting integration, p-adic and ultrametric space. All three have applications in both electrical engineering and physics. Once, complex numbers were equally outlandish, but they frequently proved the shortest path between 'real' results. Similarly, the first two topics named have already provided a number of 'wormhole' paths. There is no telling where all this is leading – fortunately.

Thus the original scope of the series, which for various (sound) reasons now comprises five subseries: white (Japan), yellow (China), red (USSR), blue (Eastern Europe), and green (everything else), still applies. It has been enlarged a bit to include books treating of the tools from one subdiscipline which are used in others. Thus the series still aims at books dealing with:

- a central concept which plays an important role in several different mathematical and/or scientific specialization areas;
- new applications of the results and ideas from one area of scientific endeavour into another;
- influences which the results, problems and concepts of one field of enquiry have, and have had, on the development of another.

It is probably impossible to overestimate the importance of the inner product for the study of problems and phenomena which take place in a Hilbert space. However, many, and probably most, mathematical objects and models do not naturally live in a Hilbert space. One of the main roles of the inner product is to enable us to interpret an element $x \in H$ as a functional $\tilde{x}$ on H, i.e. an element of the dual Hilbert space H^*. It then has the properties $\|x\|^2 = \langle x, \tilde{x}\rangle = \|\tilde{x}\|^2$.

The normalized duality mapping on a Banach space E, which is the multivalued mapping $J: E \to \{x^* \in E^*: \|x^*\|^2 = \|x\|^2 = \langle x^*, x\rangle\}$, serves as a replacement for the isomorphism $H \simeq H^*$ in the case of Banach spaces. More generally, one also considers duality mappings associated with a weight function. A selection (not necessarily continuous) j of J, i.e. a map $J: E \to E^*$ such that $\|x\|^2 = \|jx\|^2 = \langle jx, x\rangle$
a duality mapping in the literature.

Duality mappings have become a most important tool in *nonlinear* functional analysis, in particular for questions involving monotone, accretive, and dissipative nonlinear operators. But, although they appear as a tool in various books, no systematic treatment of duality mappings has been published with the exception of the book in Romanian written 16 years ago by the present author.

It is therefore a real pleasure to welcome this completely rewritten, expanded and updated English version in this series.

The shortest path between two truths in the real domain passes through the complex domain.

J. Hadamard

La physique ne nous donne pas seulement l'occasion de résoudre des problèmes ... elle nous fait pressentir la solution.

H. Poincaré

Never lend books, for no one ever returns them; the only books I have in my library are books that other folk have lent me.

Anatole France

The function of an expert is not to be more right than other people, but to be wrong for more sophisticated reasons.

David Butler

Amsterdam, July 1990

Michiel Hazewinkel

CONTENTS

PREFACE

With this book we intend to familiarize the reader with nonlinear operators related to nonlinear functional and evolution equations; that is, with monotone and accretive operators. Our approach to this subject will be by means of duality mappings and to this purpose we develop some convex analysis and a lot of geometry of Banach space.
Why such an extended treatment of duality mappings? Early in the 1960's Lumer and Phillips developed the theory of semigroups of linear contractions on general Banach spaces by means of the notion of dissipativity defined by a "semi-linear" product which was used to fill the gap caused by a back of a scalar product. This product was introduced using the normalized duality mapping.
Soon after, the dissipativity (and monotonicity) concept was extended from Hilbert to general Banach spaces and the classes of monotone and accretive operators appeared in connection with functional and partial differential equations. There are a large number of excellent monographs on these topics; in all of them although duality mapping is the principal tool, its properties are presented summarily, most of the times in appendices, on the other hand, in books on the geometry of Banach spaces, duality mappings as a tool for nonlinear problems are rather neglected.

Sixteen years ago I tried to close this gap with my book in Romanian "Duality mappings in nonlinear functional analysis". The present work is a completely rewritten, updated and improved version of this book in which only the main idea of using the duality map as Ariadna's thread in the complicated labyrinth of problems of nonlinear functional analysis has been preserved.

The monograph contains six Chapters; the first three Chapters are conceived as a survey on the properties of duality mappings. In the remaining Chapters, dedicated to nonlinear operators, we tried to underline the principle that in all monotonicity problems the duality map takes the place of the identity in Hilbert spaces, moreover, due to Kato's formula, it is the key technique in the study of evolution equations.

Exercises and bibliographical comments are included in order to complete the exposition. I was guided by the intention to make the presentation as self-contained as possible; nothing is assumed except the elementary theory of Banach spaces and some acquaintance with local convex topologies.
The material was the subject of lectures at the University of Santiago de Chile during the 1987–1988 academic year where the first version was written.

ACKNOWLEDGEMENTS

I am glad to acknowledge my thanks to the Mathematical Department of the University of Puerto Rico where the monograph reached its final form.

My very special thanks to Mrs. Ana Rebecca Velázquez and Isis V. Reyes León for typing the manuscript very professionally and with enthusiasm. I am also indebted to Dan Pascali for helpful suggestions.

Finally I express my gratitude to my husband Matei for his encourage-

SYMBOLS AND NOTATIONS

CHAPTER I

SUBDIFFERENTIABILITY AND DUALITY MAPPINGS

Our aim in this Chapter is to introduce the notion of duality mapping on a Banach space X; to this purpose we survey basic results of convex analysis and connect them with the differentiability properties of the norm on X.

§.1. GENERALITIES ON CONVEX FUNCTIONS

Let X be a linear topological space and $f:X \to \tilde{R} =]-\infty, +\infty]$; then the sets

$$D(f) = \{x \in X; f(x) < +\infty\} \text{ and } \mathrm{Epi} f = \{(x,t) \in X \times R;\ f(x) \le t\}$$

are called the effective domain and the epigraph of f. The function f is called proper if $D(f) \neq \phi$.

1.1. DEFINITION. The function $f:X \to \tilde{R}$ is said to be convex provided that:

$$f(\lambda x + (1-\lambda)y) \le \lambda f(x) + (1-\lambda) f(y) \text{ for all } x, y \in X \text{ and } \lambda \in [0,1].$$

If this inequality is strict whenever $x \neq y$, then the function f is called strictly convex.

1.2. REMARK. One can easily verify that f is convex if and only if for each $\lambda_1, \ldots, \lambda_n \in [0,1]$ with $\sum_{i=1}^{n} \lambda_i = 1$ and $x_1, \ldots, x_n \in X$, we have

$$f\left(\sum_{i=1}^{n} \lambda_i x_i\right) \le \sum_{i=1}^{n} \lambda_i f(x_i). \tag{1.1}$$

Moreover, if $f: R \to R$ is convex, then for any reals $t_1 < t_2 < t_3$

$$\frac{f(t_2) - f(t_1)}{t_2 - t_1} \le \frac{f(t_3) - f(t_1)}{t_3 - t_1} \le \frac{f(t_3) - f(t_2)}{t_3 - t_2}. \tag{1.2}$$

(Exercise 1)

1.3. REMARK. It is obvious that for a convex function f, D(f) is convex; moreover its restriction to D(f) is a convex R-valued function.

1.4. REMARK. We can associate to any set $A \subset X$ the function

$$I_A(x) = \begin{cases} 0 & x \in A \\ +\infty & x \notin A \end{cases}$$

which is called the indicator function of A.

Then $D(I_A) = A$ and the set A is convex if and only if the function I_A is convex.

1.5. PROPOSITION. A function $f: X \to \tilde{R}$ is convex if and only if its epigraph is a convex set in X x R.

PROOF. Suppose f convex and $[x_1, t_1], [x_2, t_2] \in$ Epif; then, for $\lambda \in [0,1]$, we have

$$f(\lambda x_1 + (1-\lambda) x_2) \le \lambda f(x_1) + (1-\lambda) f(x_2) \le \lambda t_1 + (1-\lambda) t_2$$

hence $\quad [\lambda x_1 + (1-\lambda) x_2, \lambda t_1 + (1-\lambda) t_2] \in$ Epif.

Conversely, assume Epif is convex; since for any $x_1, x_2 \in D(f)$, $[x_1, f(x_1)], [x_2, f(x_2)] \in$ Epif, it follows that

$$\lambda[x_1, f(x_1)] + (1-\lambda)[x_2, f(x_2)] \in \text{Epi}f, \ \forall \lambda \in [0,1].$$

It is now an easy matter to see that f is convex on D(f). ■

1.6. DEFINITION. A function $f:X \to \tilde{R}$ is said to be lower-semi continuous at $x_0 \in X$, in short *l.s.c.*, if

$$f(x_0) = \liminf_{x \to x_0} f(x) = \sup_{V \in \mathcal{V}_{x_0}} \inf_{x \in V} f(x)$$

where $\mathcal{V}_{x_0}$ is the set of all neighborhoods of the point x_0.

1.7. PROPOSITON. For $f:X \to \tilde{R}$ the following statements are equivalent

i) f is l.s.c on X;
ii) for every $\alpha \in R$ the level set $\{x \in X;\ f(x) \le \alpha\}$ is closed in X;
iii) Epif is closed in $X \times R$.

PROOF. *i)* $\Rightarrow$ *ii)* Let $\alpha \in R$; first we observe that the set $\{x \in X; f(x) \le \alpha\}$ is closed if and only if the set $\{x \in X; f(x) > \alpha\}$ is open.
Consider $x_0 \in X$ with $f(x_0) = \sup_{V \in \mathcal{V}_{x_0}} \inf_{x \in V} f(x) > \alpha$; then there exists a $V_0 \in \mathcal{V}_{x_0}$ such that $\inf_{x \in V_0} f(x) > \alpha$. Hence $V_0 \subset \{x \in X;\ f(x) > \alpha\}$ so that $\{x \in X;\ f(x) > \alpha\}$ is open.
ii $\Rightarrow$ *i)* Let $x_0 \in D(f), \varepsilon > 0$ and $V_\varepsilon = \{x \in X;\ f(x) > f(x_0) - \varepsilon\}$; then $V_\varepsilon \in \mathcal{V}_{x_0}$ and $\inf_{x \in V_\varepsilon} f(x) \ge f(x_0) - \varepsilon$. Thus $\liminf_{x \to x_0} f(x) \ge f(x_0) - \varepsilon$. Since ε is arbitrarily chosen, we get $\liminf_{x \to x_0} f(x) \ge f(x_0)$. The converse inequality is trivial and thus we conclude that f is l.s.c at x_0.
If $f(x_0) = +\infty$, then for every $n \in N$, $V_n = \{x \in X; f(x) > n\} \in \mathcal{V}_{x_0}$ and $\inf_{x \in V_n} f(x) \ge n$; hence $\liminf_{x \to x_0} f(n) = +\infty$.

i) $\Leftrightarrow$ *iii)* Define $F: X \times R \to \tilde{R}$ by $F(x, \alpha) = f(x) - \alpha$; then f is l.s.c. on X if and only if F is l.s.c. on $X \times R$; moreover Epif is a level set for F. Thus the present equivalence is a consequence of the above one. ■

1.8. COROLLARY. Every l.s.c. function f on a compact topological space is bounded from below and attains its infimum on X.

PROOF. Suppose that f is not bounded from below and let $X_n = \{x \in X; f(x) \le -n\}$, $n \in N$; then the level sets X_n are nonvoid, closed and increasing. The compacity hypothesis implies the existence of an $x_0 \in \bigcap_{n\in N} X_n$ such that $f(x_0) = -\infty$, which is impossible.

Let now $-\infty < \alpha = \inf_{x\in X} f(x)$ and $Y_n = \left\{x \in X; f(x) \le \alpha + \frac{1}{n}\right\}$, $n \in N$; then by a similar argument as above, there exists $y_0 \in \bigcap_{n\in N} Y_n$ for which we get $f(y_0) = \alpha$. ■

1.9. COROLLARY. Any convex function on a locally convex space is l.s.c. if and only if it is weakly l.s.c.

PROOF. Since Epif is convex in $X \times R$, then it is closed if and only if it is weakly closed; therefore the result is a consequence of Proposition 1.7. ■

1.10. THEOREM. Let $f: X \to \tilde{R}$ be a convex function bounded from above on a neighborhood of an interior point $x_0 \in D(f)$; then f is continuous on Int D(f).

PROOF. First we prove the continuity of f at x_0. Let $V \in \mathcal{V}_{x_0}$ and M>0 be such that $f(x) \le M, \forall x \in V$. Passing if necessary to the neighborhood $V \cap (-V)$, we can assume V symmetric.
We also can suppose $x_0=0$ and $f(x_0) = 0$; indeed, f is continuous at x_0 if and only if the function $F(x) = f(x + x_0) - f(x_0)$, $x \in X$, is continuous in the origin; moreover $F(0) = 0$ and F is convex.
For every $\varepsilon \in (0,1)$ and $x \in \varepsilon V$, we

$$f(x) = f\left(\varepsilon \cdot \frac{x}{\varepsilon} + (1-\varepsilon)\cdot 0\right) \le \varepsilon f\left(\frac{x}{\varepsilon}\right) \le \varepsilon \cdot M$$

On the other hand

$$0 = f(0) = f\left(\frac{1}{1+\varepsilon}\cdot x + \left(1 - \frac{1}{1+\varepsilon}\right)\left(-\frac{1}{\varepsilon}x\right)\right)$$

$$\le \frac{1}{1+\varepsilon} f(x) + \frac{\varepsilon}{1+\varepsilon} f\left(-\frac{1}{\varepsilon}x\right) \qquad ,$$

that is

$$f(x) \geq -\varepsilon \cdot f\left(-\tfrac{1}{\varepsilon}x\right) \geq -\varepsilon \cdot M \cdot$$

Hence $|f(x)| \leq \varepsilon M$, for $x \in \varepsilon V$ and thus f is continuous in 0. To end the proof it suffices to show that for any y ∈ Int D(f), there exists a neighborhood on which f is bounded from above.

Let $\rho > 1$ be such that $\rho y \in D(f)$. If V is the neighborhood of the origin used above, then any $x \in V_y = y + \left(1 - \frac{1}{\rho}\right)V$ can be written as $x = y + (1 - \frac{1}{\rho})z = \frac{1}{\rho}(\rho y) + (1 - \frac{1}{\rho})z$, for some $z \in V$. Since D(f) is convex, $x \in D(f)$ and thus $V_y \subseteq D(f)$. Finally, the convexity of f implies

$$f(x) \leq \frac{1}{\rho}\, f(\rho y) + (1 - \tfrac{1}{\rho})f(z) \leq \frac{1}{\rho}\, f(\rho y) + (1 - \tfrac{1}{\rho})M, \ \forall x \in V_y .$$ ■

1.11. COROLLARY. Any proper convex function on a finite dimensional space X is continuous on Int D(f).

PROOF. Let $x_0 \in \text{Int } D(f)$; by a similar argument used in the proof of Theorem 1.10. we can reduce the problem to the case $x_0 = 0$.

Let n be the dimension of X and $\alpha > 0$ small enough such that the cube $V = \{x = (x_i)_{1 \leq i \leq n};\ 0 < x_i < \alpha/n, 1 \leq i \leq n\}$ be contained in D(f).

If $\{e_i; 1 \leq i \leq n\}$ is the canonical base of the space X, each x ∈ V can be written as

$$x = \sum_{i=1}^{n} x_i e_i = \sum_{i=1}^{n} \frac{x_i}{\alpha} \cdot \alpha e_i + \left(1 - \sum_{i=1}^{n} \frac{x_i}{\alpha}\right) \cdot 0$$

where $0 < \sum_{i=1}^{n} \frac{x_i}{\alpha} < 1$. Now, the convexity of f provides

$$f(x) \leq \sum_{i=1}^{n} \frac{x_i}{\alpha} f(\alpha e_i) + \left(1 - \sum_{i=1}^{n} \frac{x_i}{\alpha}\right) f(0)$$

$$\leq \frac{1}{n} \sum_{i=1}^{n} |f(\alpha e_i)| + |f(0)|.$$

Thus f is bounded from above on V. ■

1.12. COROLLARY. Any proper convex l.s.c. function on a barrelled locally convex space X (in particular a Banach space) is continuous on Int D(f).

PROOF. We can suppose again, without loss of generality, that $0 \in \text{Int } D(f)$. Let $M \in R$ with $f(0) < M$; the level set $V = \{x \in f(x); f(x) \leq M\}$ is closed and convex; moreover it is also absorbent. Indeed, let $x \in X$ and $X_0 = \{\lambda x; \lambda \in R\}$; then, by Corollary 1.11., the restriction of f to X_0 is continuous. Hence for ε positive so that $\varepsilon \leq M - F(0)$, there exists $\lambda_\varepsilon \in R$ with $f(\lambda x) - f(0) \leq \varepsilon$ for $|\lambda| \leq \lambda_\varepsilon$. Hence $f(\lambda x) \leq M$, for any $|\lambda| \leq \lambda_\varepsilon$, i.e. V is absorbent. The result follows now from Theorem 1.10. ■

In the remainder of this paragraph X and Y will be real Banach spaces, X*, Y* their normed duals and $\mathcal{L}(X,Y)$ the space of all linear and continuous mappings of X in Y.

1.13. DEFINITION. *i)* Let $D \subseteq X$ be open, $F: D \to Y$ and $x \in D$; we call the directional derivative of F at x in the direction $y \in X$, the limit

$$\lim_{t \to 0_+} \frac{F(x+ty)}{t} = F'_+(x,y)$$

when it exists.

ii) If there exists an operator in $\mathcal{L}(X; Y)$, denoted by F'(x) such that

$$\lim_{t \to 0} \frac{F(x+ty) - F(x)}{t} = F'(x)y \quad \text{for every } y \in X,$$

then we say that F is Gateaux-differentiable (in short G-differentiable) at x.

(We note that some authors call F'(x) the gradient of F at x).

iii) We say that $F: D \to Y$ is Fréchet-differentiable (in short F-differentiable) at x if it is G-differentiable at x and

$$\lim_{t \to 0} \sup_{\|y\|=1} \left\| \frac{F(x+ty) - F(x)}{t} - F'(x)y \right\| = 0.$$

We have the following chain rule for the Fréchet derivatives:

1.14. THEOREM. Let X, Y, Z be Banach spaces and $f: X \to Y$, $g: Y \to Z$ be Fréchet-differentiables on X, respectively on Y; then $g \circ f: X \to Z$ is Fréchet differentiable on X and we have

$$(g \circ f)'(x) = g'(f(x)) \cdot f'(x), x \in X.$$

PROOF. First we remark that for ever $x \in X$, $u \in Y$, $f'(x) \in \mathcal{L}(X,Y)$, and $g'(u) \in \mathcal{L}(Y,Z)$ such that the right side in the above formula is understood as the product of the respective operators.
For $x, y \in X$, $u, v \in Y$ and $t \in R$ we have

$$f(x + ty) - f(x) = tf'(x)y + t\, w_1(x, y, t),$$
$$g(u + tv) - g(u) = tg'(u)v + t\, w_2(u, v, t),$$

where

$$\lim_{t \to 0} \sup_{\|y\| = 1} \|w_1(x,y,t)\| = 0 \text{ and } \lim_{t \to 0} \sup_{\|v\| = 1} \|w_2(u,v,t)\| = 0.$$

Then

$$\begin{aligned} g(f(x + ty)) &= g(f(x) + tf'(x)y + tw_1(x,y,t)) \\ &= g[f(x) + t(f'(x)y + w_1(x,y,t))] \\ &= g(f(x)) + t\, g'(f(x))\,(f'(x)y + w_1(x,y,t)) \\ &\qquad + t\, w_2(f(x), f'(x)y + w_1(x,y,t), t) \\ &= g(f(x)) + tg'(f(x)) \cdot f'(x)y + tw_3(x,y,t) \end{aligned}$$

where

$$w_3(x,y,t) = g'(f(x))w_1(x,y,t) + w_2(f(x), f'(x)y + w_1(x,y,t), t).$$

It is not difficult to see that

$$\lim_{t \to 0} \sup_{\|y\| = 1} \|w_3(x,y,t)\| = 0 \bullet$$

Thus $g \circ f$ is Fréchet differentiable and the chain rule is proved. ■

1.15. REMARK. It is clear that if F is G-differentiable at $x \in D$, then F has directional derivatives at x in any direction $y \in X$ and that we have

$$F'_+(x,y) = F'_+(x,-y), \forall y \in X.$$

As in general the above two vectors are distinct, we put

$$F'_-(x,y) = -F'_+(x,-y).$$

1.16. REMARK. Consider $f:D \subseteq X \to R$, G-differentiable at $x \in D$; then $f'(x) \in X^*$ and

$$<f'(x),y> = \frac{d}{dt} f(x+ty)|_{t=0}, \forall y \in X.$$

If f is twice G-differentiable on D, (i.e. f ' (x) exists, $\forall x \in D$ and $f':D \subseteq X \to X^*$ is G-differentiable) then $f''(x) \in L(X;X^*)$ and we have

$$<f''(x)y,y> = \frac{d^2}{dt^2} f(x+ty)|_{t=0}, \quad \forall x \in D, y \in X.$$

(Exercise 4).

The following result on scalar convex functions is crucial in what follows

1.17. LEMMA . Let $\varphi:R \to \tilde{R}$ be convex; then φ has one-sided derivatives at any $t \in \text{Int } D(\varphi)$ and

$$\varphi'_-(t) \le \varphi'_+(t). \tag{1.3}$$

Moreover for any $t_1, t_2 \in \text{Int } D(\varphi)$ so that $t_1 < t_2$, $t_1, t_2 \in \text{Int } D(\varphi)$ we have

$$\varphi'_+(t_1) \le \frac{\varphi(t_2) - \varphi(t_1)}{t_2 - t_1} \le \varphi'_-(t_2). \tag{1.4}$$

PROOF. Let $0 < s_1 < s_2$; then for $t \in \text{Int } D(\varphi)$, $t-s_2 < t-s_1 < t < t+s_1 < t+s_2$ and the inequalities (1.2) imply

$$\frac{\varphi(t) - \varphi(t-s_2)}{s_2} \le \frac{\varphi(t) - \varphi(t-s_1)}{s_1} \le \frac{\varphi(t+s_1) - \varphi(t)}{s_1} \le \frac{\varphi(t+s_2) - \varphi(t)}{s_2}.$$

Thus the function $s \to \frac{\varphi(t+s) - \varphi(t)}{s}$ does not increase as $s \downarrow 0_+$ and is bounded from bellow; hence it has a limit $\varphi'_+(t)$.

Similarly, the function $s \to \frac{\varphi(t) - \varphi(t-s)}{s}$ does not decrease as $s \downarrow 0_+$ and is bounded from above; hence $\varphi'_-(t)$ exists. Moreover, it is clear that (1.3) is true.

Further, for $s > 0$ and $t_1, t_2 \in \text{Int } D(f), t_1 < t_2$ we get

$$\varphi'_+(t_1) \le \frac{\varphi(t_1 + s) - \varphi(t_1)}{s} \quad \text{and} \quad \varphi'_-(t_2) \ge \frac{\varphi(t_2) - \varphi(t_2 - s)}{s}.$$

Taking $s = t_2 - t_1$, we obtain

$$\varphi'_+(t_1) \le \frac{\varphi(t_2) - \varphi(t_1)}{t_2 - t_1} \le \varphi'_-(t_2).$$ ■

1.18. REMARK. We note that the first inequality in (1.4) is valid for $t_1 \in \operatorname{Int} D(\varphi)$ and any t_2 in R, with $t_1 < t_2$.

1.19. PROPOSITION. Let $f:X \to \tilde{R}$ be a proper convex function; then for each $x \in \operatorname{Int} D(f)$ the directional derivative $f'_+(x,y)$ exists in any direction $y \in X$; the function $y \to f'_+(x,y)$ is convex, positive homogeneous and satisfies

$$f'_-(x,y) \le f'_+(x,y), \quad \forall y \in X. \tag{1.5.}$$

Moreover, if f is continuous in an interior point x_0 of its effective domain, then $y \to f'_+(x_0,y)$ is continuous on X.

PROOF. Let $x \in \operatorname{Int} D(f), y \in X$ and define $\varphi(t) = f(x+ty)$; then φ is convex and $0 \in \operatorname{Int} D(\varphi)$. Hence by Lemma 1.17., $\varphi'_+(0)$ and $\varphi'_-(0)$ exist and $\varphi'_-(0) \le \varphi'_+(0)$. But $\varphi'_+(0) = f'_+(x,y)$ and $\varphi'_-(0) = f'_-(x,y)$, so that (1.5) holds.

It is obvions that $f'_+(x,\lambda y) = \lambda f'_+(x,y)$, $\lambda > 0$. We shall prove that $y \to f'_+(x,y)$ is a subadditive function and this would imply its convexity.

For $y_1, y_2 \in X$, we have

$$f'_+(x,y_1+y_2) = \lim_{t \to 0_+} \frac{f\left(x+\frac{t}{2}(y_1+y_2)\right) - f(x)}{\frac{t}{2}}$$

$$= \lim_{t \to 0_+} \frac{2f\left(\frac{x+ty_1}{2} + \frac{x+ty_2}{2}\right) - 2f(x)}{t}$$

$$\le \lim_{t \to 0_+} \frac{f(x+ty_1) - f(x)}{t} + \lim_{t \to 0_+} \frac{f(x+ty_2) - f(x)}{t}$$

$$= f'_+(x,y_1) + f'_+(x,y_2).$$

Suppose now f continuous at $x_0 \in \text{Int } D(f)$ and let $\varepsilon > 0$ and $M > 0$ be such that

$$\bar{S}_\varepsilon(x_0) = \{x \in X; \|x - x_0\| \le \varepsilon\} \subset \text{Int } D(f) \text{ and } |f(x)| \le M, \forall\, x \in \bar{S}_\varepsilon(x_0).$$

For $y \in X$, consider the function $\varphi(t) = f(x_0 + ty)$; then $\left[-\frac{\varepsilon}{\|y\|}, \frac{\varepsilon}{\|y\|}\right] \subseteq \text{Int } D(\varphi)$. We use now the inequality (1.4) from Lemma 1.17. with $t_1 = 0$, $t_2 = \frac{\varepsilon}{\|y\|}$ to get

$$f'_+(x_0,y) = \varphi'_+(0) \le \frac{\varphi\left(\frac{\varepsilon}{\|y\|}\right) - \varphi(0)}{\frac{\varepsilon}{\|y\|}} = \frac{f\left(x_0 + \frac{\varepsilon}{\|y\|} y\right) - f(x_0)}{\varepsilon} \|y\| \le \frac{2M}{\varepsilon} \|y\|.$$

Similarly

$$f'_+(x_0,y) \ge f'_-(x_0,y) = \varphi'_-(0) \ge \frac{\varphi(0) - \varphi\left(\frac{\varepsilon}{\|y\|}\right)}{\varepsilon} \|y\|$$

$$= \frac{f(x_0) - f\left(x_0 - \frac{\varepsilon}{\|y\|} y\right)}{\varepsilon} \|y\| \ge -\frac{2M}{\varepsilon} \cdot \|y\|.$$

Hence

$$|f'_+(x_0,y)| \le \frac{2M}{\varepsilon} \cdot \|y\|, \quad \forall y \in X$$

and this means that $y \to f'_+(x_0,y)$ is continuous at $y = 0$. But then, by Theorem 1.10. it is continuous at any point $y \in X$. ■

1.20. COROLLARY. A convex continuous function is G-differentiable at $x \in \text{Int } D(f)$ if and only if

$$f'_+(x,y) = f'_-(x,y), \quad \forall\, y \in X.$$

PROOF. Suppose $f'_+(x,-y) = f'_+(x,y)$, $\forall\, y \in X$; then the functions $X \ni y \longrightarrow \pm f'_+(x,y)$ are positive homogeneous and convex; hence the function $y \to f'_+(x,y)$ is linear. By the above Proposition, it is also continuous on X, hence is an element of X*. Thus f is G-differentiable at x. ■

§2. THE SUBDIFFERENTIAL AND THE CONJUGATE OF A CONVEX FUNCTION

In all this section X will be a real Banach space.

2.1. DEFINITION. We say that $f:X \to \tilde{R}$ is subdifferentiable at a point $x \in X$ if there exists a functional $x^* \in X^*$, called subgradient of f at x, such that

$$f(y) - f(x) \geq \langle x^*, y - x \rangle, \ \forall y \in X. \tag{2.1}$$

The set of all subgradients of f at x is denoted by $\partial f(x)$ and the mapping $\partial f : X \to 2^{X^*}$ is called the subdifferential of f.

2.2. REMARK. As a direct consequence of the definition we get that $\partial f(x)$ is a weak* – closed convex subset of X*; moreover f has a minimum value at x if and only if $0 \in \partial f(x)$. Finally, if f is proper and $\partial f(x) \neq \phi$, then $x \in D(f)$.

2.3. REMARK. Let us recall that a closed hyperplane in $X \times R$ has the form

$H = \{(x,t) \in X \times R; \langle x^*, x \rangle + \alpha t = \beta\}$ for some α, $\beta \in R$ and $x^* \in X^*$.

If $\alpha \neq 0$, then the hyperplane is called non vertical and has the form

$H = \{(x,t) \in X \times R; \langle x^*, x \rangle + \alpha = t\}$ (for some other $\alpha \in R$ and $x^* \in X^*$).

A non vertical hyperplane H in $X \times R$ is called supporting hyperplane for Epif at the point $(x_0, f(x_0))$ if $(x_0, f(x_0)) \in H$ and Epif is contained in one of the two closed subspaces determined by H. In this case H has the form

$$H = \{(x,t) \in X \times R; \langle x^*, x - x_0 \rangle + f(x_0) = t\} \tag{2.2}$$

and moreover

$$\text{Epi}\, f \subseteq \{(x,t) \in X \times R; \langle x^*, x - x_0 \rangle + f(x_0) \leq t\} \tag{2.3}$$

Let us remark that the inequality $<x^*, x - x_0> + f(x_0) \geq t$, for every $(x,t) \in \text{Epi}f$ is impossible because $(x,t) \in \text{Epi}f$ implies that for all n in N, $(x, t+n) \in \text{Epi}f$.

2.4. PROPOSITION. $\partial f(x_0) \neq \phi$ if and only if there exists a supporting hyperplane for Epif at $(x_0, f(x_0))$.

PROOF. Suppose $\partial f(x_0) \neq \phi$; then there is $x^* \in X^*$ such that

$$f(x) - f(x_0) \geq <x^*, x - x_0>, \quad \forall x \in X ;$$

then

$$H = \{(x,t) \in X \times R;\ <x^*, x - x_0> + f(x_0) = t\}$$

is obviously a supporting hyperplane for Epif at $(x_0, f(x_0))$.
Conversely, if H is a supporting hyperplane for Epif at $(x_0, f(x_0))$ then it has the form (2.2) for some $x^* \in X^*$ and (2.3) holds.
Since for any $x \in D(f)$, $(x, f(x)) \in$ Epif, we obtain

$$<x^*, x - x_0> + f(x_0) \leq f(x) \quad \text{i.e.} \quad x^* \in \partial f(x_0). \qquad \blacksquare$$

2.5. PROPOSITION. Let $f: X \to \tilde{R}$ be a proper convex function and $x \in \text{Int}\, D(f)$; then $x^* \in \partial f(x)$ if and only if

$$f'_-(x,y) \leq <x^*, y> \leq f'_+(x,y),\ \forall y \in X. \tag{2.4}$$

PROOF. Let $x^* \in \partial f(x)$; then for any $y \in X$ we have

$$\frac{f(x) - f(x - ty)}{t} \leq <x^*, y> \leq \frac{f(x + ty) - f(x)}{t}, \quad t > 0$$

and this yields when $t \to 0_+$ the inequality (2.4).

Conversely, let $y \in X$, $z = y - x$ and $\varphi(t) = f(x + tz)$, $t \in R$; then from Lemma 1.17. for $t_1 = 0$ and $t_2 = 1$ we obtain

$$\varphi'_+(0) \leq \frac{\varphi(1) - \varphi(0)}{1 - 0}.$$

Furthermore (2.4) yields
$f(x + z) - f(x) \geq f'_+(x,y) \geq <x^*, z>$ i.e. $f(y) - f(x) \geq <x^*, y - x>$.
It follows that $x^* \in \partial f(x)$. $\blacksquare$

2.6. THEOREM. *i)* A subdifferentiable function f is convex and *l. s. c.* on any open convex set $D \subset D(f)$.
ii) A proper convex *l. s. c.* function is subdifferentiable on Int D(f).

PROOF. *i)* Let $x_1, x_2 \in D$ and $\lambda \in [0,1]$; then $x = \lambda x_1 + (1-\lambda)x_2 \in D$ and for $x^* \in \partial f(x)$ we have

$$f(x_1) \geq f(x) + \langle x^*, x_1 - x \rangle \text{ and } f(x_2) \geq f(x) + \langle x^*, x_2 - x \rangle.$$

This implies

$$\lambda f(x_1) + (1-\lambda) f(x_2) \geq f(x) = f(\lambda x_1 + (1-\lambda)x_2),$$

i.e. f is convex.
Now consider $x \in D, V \in \mathcal{V}_x$ such that $V \subset D$ and $x^* \in \partial f(x)$; then we have

$$f(y) \geq f(x) + \langle x^*, y - x \rangle; \quad \forall y \in V$$

and this implies $\liminf\limits_{y \to x} f(y) \geq f(x)$. Since the other inequality is trivial, f is l. s. c.
ii) It is sufficient to prove that for any $x \in \operatorname{Int} D(f)$ there is $x^* \in X^*$ which satisfies (2.4). By the Corollary 1.12, f is continuous on Int D(f). Fix $y_0 \in X$ and consider on the subspace $Y_0 = \{\lambda y_0; \lambda \in R\}$ a linear functional F such that $F(y_0) = f'_+(x, y_0)$.
By Proposition 1.19., $y \to f'_+(x,y)$ is positively homogeneous; hence $F(\lambda y_0) = f'_+(x, \lambda y_0)$ for $\lambda \geq 0$.
For $\lambda < 0$, we have

$$F(\lambda y_0) = \lambda F(y_0) = \lambda f'_+(x, y_0) = |\lambda| f'_-(x, -y_0)$$

$$\leq |\lambda| f'_+(x, -y_0) = f'_+(x, \lambda y_0).$$

Thus $F(y) \leq f'_+(x,y)$, $\forall y \in Y_0$.
Then, by the Hahn - Banach Theorem, F can be extended to a linear functional x* on X such that

$$\langle x^*, y \rangle \leq f'_+(x,y), \ \forall y \in X \text{ and } \langle x^*, y_0 \rangle = f'_+(x, y_0). \tag{2.5.}$$

Therefore

$$<x^*, y> = -<x^*, -y> \geq -f'_+(x, -y) = f'_-(x,y), \quad \forall y \in X$$

i.e.

$$f'_-(x,y) \leq <x^*, y> \leq f'_+(x,y), \quad \forall y \in X.$$

Since the functions $y \to f'_+(x,y)$ and $y \to f'_-(x,y)$ are continuous it follows that $x^* \in X^*$ and thus the proof is complete. ■

2.7. COROLLARY. A proper convex continuous function f is G-differentiable at $x \in \text{Int}\, D(f)$ if and only if it has a unique subgradient at x; in this case $\partial f(x) = f'(x)$.

PROOF. If f is G-differentiable at x, then $f'_+(x,y) = f'_-(x,y)$ for all $y \in X$ and the assertion is a consequence of Proposition 2.4 and the existence result given by Theorem 2.1. (ii).
Conversely, if the subgradient x^* at x is unique, then by the construction of Theorem 2.6. ii) we obtain the same functional x^* starting with arbitrary $y_0 \in X$. By (2.5) we obtain

$$<x^*, y> = f'_+(x,y), \quad y \in X.$$

Hence

$$<x^*, -y> = f'_+(x, -y) = f'_-(x,y)$$

and this yields

$$<x^*, y> = f'_-(x,y) = f'_+(x,y).$$

The G-differentiability of f at x is now a consequence of the Corollary 1.18. ■

2.8. THEOREM. If f_1, f_2 are two convex functions on X such that there is a point $x_0 \in D(f_1) \cap D(f_2)$ where f_1 is continuous, then we have

$$\partial(f_1 + f_2)(x) = \partial f_1(x) + \partial f_2(x), \quad \text{for all } x \in X.$$

PROOF. It is easy to verify that we always have

$$\partial(f_1 + f_2)(x) \supseteq \partial f_1(x) + \partial f_2(x), \quad x \in X.$$

We shall prove that also the inverse inclusion is true, i.e. that each $x^* \in \partial(f_1 + f_2)(x)$ can be decomposed as $x^* = x_1^* + x_2^*$, with $x_1^* \in \partial f_1(x)$ and $x_2^* \in \partial f_2(x)$.
Let be $x^* \in \partial(f_1 + f_2)(x)$; then $x \in D(f_1) \cap D(f_2)$ and

$$f_1(y) + f_2(y) - <x^*, y - x> \geq f_1(x) - f_2(x), \quad y \in X. \tag{2.6}$$

Consider the convex sets in $X \times R$

$$A_1 = \{(y,t) \in X \times R;\ f_1(y) - f_1(x) - x^*, y - x> \leq t\}$$

$$A_2 = \{(y,t) \in X \times R;\ f_2(x) - f_2(y) \geq t\}.$$

It is not difficult to see that A_1 is the epigraph of the convex function $f(y) = f_1(y) - f_1(x) - <x^*, y - x>, y \in X$.
Since f is by hypotesis continuous at x_0, then $(x_0, f(x_0)) \in$ Int Epif. Indeed, for $\varepsilon > 0$, there is $V_\varepsilon \in \mathcal{V}_{x_0}$ with $|f(x) - f(x_0)| \leq \varepsilon$, for all $x \in V$ and this implies $(x, f(x_0) + \varepsilon)_{x \in V_\varepsilon} \subseteq \text{Epif}$.
Thus Int $A_1 \neq \phi$. Moreover by (2.6) the intersection of the sets A_1 and A_2 consists only of boundary points. Then we can separate A_1 from A_2 by a closed hyperplane H, which moreover strictly separates Int A_1 form A_2.
The hyperplane H can not be vertical since in this case its projection on X would strictly separate Int $D(f_1)$ from $D(f_2)$ and this is impossible because $x_0 \in \text{Int}\, D(f_1) \cap D(f_2)$.
Therefore there exists $x_0^* \in X^*$ and $\alpha \in R$ such that $H = \{(y,t) \in X \times R; <x_0^*, y> + \alpha = t\}$.
By the Remark 2.3. the separation relation can be written as

$$\text{Epif} = A_1 \subseteq \{(y,t) \in X \times R; <x_0^*, y> + \alpha \leq t\}$$

and

$$A_2 \subseteq \{(y,t) \in X \times R; <x_0^*, y> + \alpha \geq t\}.$$

In particular, for $(y, f(y)) \in A_1$ and $(y, f(x) - f_2(y)) \in A_2$ we have

$$f_2(x) - f_2(y) \leq <x_0^*, y> + \alpha \leq f_1(y) - f_1(x) - <x^*, y - x>, \qquad \forall y \in X.$$

For $y = x$ we get $\alpha = -<x_0^*, x>$; hence for every $y \in X$ we obtain

$$f_1(y) - f_1(x) \geq <x_0^* + x^*, y - x> \qquad \text{and} \qquad f_2(y) - f_2(x) \geq <-x_0^*, y - x>$$

that is

$x_0^* + x \in \partial f_1(x), - x_0^* \in \partial f_2(x)$ and $x^* = (x^* + x_0^*) + (-x_0^*)$. ∎

2.9. EXAMPLE. Let $f(x) = \frac{1}{2}\|x\|^2$; then, for $x \neq 0$

$$\partial f(x) = \{x^* \in X^*; \langle x^*, x\rangle = \|x\|^2 = \|x^*\|^2\}.$$

Indeed, if $x^* \in X^*$ is such that $\langle x^*, x\rangle = \|x\|^2 = \|x^*\|^2$, then for every $y \in X$ we have

$$\langle x^*, y - x\rangle = \langle x^*, y\rangle - \|x\|^2 \leq \|x\|y\| - \|x\|^2$$
$$\leq \frac{1}{2}\left(\|x\|^2 - \|x\|^2\right) = f(y) - f(x),$$

i.e. $x^* \in \partial f(x)$.
Conversely, if $x^* \in \partial f(x)$., then

$$\langle x^*, y - x\rangle \leq \frac{1}{2}\left(\|y\|^2 - \|x\|^2\right), \qquad \forall\, y \in X.$$

Take in the above inequality $y = x + \lambda z, \lambda \in R; z \in X$, we obtain

$$\lambda \langle x^*, z\rangle \leq \frac{1}{2}\left(\|x + \lambda z\|^2 - \|x\|^2\right) \leq \frac{1}{2}\left(\lambda^2\|z\|^2 + 2|\lambda|\|x\|\,\|z\|\right). \tag{2.7}$$

Letting now $\lambda \to 0_+$, we have that $\langle x^*, z\rangle \leq \|x\|\,\|z\|$ for every $z \in X$. Therefore $|\langle x^*. z\rangle| \leq \|x\|\,\|z\|$, $\forall\, z \in X$, so that for $z = x$ we have

$$|\langle x^*, x\rangle| \leq \|x\|^2 \quad \text{and} \quad \|x^*\| \leq \|x\| \tag{2.8}$$

Take now in the first inequality of (2.7) z=x and $\lambda < 0$; we obtain

$$\langle x^*, x\rangle \geq \frac{\lambda + 2}{2}\|x\|^2.$$

Letting now $\lambda \to 0_-$, we have

$$\langle x^*, x\rangle \geq \frac{\lambda + 2}{2}\|x\|^2. \tag{2.9}$$

Combining (2.8) and (2.9) we finally obtain $\langle x^*, x\rangle = \|x\|^2 = \|x^*\|^2$.

2.10. REMARK. The mapping $X \ni x \to \partial(\frac{1}{2}\|x\|^2) \in 2^{X^*}$ is called the normalized duality mapping on X and will play a basic role in our further considerations.

2.11. DEFINITION. Let $f:X \to \tilde{R}$; the function $f^*:X^* \to \bar{R} = [-\infty,+\infty]$ defined by

$$f^*(x^*) = \sup_{x^* \in X^*} (<x^*,x> - f^*(x^*))$$

is called the conjugate function of f.

The biconjugate function of $f, f^{**}:X \to \bar{R}$ is defined as

$$f^{**}(x) = \sup_{x^* \in X^*} (<x^*,x> - f^*(x^*)) \quad .$$

2.12. REMARK. If $f:X \to \tilde{R}$ is proper then we can consider in the definition of f^* only the supremum over D(f).

The following properties can be directely derived from the definition:

$f^*(0) = -\inf_{x \in X} f(x)$.

$f \le g$ implies $f^* \ge g^*$.

$(\inf_{\iota \in I} f_\iota)^* = \sup_{\iota \in I} f_\iota^*$ and $(\sup_{\iota \in I} f_\iota)^* \le \inf_{\iota \in I} f_L$.

$(\lambda f)^*(x)^* = \lambda f^*(x^*/\lambda), \lambda > 0$.

$(f + \lambda)^* = f^* - \lambda, \quad \lambda \in R$.

$(\tau_y f)^*(x^*) = f^*(x^*) + <x^*,y>, y \in x$ where $(\tau_y f)(x) = f(x-y), \quad \forall x \in X$.

$f^{**} \le f$ (see Exercise 6).

f^* is convex on $D(f^*)$.

2.13. PROPOSITION. If $f:X \to \tilde{R}$ is proper, convex and l.s.c. then also f^* is proper.

PROOF. Let be $x_0 \in D(f)$ and $\varepsilon > 0$; then $(x_0, f(x_0) - \varepsilon) \notin Epif$. Since Epif is a convex closed subset of $X \times R$, there exists $(x_0^*, \alpha) \in X^* \times R$ such that

$$\sup_{(x,t) \in Epif} (<x_o^*,x> + \alpha t) < <x_o^*,x_o> + \alpha(f(x_0) - \varepsilon). \qquad (2.10)$$

Since $x_0 \in D(f)$, it follows that $\alpha \neq 0$; moreover $\alpha<0$, since in the case $\alpha>0$, we would obtain that the left side in (2.10.) is $+\infty$.

Therefore we can suppose $\alpha=-1$. Since $(x,f(x)) \in Epif$, then by (2.10) we may write

$$f^*(x_0^*) = \sup_{x \in D(f)} (<x_o^*,x> - f(x)) \le <x_o^*,x_o> - f(x_0) + \varepsilon$$

i.e. $x^*_o \in D(f^*)$. ■

2.14. PROPOSITION. Let $f:X \to \tilde{R}$ be proper; then $x^* \in \partial f(x)$ if and only if

$$f(x) + f^*(x^*) = \langle x^*, x \rangle \tag{2.11}$$

PROOF. Let $x^* \in \partial f(x)$; we have $f(y) \geq f(x) + \langle x^*, y - x \rangle \quad , \quad \forall y \in X$.
Hence

$$f^*(x^*) = \sup_{y \in X} (\langle x^*, y \rangle - f(y))$$
$$\leq \sup_{y \in X} (\langle x^*, y \rangle - f(x) - \langle x^* - y - x \rangle) = \langle x^*, x \rangle - f(x).$$

Since by definition we have that $f^*(x^*) \geq \langle x^*, x \rangle - f(x)$, it follows that (2.11) is satisfied.
Conversely, if (2.11) is satisfied, then

$$\langle x^*, x \rangle - f(x) = f^*(x^*) \geq \langle x^*, y \rangle - f(y), \qquad \forall y \in X$$

and this means that $x \in \partial f(x)$. ■

2.15. LEMMA. (Fenchel's duality) Let be $f,g:X \to R$ two convex functions; if g is continous then

$$\inf_{x \in X} [f(x) + g(x)] = \max_{x^* \in X^*} [-f^*(x^*) - g^*(-x^*)].$$

PROOF. It is an easy matter to see that

$$-f^*(x^*) - g^*(-x^*) \leq f(x) + g(x), \qquad \forall x \in X, x^* \in X^*.$$

Hence, denoting $m = \inf_{x \in X} [f(x) + g(x)]$ we can suppose $m \neq -\infty$.
Consider the sets

$$D_1 = \{(x,t); t \geq f(x)\}, \quad D_2 = \{(x,t); t < m - g(x)\}.$$

Then $D_1 \cap D_2 = \phi$, D_1 and D_2 are convex and D_2 is open; hence there exists $(x^*_o, \alpha) \in X^* \times R$ so that the non-vertical hyperplane $H = \{(x,t) \in X \times R; \langle x^*_o, x \rangle + \alpha = t\}$ separes D_1 and D_2, i.e.

$$\langle x^*_o, x \rangle + \alpha \leq f(x), x \in D_1 \text{ and } \langle x^*_o, x \rangle + \alpha \geq m - g(x), x \in D_2.$$

This yields

$$f^*(x^*) \leq -\alpha \quad \text{and} \quad -g^*(-x^*_o) \geq m - \alpha.$$

Hence

$$m = \alpha + (m - \alpha) \leq -f^*(x^*_o) - g^*(-x^*_o)$$
$$\leq \sup_{x^* \in X^*} [-f^*(x^*) - g^*(-x^*)] \leq \inf_{x \in X} [f(x) + g(x)] = m.$$ ■

2.16. PROPOSITION. Let $f_1, f_2 : X \to R$ be convex and f_1 continuous; then

$$(f_1 + f_2)^*(x^*) = \min_{z^* \in X^*}\left[f_1^*(x^* - z^*) + f_2^*(z^*)\right].$$

PROOF. For a fixed $x^* \in X^*$ we define on X the functions

$$f(x) = f_2(x) \text{ and } g(x) = f_1(x) - \langle x^*, x\rangle;$$

then g is convex and continuous and we have

$$-g^*(-z^*) = -\sup_{x \in X}\left[\langle -z^*, x\rangle + \langle x^*, x\rangle - f_1(x)\right] = -f_1^*(x^* - z^*).$$

Hence, using the above Lemma we obtain

$$(f_1 + f_2)^*(x^*) = -\inf_{x \in X}\left[f_1(x) + f_2(x) - \langle x^*, x\rangle\right] - \max_{z^* \in X^*}\left[-f^*(z^*) - g^*(-z^*)\right]$$

$$= -\max_{z^* \in X^*}\left[-f_1^*(x^* - z^*) - f_1^*(z^*)\right] = \min_{z^* \in X^*}\left[f_1^*(x^* - z^*) + f_2^*(z^*)\right] \qquad \blacksquare$$

2.17. THEOREM. Let $f : X \to \tilde{R}$ be proper; then $f^{**} = f$ if and only if f is convex and l.s.c.

PROOF. If $f^{**} = f$, then $f(x) = \sup_{x^* \in X^*}\left(\langle x^*, x\rangle - f^*(x^*)\right)$, $\quad \forall x \in X$.

But for every $x^* \in X^*$, the function $x \to \langle x^*, x\rangle - f^*(x^*)$ is linear and weakly continuous, in particular, weakly l.s.c. Then by Corrollary 1.9 it is l.s.c.

Hence f is convex and l.s.c. as upper envelope of convex and l.s.c. functions (Exercise 2).

Conversely, suppose that f is l.s.c. and convex. It is clear that $f^{**} \le f$. Suppose that there is $x_o \in X$ such that $f^{**}(x_o) < f(x_o)$, i.e. $(x_o, f^{**}(x_o)) \notin \operatorname{Epi} f$.

Since Epif is a nonvoid convex closed set in $X \times R$, there exists $(x_o^*, \alpha) \in X^* \times R$ such that

$$\sup_{(x,t) \in \operatorname{Epi} f}\left(\langle x_o^*, x\rangle + \alpha t\right) < \langle x_o^*, x_o\rangle + \alpha f^{**}(x_o). \tag{2.12}$$

We note that $\alpha \le 0$ since in the case $\alpha > 0$ the property of the epigraph would imply that the left side in (2.12) is $+\infty$.

Suppose now $\alpha = 0$; then (2.12) may be written

$$\langle x_o^*, x\rangle > \sup_{x \in D(f)} \langle x_o^*, x_o\rangle. \tag{2.13}$$

By Proposition 2.13. f^* is proper; then for $y^* \in D(f^*)$ and $\lambda > 0$ we have

$$f^*(y^* + \lambda x_0^*) = \sup_{x \in D(f)} (< y^*, x > + \lambda < x_o^*, x > - f(x)) \leq$$

$$\sup_{x \in D(f)} (< y^*, x > - f(x)) + \lambda \sup_{x \in D(f)} < x_o^*, x > = f^*(y^*) + \lambda \sup_{x \in D(f)} < x_o^*, x >.$$

Further, using the definition of f^{**} at x_0, we obtain

$$f^{**}(x_o) \geq < y^* + \lambda x_o^*, x_o > - f^*(y^* + \lambda x_o^*) \geq$$

$$< y^*, x_o^* > - f^*(y^*) + \lambda(< x_o^*, x_o > - \sup_{x \in D} < x_o^*, x >).$$

Letting $\lambda \to \infty$ and using (2.13) it follows that $f^{**}(x_0) += \infty$. Since this contradicts the fact that $f^{**}(x_0) < f(x_0)$, we obtain $\alpha < 0$.

Dividing (2.12) by $-\alpha$ and denoting by $y_0^* = -x_0^*/\alpha$, we finally obtain:

$$< y_o^*, x_o > - f^{**}(x_0) > \sup_{(x,t) \in \operatorname{Epif}} (< y_o^*, x > - t) \geq \sup_{x \in D(f)} (< y_o^*, x > - f(x)) = f^*(y_0^*)$$

i.e. $f^{**}(x_o) < < y_o^*, x_o > - f^*(y_0^*)$, which is impossible. ■

2.18. EXAMPLE. For $f(x) = \frac{1}{2}\|x\|^2$ the conjugate function f^* has the form $f^*(x^*) = \frac{1}{2}\|x^*\|^2$. Indeed, for any $x \in X$ and $x^* \in X$, we have $< x^*, x > \leq \|x^*\| \|x\| \frac{1}{2}(\|x^*\|^2 + \|x\|^2)$ hence $< x^*, x > - \frac{1}{2}\|x\|^2 \leq \frac{1}{2}\|x^*\|^2$.
This yields

$$f^*(x^*) = \sup_{x \in X} (< x^*, x > - \frac{1}{2}\|x\|^2) \leq \frac{1}{2}\|x^*\|^2$$

Let be $y_n \in X, \|y_n\| = 1$ such that $< x^*, y_n > \xrightarrow[n]{} \|x^*\| = \sup_{\|y\|=1} |< x^*, y >|$.

Then for $x_n = \|x^*\| . y_n$ we obtain

$$< x^*, x_n > - \frac{1}{2}\|x_n\|^2 = \|x^*\| . < x^*, y_n > - \frac{1}{2}\|x_n\|^2 \xrightarrow[n \to \infty]{} \frac{1}{2}\|x^*\|^2$$

This yields the statement.

§ 3. SMOOTH BANACH SPACES

We start our considerations with the

3.1. DEFINITION. A Banach space X (real or complex) is called smooth if for every $x \neq 0$ there is a unique $x^* \in X^*$ such that $\|x^*\| = 1$ and $< x^*, x > = \|x\|$.

3.2. REMARK. In the case X is a complex Banach space, we shall denote by X_R the space X considerated as a real space.
For each $x^* \in X^*$ we denote by $\mathrm{Re}\, x^* \in X_R^*$, the real functional defined by $< \mathrm{Re}x^*, x > = \mathrm{Re} < x^*, x >$, $x \in X$.
Then it is easy to verify that $\|x^*\| = \|\mathrm{Re}\, x^*\|$ and that for each $x^* \in X$, we have:

$$< x^*, x > = < \mathrm{Re}\, x^*, x > - i < \mathrm{Re}\, x^*, i\, x >$$

Thus $X^* \ni x^* \longrightarrow \mathrm{Re}\ x^* \in X_R^*$ is a R-linear isometric isomorphism.
These considerations lead to the conclusion that a complex Banach space X is smooth if and only if X_R is smooth; therefore we may assume in what follows that X is a real Banach space.

3.3. NOTATION. For $r > 0$ and $x_o \in X$ we denote by

$$S_r(x_o) = \{x \in X; \|x - x_o\| < r\}; \bar{S}_r(x_o) = \{x \in X; \|x - x_o\| \leq r\}.$$

We call sphere at the boundary of $S_r(x_o)$.

3.4. PROPOSITION. For every $x \neq 0$ we have

$$\partial\|x\| = \{x^* \in X^*; < x^*, x > = \|x\|, \|x^*\| = 1\}.$$

PROOF. We note that if $x^* \in X^*$ is such that $\|x^*\| = 1$ and $< x^*, x > = \|x\|$, then $< x^*, y - x > = < x^*, y > - \|x\| \leq \|y\| - \|x\|$, for all $y \in X$, i.e. $x^* \in \partial\|x\|$.
Conversely, if $x^* \in \partial\|x\|$, then

$$< x^*, y - x > \leq \|y\| - \|x\| \leq \|y - x\|, \qquad \forall\ y \in X. \tag{3.1}$$

It follows that $x^* \in X^*$ and $\|x^*\| \leq 1$. Taking now y=0 in (3.1) we obtain $< x^*, x > \geq \|x\|$ i.e. $< x^*, x > \|x\|$ and $\|x^*\| = 1$. ■

From this Proposition and Corollary 2.1 we directly obtain the

3.5. THEOREM. X is smooth is and only if the norm is G-differentiable on $X\setminus\{0\}$.

This result can also be expressed in the following geometric form:

3.6. PROPOSITION. X is smooth if and only if for every $x \neq 0$ there is a unique supporting hyperplane for the ball $\bar{S}_{\|x\|}(0)$ at x.

PROOF. Let $x^* \in \partial\|x\|$; then for every $y \in \bar{S}_{\|x\|}(0)$ we may write:

$$<x^*, y - x> \leq \|y\| - \|x\| \leq 0, \text{ i.e } \quad <x^*, y> \leq \|x\|.$$

Then $H = \{y \in X; <x^*, y> = \|x\|\}$ is a supporting hyperplane for $\bar{S}_{\|x\|}(0)$ at x.
Conversely, if for some $x^* \in X^*$ the set $H = \{y \in X; <x^*, y> = c\}$ is a supporting hyperplane for $\bar{S}_{\|x\|}(0)$ at x, then $c = <x^*, x>$. We can suppose $\|x^*\| = 1$ and $<x^*, x> \geq 0$; hence we have

$$<x^*, y> \leq <x^*, x> \qquad \forall\, y \in S_{\|x\|}(0). \tag{3.2}$$

Let $x_n \in X, \|x_n\| = 1$ and $<x^*, x_n> \to 1$; put $y_n = x_n . \|x\|$; then (3.2) yields

$$<x^*, y_n> = <x^*, x_n> \|x\| \leq <x^*, x>, \; n \in N,$$

Letting $n \to \infty$, we obtain $\|x\| \leq <x^*, x>$ and this implies $<x^*, x> = \|x\|$. Hence by Proposition 3.1, $x^* \in \partial\|x\|$. ∎

3.7. EXAMPLE. A simple computation shows that the norm on a real Hilbert space is G-differentiable and that for $f(x) = \|x\|$, we have $f'(x) = \frac{x}{\|x\|}$ for every $x \neq 0$.

3.8. NOTATIONS. For $\tau > 0$ we define

i) $$\rho(\tau) = \frac{1}{2} \sup_{\|x\| = \|y\| = 1} (\|x + \tau y\| + \|x - \tau y\| - 2);$$

ii) $$\rho(\tau, x) = \frac{1}{2} \sup_{\|y\| = 1} (\|x + \tau y\| + \|x - \tau y\| - 2\|x\|) \qquad \text{for } x \in X.$$

The functions $R_+ \ni \tau \to \rho(\tau)$ and $R_+ \ni \tau \to \rho(\tau, x)$ are called modulus of smoothness of X, respectively modulus of smothness at x.
Sometimes, in order to avoid a possible confusion we shall use the notation ρ_X for the modulus of smoothness of the space X.

3.9. DEFINITION. A Banach space X is said to be uniformly smooth, respectively locally uniformly smooth if

$$\lim_{\tau\to 0}\frac{\rho(\tau)}{\tau}=0, \text{ respectively } \lim_{\tau\to 0}\frac{\rho(\tau,x)}{\tau}=0, \text{ for all } x\in X\setminus\{0\}.$$

It is clear that the uniform smoothness implies the locally uniform smoothness.

3.10. REMARK. Observe that $\rho(\tau) \geq 0$ and $\rho(\tau, x) \geq 0$, $\forall\, \tau > 0$.
Indeed, for fixed $x, y \in X$ we apply Lemma 1.17. to the convex function $\varphi(\tau) = \|x + \tau y\|$, $\tau \in R_+$ to obtain

$$\frac{\|x\| - \|x - \tau y\|}{\tau} \leq \varphi'_-(0) \leq \varphi'_+(0) \leq \frac{\|x + \tau y\| - \|x\|}{\tau} \tag{3.3}$$

This yields $\|x + \tau y\| + \|x - \tau y\| - 2 \geq 0$.

3.11. PROPOSITION. If X is locally uniformly smooth, then X is smooth.

PROOF. Suppose that X is not smooth; then there exists $x_0 \in X, \|x_0\| = 1$ and $x_1^*, x_2^* \in X_1^*, x_1^* \neq x_2^*$, with
$\|x_1^*\| = \|x_2^*\| = 1$ and $<x_1^*, x_o> = <x_2^*, x_o> = 1$.
Let be $y_0 \in X$ such that $\|y_0\| = 1, <x_1^*, y_o> > 0$ and $y_0 \in \mathrm{Ker}(x_1^* + x_2^*)$.
Then for every $\tau > 0$ we have

$$\tfrac{1}{2}(\|x_0 + \tau y_0\| + \|x_0 - \tau y_0\| - 2)$$

$$\geq \tfrac{1}{2}[(<x_1^*, x_o> + \tau <x_1^*, y_o>) + (<x_2^*, x_o> - \tau <x_2^*, y_o>) - 2]$$

$$= \tfrac{\tau}{2}(<x_1^*, y_o> - <x_2^*, y_o>) = \tau <x_1^*, y_o>$$

where in the last equality we used the fact that $<x_2^*, y_o> = - <x_1^*, y_o>$.
Hence $\frac{\rho(\tau, x_0)}{\tau} \geq <x_1^*, y_o> > 0$, which is a contradiction. ■

3.12. THEOREM. i) The norm is F-differentiable on $X\setminus\{0\}$ if and only if X is locally uniformly smooth.
ii) The norm is uniformly F-differentiable on the unit sphere, i.e. $f(x)=\|x\|$ is F-differentiable and $\lim_{\tau\to 0}\sup_{\|x\|=\|y\|=1}\left|\frac{\|x+\tau y\|-\|x\|}{\tau}-<f'(x),y>\right|=0$ if and only if X is uniformly smooth.

PROOF. i) Suppose that the function $f(x)=\|x\|$ is Fréchet differentiable on $X\setminus\{0\}$ and let $x\neq 0$; then by (3.3) we have

$$\frac{\|x\|-\|x-\tau y\|}{\tau}\leq <f'(x),y>\leq\frac{\|x+\tau y\|-\|x\|}{\tau},\qquad \forall\, y\in X. \tag{3.4}$$

Moreover we may write

$$\|x+\tau y\|+\|x-\tau y\|-2\|x\|$$
$$=(\|x+\tau y\|-\|x\|-\tau<f'(x),y>)+(\|x-\tau y\|-\|x\|+\tau<f'(x),y>).$$

Then (3.4) yields for $\tau>0$

$$0\leq\frac{\rho(\tau,x)}{\tau}\leq\frac{1}{2}\sup_{\|y\|=1}\left(\frac{\|x+\tau y\|-\|x\|}{\tau}-<f'(x),y>\right)$$
$$+\frac{1}{2}\sup_{\|y\|=1}\left(\frac{\|x+\tau(-y)\|-\|x\|}{\tau}-<f'(x),-y>\right)\underset{\tau\to 0}{\longrightarrow}0.$$

Conversely, suppose X locally uniformly smooth and let $x\neq 0$; by Proposition 3.11 and Theorem 3.5 the norm is G-differntiable on $X\setminus\{0\}$ and (3.4) holds; hence, for every $x\neq 0$ and $y\in X$ we have

$$0\leq\frac{\|x+\tau y\|-\|x\|}{\tau}-<f'(x),y>\leq\left(\frac{\|x+\tau y\|+\|x-\tau y\|-2\|x\|}{\tau}\right).$$

Hence

$$\sup_{\|y\|=1}\left(<f'(x),-y>-\frac{\|x\|-\|x-\tau y\|}{\tau}\right)\leq 2\frac{\rho(\tau,x)}{\tau}\underset{\tau\to 0}{\longrightarrow}0.$$

Analogously

$$\sup_{\|y\|=1}\left(<f'(x),y>-\frac{\|x\|-\|x-\tau y\|}{\tau}\right)\leq 2\frac{\rho(\tau,x)}{\tau}\underset{\tau\to 0}{\longrightarrow}0.$$

Then f is Fréchet differentiable at x.
ii) The proof is similar. ■

§.4. DUALITY MAPPINGS ON BANACH SPACES

Let X be a Banach space.

4.1. DEFINITION. i) A continuous and strictly increasing function $\varphi: R_+ \to R_+$, such that $\varphi(0) = 0$ and $\lim_{t \to +\infty} \varphi(t) = +\infty$ is called a weight function.
ii) We call duality mapping of weight φ at the mapping $J: X \to 2^{X^*}$ defined by

$$Jx = \{x^* \in X^*_R; \langle x^*, x \rangle = \|x^*\| \|x\|, \|x^*\| = \varphi(\|x\|)\} .$$

The duality mapping corresponding to the weight $\varphi(t) = t$ is called normalized duality mapping.
iii) A selection of the duality mapping J is a single-valued mapping $\tilde{J}: X \to X^*$ satisfying $\tilde{J} x \in Jx$ for each $x \in X$.

4.2. REMARK. We note that $Jx \neq 0$ for every $x \in X$. Indeed, consider $y = x.\varphi(\|x\|)$; by the Hahn-Banach Theorem, there exists $y^* \in X^*$ with $\|y^*\| = 1$ and $\langle y^*, y \rangle = \|y\|$; then obviously $x^* = y^*.\varphi(\|x\|) \in Jx$.

We shall further need the

4.3 LEMMA. Let φ be a weight on R_+ and $\psi(t) = \int_0^t \varphi(s)ds$; then Ψ is a convex function on R_+.

PROOF. For $h > 0$ and $t > 0$ we have

$$\frac{\psi(t+h) - \psi(t)}{h} = \frac{1}{h}\int_t^{t+h} \varphi(s)ds \geq \varphi(t)$$

and

$$\frac{\psi(t) - \psi(t-h)}{h} = \frac{1}{h}\int_{t-h}^{t} \varphi(s)ds \leq \varphi(t) .$$

Let be $0 \le t_1 < t_2$, $\lambda \in [0,1]$ and $t = \lambda t_1 + (1-\lambda)t_2$; then $\lambda = \dfrac{t_2 - t}{t_2 - t_1}$ and

$1 - \lambda = \dfrac{t - t_1}{t_2 - t_1}$ and the above two inequalities yield

$$\psi(t_2) - \psi(t) \ge (t_2 - t)\varphi(t) \quad \text{and} \quad \psi(t) - \psi(t_1) \ge (t - t_1)\varphi(t).$$

Multiplying the first inequality by $(1-\lambda)$, the second one by $(-\lambda)$ and summing, we obtain

$$\lambda\psi(t_1) + (1-\lambda)\psi(t_2) - \psi(t) \ge 0$$

i.e.

$$\psi(\lambda t_1 + (1-\lambda))t_2 \le \lambda\psi(t_1) + (1-\lambda)\psi(t_2).$$

■

4.4. THEOREM. (Asplund). If J is a duality mapping of weight φ, then

$$Jx = \partial\psi(\|x\|) \quad \text{for each } x \in X.$$

PROOF. Let be $x^* \in Jx$ and $y \in X$ with $\|y\| > \|x\|$; by Lemma 4.3 Ψ is a convex function to which we may apply Lemma 1.17 to get

$$\|x^*\| = \varphi(\|x\|) = \psi'(\|x\|) \le \frac{\psi(\|y\|) - \psi(\|x\|)}{\|y\| - \|x\|}.$$

Hence

$$\psi(\|y\|) - \psi(\|x\|) \ge \|x^*\|(\|y\| - \|x\|) = \|x^*\|\,\|y\| - \langle x^*, x\rangle \ge \langle x^*, y - x\rangle.$$

In the case $\|y\| < \|x\|$ we obtain the same result using the inequality

$$\psi'(\|x\|) \ge \frac{\psi(\|x\|) - \psi(\|y\|)}{\|x\| - \|y\|}.$$

Thus, for every $y \in X, \psi(\|y\|) - \psi(\|x\|) \ge \langle x^*, y - x\rangle$ i.e. $x^* \in \partial\psi(\|x\|)$.

Conversely, consider $x^* \in \partial\psi(\|x\|)$ and $y \in X$ with $\|y\| = \|x\|$; then $\langle x^*, y\rangle \le \langle x^*, x\rangle$. Hence

$$\|x^*\|.\|x\| = \sup_{\|y\| = \|x\|} \langle x^*, y\rangle \le \langle x^*, x\rangle \quad \text{that is} \quad \langle x^*, x\rangle = \|x^*\|.\|x\|.$$

Denote $z = x/_{\|x\|}$ and $y = t.z, t > 0$; we have

$$\psi(\|y\|) - \psi(\|x\|) = \psi(t) - \psi(\|x\|) \ge \langle x^*, y - x\rangle = \frac{t - \|x\|}{\|x\|}\langle x^*, x\rangle = (t - \|x\|)\|x^*\|.$$

Also

$\|x^*\| \le \dfrac{\psi(t) - \psi(\|x\|)}{t - \|x\|}$ for $t > \|x\|$ and $\|x^*\| \ge \dfrac{\psi(t) - \psi(\|x\|)}{t - \|x\|}$ for $t < \|x\|$.
Now letting $t \to \|x\|$ we obtain $\|x^*\| = \psi'(\|x\|) = \varphi(\|x\|)$. ■

4.5. COROLLARY. The Banach space X is smooth if and only if each duality mapping J of weight φ is single valued; in this case

$$< Jx,y > = \frac{d}{dt}\psi(\|x + ty\|)\Big|_{t=0}, \qquad \forall\, x,y \in X. \tag{4.1}$$

PROOF. There is a unique $x^* \in X^*$ satisfying
$< x^*, x.\varphi(\|x\|) > = \|x\|\varphi(\|x\|)$ and $\|x^*\| = 1$
if and only if X is smooth; in this case $x^*.\varphi(\|x\|) = Jx = \partial\psi(\|x\|)$ and by Corollary 2.7 the function $x \to \psi(\|x\|)$ is G-differentiable on X. Because of these facts the formula (4.1) is clear. ■

4.6. COROLLARY. Let J be a duality mapping of weight φ; then $x^* \in Jx$ if and only if $H = \{y \in X; < x^*,y > = \varphi(\|x\|)\|x\|\}$ is a supporting hyerplane for the ball $\bar{S}_{\|x\|}(0)$ at x.

PROOF. Let $x^* \in Jx = \partial\psi(\|x\|)$; then for each $y \in S_{\|x\|}(0)$ we have $< x^*, y - x > \le \psi(\|y\|) - \psi(\|x\|) \le 0$ and hence $< x^*,y > \le \varphi(\|x\|)\|x\|$.
Consequently H has the required properties.
Conversely if for $x^* \in X^*, H = \{y \in X; < x^*,y > = \varphi(\|x\|)\|x\|\}$ is a supporting hyperplane for $\bar{S}_{\|x\|}(0)$ at x, then
$< x^*, x > = \varphi(\|x\|)\|x\|$ and $< x^*,y > \le \varphi(\|x\|)\|x\|$ for all $y \in \bar{S}_{\|x\|}$.
Also $\|x^*\| \le \varphi(\|x\|)$, so that $\|x^*\| = \varphi(\|x\|)$ and $x^* \in Jx$. ■

4.7 PROPOSITION. Let J be a duality mapping associated with a weight φ; then

a) For every $x \in X$ the set Jx is convex and weakly*-closed in X^*.
b) The mapping J is monotone, that is

$$< x^* - y^*, x - y > \ge 0, \text{ for all } x,y \in X \text{ and } x^* \in Jx,\ y^* \in Jy\,.$$

c) $J(-x) = -Jx, \qquad x \in X.$
d) $J(\lambda x) = \dfrac{\varphi(\lambda\|x\|)}{\varphi(\|x\|)}Jx, \qquad x \in X, \qquad \lambda > 0.$

In particular each selection of the normalized duality mapping is homogeneous.

e) If φ^{-1} is the inverse of the weight φ, then φ^{-1} is a weight function and if we denote by J_* the duality mapping on X^* of weight φ^{-1}, then $x^* \in Jx$ whenever $x \in J_* x^*$.
f) If J_1 and J_2 are two duality mappings with weights φ_1 and φ_2, then

$$\varphi_2(\|x\|)J_1 x = \varphi_1(\|x\|)J_2 x, \qquad x \in X.$$

PROOF. a) Jx has the mentioned properties because $Jx = \partial\psi(\|x\|)$.
b) Let be $x,y \in X, x^* \in Jx$ and $y^* \in Jy$; then we may write

$$< x^* - y^*, x - y > = < x^*, x > + < y^*, y > - < x^*, y > - < y^*, x >$$

$$\geq \varphi(\|x\|)\|x\| + \varphi(\|y\|)\cdot\|y\| - \varphi(\|x\|)\|y\| - \varphi(\|y\|)\|x\| = (\varphi(\|x\|) - \varphi(\|y\|))(\|x\| - \|y\|) \geq 0$$

Thus J is monotone as asserted.
c) Let be $x^* \in Jx$; then we have
$$< -x^*, -x > = < x^*, x > = \|x^*\|\,\|x\| = \|-x^*\|\,\|-x\|$$
and
$$\|-x^*\| = \|x^*\| = \varphi(\|x\|) = \varphi(\|-x\|).$$
Hence $-x^* \in J(-x)$, that is $-Jx \subset J(-x)$.
The converse inclusion can be proved similarly.

d) For $x^* \in Jx$ and $\lambda > 0$, we shall prove that if $\alpha = \dfrac{\varphi(\lambda\|x\|)}{\varphi(\|x\|)}$ then $\alpha x^* \in J(\lambda x)$. Indeed, we have

$$< \alpha x^*, \lambda x > = \alpha\lambda\|x^*\|\,\|x\| = \|\alpha x^*\|\,\|\lambda x\|$$

and

$$\|\alpha x^*\| = \alpha\|x^*\| = \frac{\varphi(\lambda\|x\|)}{\varphi(\|x\|)}\varphi(\|x\|) = \varphi(\lambda\|x\|).$$

One can prove the converse inclusion analogously
If $\varphi(t) = t$, then $J(\lambda x) = \lambda Jx$, for all $x \in X, \lambda > 0$; hence by c) any selection of J is homogeneous.
e) It is clear that φ^{-1} is a weight function. If $x^* \in Jx$, then
$$< x^*, x > = \|x^*\|\,\|x\| \text{ and } \|x\| = \varphi^{-1}(\|x^*\|).$$
Hence $x \in Jx^*$.
f) Let be $x^* \in J_1 x$, then $< x^*, x > = \|x^*\|\,\|x\|$ and $\|x^*\| = \varphi_1(\|x\|)$.
We shall prove that $\alpha x^* \in J_2 x$ where $\alpha = \varphi_2(\|x\|)/\varphi_1(\|x\|)$.

In effect we have
$$\|\alpha x^*\| = \varphi_2(\|x\|) \text{ and } < \alpha x^*, x > = \alpha\|x^*\|\,\|x\| = \|\alpha x^*\|\,\|x\|.$$
The converse inclusion can be proved analogously. ■

4.8. PROPOSITION. i) In a real Hilbert space the normalized duality mapping is the identity operator.
ii) The normalized duality mapping on a real Banach space X is linear if and only if X is a Hilbert space.

PROOF. i) Since a Hilbert space is smooth, we can apply (4.1) to calculate J. Let be $\varphi(t) = t$, $\psi(t) = t^2/2$ and $x, y \in X$; we have

$$< Jx, y > = \frac{1}{2}\frac{d}{dt}\|x + ty\|^2 \Big|_{t=0} = \frac{1}{2}\frac{d}{dt} < x + ty, x + ty > \Big|_{t=0}$$

$$= \frac{1}{2}\frac{d}{dt}\left(\|x\|^2 + 2t < x, y > + t^2\|y\|^2\right)\Big|_{t=0} = < x, y >.$$

Hence $Jx = x$, for all $x \in X$.
ii) Suppose that J is linear and let $x^* \in Jx$ and $y^* \in Jx$; then $x^* \pm y^* \in J(x \pm y)$ and we have

$$\|x \pm y\|^2 = < x^* \pm y^*, x \pm y > = \|x\|^2 \pm < x^*, y > \pm < y^*, x > + \|y\|^2.$$

It is an easy matter to obtain now

$$\|x + y\|^2 + \|x - y\|^2 = 2\left(\|x\|^2 + \|y\|^2\right)$$

relation which characterizes a Hilbert space.
Conversely, if X is Hilbert, then by i) Jx=x, hence J is linear. ■

4.9. PROPOSITION. (Kato) Let $R \ni t \to f(t) \in X$ be a function which satisfy:
a) the function $t \to \|f(t)\|$ is a.e. differentiable on R;
b) the weak derivative f' of f exists a.e. on R.
Then, if J is the normalized duality mapping, we have

$$\|f(t)\|\frac{d}{dt}\|f(t)\| = < x^*, f'(t) >, \quad \text{for all} \quad x^* \in Jf(t), \text{ a.e. on } R.$$

PROOF. Since for every $t, s \in R$ and $x^* \in Jf(t)$ we have

$$< x^*, f(t) > = \|f(t)\|^2 \quad \text{and} \quad < x^*, f(s) > \le \|f(t)\| \, \|f(s)\|$$

then

$$< x^*, f(s) - f(t) > \le \|f(t)\| \left(\|f(s)\| - \|f(t)\|\right). \tag{4.2}$$

In the case s>t, dividing (4.2) by s-t and letting s→t, we obtain

$$< x^*, f'(t) > = \lim_{s \to 0} < x^*, \frac{f(t + s) - f(t)}{s} > \le \|f(t)\| \frac{d}{dt}\|f(t)\|.$$

The converse inequality can be obtained in a similar way. ■

4.10. PROPOSITION. Let J be the normalized duality mapping on X; then for every $x, y \in X$ the following statements are equivalent:

i) $\|x\| \le \|x + \lambda y\|, \quad \forall \lambda > 0$

ii) there exists $x^* \in Jx$ such that $< x^*, y > \ge 0$.

PROOF. $i \Rightarrow ii)$ Consider $\lambda > 0$, $x^*_\lambda \in J(x + \lambda y)$ and $y^*_\lambda = x^*_\lambda / \|x^*_\lambda\|$; then we have

$$\|x\| \le \|x + \lambda y\| = < x^*_\lambda, x + \lambda y > / \|x^*_\lambda\| \tag{4.3}$$

$$= < y^*_\lambda, x + \lambda y > = < y^*_\lambda, x > + \lambda < y^*_\lambda, y > \le \|x\| + \lambda < y^*_\lambda, y >.$$

Since the unit ball is weakly*-compact in X^*, the net $\{y^*_\lambda\}_{\lambda \in R_+}$ has a limit point y^* which by (4.3) satisfies

$$\|y^*\| \le 1, < y^*, x > \ge \|x\| \text{ and } < y^*, y > \ge 0.$$

But then $\|x\| \le < y^*, x > \le \|x\| \, \|y^*\| \le \|x\|$ i.e. $< y^*, x > = \|x\|$. Therefore $\|y^*\| = 1$.
Now it is clear that $x^* = y^* \|x\| \in Jx$ and $< x^*, y > \ge 0$.
Conversely, suppose that for $x, y \in X$ there is $x^* \in Jx$ such that $< x^*, y > \ge 0$. Then for $\lambda > 0$ we may write

$$\|x\|^2 = < x^*, x > \le < x^*, x > + \lambda < x^*, y >$$
$$= < x^*, x + \lambda y > \le < \|x^*\| \, \|x + \lambda y\| = \|x + \lambda y\|.$$

Hence we have $\|x\| \le \|x + \lambda y\|$. ■

4.11. DEFINITION. Let X, Y be two linear topological spaces; a mapping $T : X \to 2^Y$ is called upper-semicontinuous at $x \in X$ if for every open set V in Y containing Tx there exists $U \in \mathcal{V}_x$ such that $T(U) = \cup \{Tx, x \in U\} \subset V$.
T is upper-semicontinuous if it is upper-semicontinuous at each point $x \in X$.
In order to avoid possible confussions we shall spicify the considered topologies on X, respectively Y as follows: norm-to-norm, norm-to-weak, norm-to-weak* upper semicontinuous.
It is clear that if T is single valued the above definition coincides with that of the continuity of T.

4.12. THEOREM. Every duality mapping J of weight φ on a Banach space X is norm-to- weak* upper-semicontinuous on X.

PROOF: Let us suppose that there is a duality mapping J of weight φ which is not norm-to- weak* upper semincntinuous.

Then there exist a $x \in X$, a weak* open set V in X^* containing Tx and $\{x_n\}_{n\in N} \subseteq X, x_n^* \in Jx_n, n \in N$ such that $x_n \xrightarrow[n\to\infty]{} x$ and $x_n^* \notin V, n \in N$.

Let F_n be the weak* closure of the set $\{x_n^*, x_{n+1}^* \ldots,\}$;then $F_1 \supseteq F_2 \supseteq \ldots$ and since the unit ball is weak* compact in X^*, there exists $x^* \in \bigcap_{n=1}^{\infty} F_n$. It is clear that $x^* \notin V$, in particular $x^* \notin Jx$.

From the definition of F_n and x^* it follows that for every $n \in N$, there exists $n' \geq n$ such that

$$|<x^*, x> - <x_{n'}^*, x>| \leq \frac{1}{n}.$$

Then we have

$$\left|<x^*, x> - \varphi(\|x_{n'}\|)\|x_{n'}\|\right|$$

$$\leq |<x^*, x> - <x_{n'}^*, x>| + |<x_{n'}^*, x> - <x_{n'}^*, x_{n'}>| \leq$$

$$\leq \frac{1}{n} + \|x_{n'}^*\| \, \|x - x_{n'}\| = \frac{1}{n} + \varphi(\|x_{n'}\|)\|x - x_{n'}\| \xrightarrow[n'\to\infty]{} 0.$$

Hence $<x^*, x> = \varphi(\|x\|)\|x\|$.

Since $x^* \in F_n \subseteq \left\{x^* \in X^*; \|x^*\| \leq \sup_{k\geq n} \|x_k^*\| = \sup_{k\geq n} \varphi(\|x_k\|)\right\}$ for all $n \in N$ then $\|x^*\| \leq \varphi(\|x\|)$.

It follows that $x^* \in Jx$ and this is a contradiction. ■

§.5. POSITIVE DUALITY MAPPINGS

Let X be a real Banach lattice and P the cone of all positive elements in X. We shall use the following notations:

$x_+ = x \vee 0$, $x_- = -x \vee 0$, $|x| = x_+ + x_-$ and $x^\perp = \{y \in X; |x| \wedge |y| = 0\}$.

5.1. DEFINITION. We call positive duality mapping on X at the map $J_+ : P \to 2^{X^*}$ defined as

$$J_+x = \{x^* \in X^*; x^* \geq 0, <x^*, x> = \|x\|^2, \|x^*\| = \|x\|, <x^*, y> = 0, \forall y \in x^\perp\}.$$

It is obvious that if J is the normalized duality map on X, then $J_+x \subseteq Jx$, $x \in P$.

5.2 PROPOSITION. For every $x \in P, J_+x \neq \phi$.

PROOF. We note that the set $x^\perp$ is a closed subset of X. By the Hahn-Banach Theorem there is $x^* \in X^*$ such that

$$<x^*, x> = \|x^*\| \, \|x\|, \|x^*\| = \|x\| \text{ and } <x^*, y> = 0, \qquad \forall\, y \in x^\perp.$$

Define $\displaystyle <x^*_+, y> = \sup_{0 \le z \le y} <x^*, z>, \qquad y \in P.$

Then x^*_+ is positive, $<x^*_+, \lambda y> = \lambda <x^*_+, y>$, $\forall \lambda \ge 0$ and $<x^*_+, y = 0$, $\forall\, y \in x^\perp$ (because $0 \le z \le y$ and $y \in x^\perp \Rightarrow z \in x^\perp$).
Moreover x^*_+ is additive; indeed, let be $y_1, y_2 \in P$ and $z \in P, z \le y_1 + y_2$.
By Riesz' decomposition result, there are $z_1, z_2 \in P$ such that

$$0 \le z_1 \le y_1, 0 \le z_2 \le y_2 \text{ and } z = z_1 + z_2.$$

Hence

$$<x^*_+, y_1 + y_2> = \sup_{\substack{0 \le z_1 \le y_1 \\ 0 \le z_2 \le y_2}} <x^*, z_1 + z_2>$$

$$= \sup_{0 \le z_1 \le y_1} <x^*, z_1> + \sup_{0 \le z_2 \le y_2} <x^*, z_2>.$$

We can extend x^*_+ to whole X by $\tilde{x}^*_+$ defined as

$$<\tilde{x}^*_+, x> = <x^*_+, x_+> - <x^*_+, x_->, \qquad \forall\, x \in X.$$

One can easily verify that $\tilde{x}^*_+$ is linear, positive and zero on $x^\perp$(indeed if $x \in x^\perp$, then $x_+, x_- \in x^\perp$).
We shall prove that $x^*_+ \in J_+(x_0)$. We have

$$<x^*_+, y> = \sup_{0 \le z \le y} <x^*, z> \le \sup_{0 \le z \le y} \|x^*\| \, \|z\| \le \|x^*\| \, \|y\|, \; y \in P.$$

Hence

$$|<\tilde{x}^*_+, x>| = |<x^*_+, x_+> - <x^*_+, x_->| \le \sup(<x^*_+, x_+>, <x^*_+, x_->)$$

$$\le \|x^*\| \max(\|x_+\|, \|x_-\|) \le \|x^*\| \, \|x\|.$$

Consequently $\|\tilde{x}^*_+\| \le \|x^*\|$.
We further have

$$<x^*, x> \le <x^*_+, x> = <\tilde{x}^*_+, x> \le \|\tilde{x}^*_+\| \, \|x^*\| \le \|x^*\|. \|x\| = \|x\|^2 = <x^*, x>.$$

This yields

$<\tilde{x}^*_+, x> = \|x\| \|\tilde{x}^*_+\|$ and $\|\tilde{x}^*_+\| = \|x^*\|$; hence $\tilde{x}^*_+ \in J_+(x)$. ■

5.3. LEMMA. For $x \in P$ and $y \in X$ we define

$$\sigma(x, y) = \inf_{z \in x^{\perp}, \alpha \geq 0} \lim_{\lambda \to 0_+} \lambda^{-1}(\|x + \lambda[(y + z) \vee (-\alpha x)]\| - \|x\|) .$$

Then

i) $\sigma(x, y) \leq \|y_+\|$;

ii) $y \to \sigma(x, y)$ is subadditive, positive and positive homogeneous;

iii) $\sigma(x, ax + y) = a\|x\| + \sigma(x, y)$, for every $a \geq 0$;

iv) if $y' \in x^{\perp}$, then $\sigma(x, y) = \sigma(x, y + y')$.

PROOF. i) Take in the definition of $\sigma, z = 0$ and $\alpha = 0$; we obtain

$$\sigma(x, y) \leq \lim_{\lambda \to 0_+} \lambda^{-1}(\|x + \lambda y_+\| - \|x\|) \leq \|y_+\|.$$

ii) Let $y, y' \in X$, $z, z' \in x^{\perp}$ and $\alpha \geq 0$; we have

$$\lim_{\lambda \to 0_+} \lambda^{-1}(\|x + \lambda[(y + y' + z + z') \vee (-\alpha x)]\| - \|x\|)$$
$$= \lim_{\gamma \to 0_+} \gamma^{-1}(\|2x + \gamma[(y + y' + z + z') \vee (-\alpha x)]\| - 2\|x\|)$$
$$\leq \lim_{\gamma \to 0_+} \gamma^{-1}(\|x + \gamma[(y + z) \vee (-\alpha x)]\| - \|x\|)$$
$$+ \lim_{\gamma \to 0_+} \gamma^{-1}(\|x + \gamma[(y' + z') \vee (-\alpha x)]\| - \|x\|), \text{ where } \gamma = \lambda/2 .$$

It follows that $\sigma(x, y + y') \leq \sigma(x, y) + \sigma(x, y')$.

Consider now $y \in P$; then for every $z \in x^{\perp}$ and $\alpha \geq 0$, we have

$$(y + z) \vee (-\alpha x) \geq z \vee (-\alpha x) = -(-z) \wedge (\alpha x)$$
$$\geq -\alpha(|z| \wedge x) \geq -\alpha(|z| \wedge |x|) = 0$$

Hence

$\|x + \lambda[(y + z) \vee (\alpha x)]\| - \|x\| \geq 0$ and this yields $\sigma(x, y) \geq 0$.

Let be a>0; then with $\gamma = a\lambda$ we obtain

$$\sigma(x, ay) = \inf_{z \in x^{\perp}, \alpha \geq 0} \lim_{\lambda \to 0_+} \lambda^{-1}(\|x + \lambda[(ay + az) \vee (-\alpha x)]\| - \|x\|)$$
$$= a \inf_{z \in x^{\perp}, \alpha \geq 0} \lim_{\lambda \to 0_+} \gamma^{-1}(\|x + \gamma[(y + z) \vee (-\alpha x)]\| - \|x\|) = a\sigma(x, y)$$

iii) It is not difficult to see that for $z \in x^{\perp}$, and $\alpha \geq 0$ we have

$$(ax + z + y) \vee (-\alpha x) = ax + (y + z) \vee [-(a + \alpha)x].$$

Then we may write

$$\lim_{\lambda \to 0_+} \lambda^{-1}(\|x + \lambda(ax + y + z) \vee (-\alpha x)]\| - \|x\|)$$

$$= \lim_{\lambda \to 0_+} \lambda^{-1}(\|x + \lambda ax + \lambda[(y + z) \vee -(a + \alpha)x] - \|x\|\|)$$

$$= \lim_{\lambda \to 0_+} \lambda^{-1}(1 + a\lambda)(\|x + (1 + a\lambda)^{-1}\lambda[(y + z) \vee -(a + \alpha)x]]\| - (1 + a\lambda)^{-1}\|x\|)$$

$$= \lim_{\gamma \to 0} \gamma^{-1}(\|x + \gamma[(y + z) \vee -(a + \alpha)x]\| - \|x\| + \gamma a\|x\|), \quad \text{for} \quad \gamma = \lambda(1 + a\lambda)^{-1}$$

We note that $\sigma(x, y)$ doesn't change if we replace in its definition $\alpha \geq 0$ by $\alpha \geq a > 0$ so that it follows that

$$\sigma(x, ax + y) = a\|x\| + \sigma(x, y).$$

iv) This last relation is a direct consequence of the definition of σ. ■

5.4. REMARK. Since by (i) $\sigma(x, 0) \leq 0$ and by ii) $\sigma(x, 0) \geq 0$ it follows that $\sigma(x, 0) = 0$. Therefore $\sigma(x, 0) = \sigma(x, y - y) \leq \sigma(x, y) + \sigma(x, -y)$, i.e.

$$\sigma(x, -y) \leq \sigma(x, y).$$

5.5. PROPOSITION. Let be $x \in P$ and $y \in X$; there exists $\tilde{x}^* \in J_+x$ such that

$$< \tilde{x}^*, y > = \sup_{x^* \in J_+x} < x^*, y > = \|x\|\sigma(x, y).$$

PROOF. We note that we have

$$< x^*, y > \leq \|x\|\sigma(x, y) \text{ for all } x^* \in J_+x. \tag{5.1}$$

Indeed, for $x^* \in J_+x$, $z \in x^{\perp}$ and $\lambda, \alpha \geq 0$, we have

$$\lambda^{-1}\|x\|(\|x + [\lambda(y + z) \vee (-\alpha)x]\| - \|x\|)$$

$$\geq \lambda^{-1}(< x^*, x + \lambda[(y + z) \vee (-\alpha)x] > - < x^*, x >)$$

$$= < x^*, (y + z) \vee (-\alpha x) > \geq < x^*, y + z > = < x^*, y >.$$

Now it is an easy matter to obtain (5.1).

Consider the subspace $X_0 = \{u \in X; u = \lambda(x + y) + z, \lambda \in R, z \in x^{\perp}\}$.

We define on X_0 the linear functional

$$< F, u > = \lambda\|x\|\sigma(x, x + y) \quad \text{for} \quad u = \lambda(x + y) + z \in X_0.$$

By ii) and iv) in the above Lemma, we have,

$$< F, u > = \|x\|\sigma(x, \lambda(x+y)) = \|x\|\sigma(x, u) \text{ for } \lambda \geq 0.$$

If $\lambda < 0$, using the Remark 5.4 we obtain

$$< F, u > = - \|x\|\sigma(x, -\lambda(x+y)) \leq \|x\|\sigma(x, \lambda(x+y)) = \|x\|\sigma(x, u).$$

Hence

$$< F, u > \leq \|x\|\sigma(x, u), \qquad \forall\, u \in X_o.$$

Since $u \to \|x\|\sigma(x, u)$ is subadditive and positive homogeneous on X, by the Hahn-Banach theorem there exists a linear functional $\tilde{x}^*$ on X extending F and such that

$$< \tilde{x}^*, u > \leq \|x\|\sigma(x, u), \text{ for all } u \in X. \tag{5.2}$$

By Lemma 5.1 (iii) we have

$$< \tilde{x}^*, x > + < \tilde{x}^*, y > = < F, x+y > = \|x\|\sigma(x, x+y) = \|x\|^2 + \|x\|\sigma(x, y). \tag{5.3}$$

Since $< \tilde{x}^*, x > \leq \|x\|\,\sigma(x, x) = \|x\|^2$ and $< \tilde{x}^*, y > \leq \|x\|\,\sigma(x, y)$, then (5.3) yields

$$< \tilde{x}^*, x > \leq \|x\|^2 \text{ and } < \tilde{x}^*, y > \leq \|x\|\,\sigma(x, y). \tag{5.4}$$

Moreover for every $u \in X$ we have

$$- \|x\|\,\|u\| \leq - \|x\|\,\|(-u)_+\| \leq - \sigma(x, -u)\|x\|$$
$$\leq - \|x\|\,\|u\| \leq - \|x\|\,\|(-u)_+\| \leq - \sigma(x, -u)\|x\|$$

hence $\tilde{x}^* \in X^*$. We also note that $\tilde{x}^*$ is positive; indeed, if $u \geq 0$, then since $u \to \sigma(x, u)$ is positive (5.2) yields

$$- < \tilde{x}^*, u > \leq \|x\|\sigma(x, -u) \leq 0; \text{ i.e. } < \tilde{x}^*, u > \geq 0.$$

Finally, it $u \in x^{\perp}$, then $< \tilde{x}^*, u > = < F, u > = 0$, hence $\tilde{x}^* \in J_+x$. The statement is now a consequence of (5.1) and (5.4). ■

5.6. PROPOSITION. For each $x \in X$, $J_+(x_+) \subseteq \partial\left(\frac{1}{2}\|x_+\|^2\right)$.

PROOF. Let $x^* \in J_+(x_+)$; for each $y \in X$ we may write

$$\|x_+\|^2 - \|y_+\|^2 + 2 < x^*, y_+ - x > = \|x_+\|^2 - \|y_+\|^2 + 2 < x^*, y_+ - y_- - x_+ + x_- > =$$

$$- \|x_+\|^2 - \|y_+\|^2 + 2 < x^*, y_+ > + 2 < x^*, x_- > - 2 < x^*, y_- > \leq - \left(\|x_+\|^2 - \|y_+\|^2\right)^2.$$

Hence

$$\tfrac{1}{2}\|y_+\|^2 - \tfrac{1}{2}\|x_+\|^2 \geq < x^*, y - x >, \qquad y \in X.$$ ■

EXERCISES

1. Let X be a linear topological space and $f: X \to R$; the following statements are equivalent:
   i) f is convex;
   ii) for any $x_1, \ldots, x_n \in X$ and $\lambda_1, \ldots, \lambda_n \in [0,1]$ with $\sum_{i=1}^{n} \lambda_i = 1$, we have:
   $$f(\sum_{i=1}^{n} \lambda_i x_i) \leq \sum_{i=1}^{n} \lambda_i f(x_i);$$
   If X=R then the above conditions are also equivalent to
   iii) $\dfrac{f(t_2)-f(t_1)}{t_2-t_1} \leq \dfrac{f(t_3)-f(t_1)}{t_3-t_1} \leq \dfrac{f(t_3)-f(t_2)}{t_3-t_2}$, for $t_1 < t_2 < t_3$.

2. i) Let $\{f_\iota\}_{\iota \in I}$ be a family of l.s.c. functions on a linear topological space X; then the function $f(x) = \sup_{\iota \in I} f_\iota(x)$, $x \in X$ is l.s.c.
   ii) Let $f: X \to \tilde{R}$ be proper and $\bar{f}(x) = \liminf_{y \to x} f(y)$, $x \in X$; then
   $$\text{Epi}\, \bar{f} = \overline{\text{Epi} f}.$$

In what follows X will be a real Banach space.

3. Let $f: X \to R$ be convex; there are equivalent:
   i) f is bounded from above on an open set of X;
   ii) Int D(f) $\neq \phi$ and f is locally Lipschitz on Int D(f).

<u>Hint</u>. Consider $x_0 \in \text{Int}\, D(f)$ and $r_0 > 0$ such that
$$m \leq f(y) \leq M, \forall\, y \in \bar{S}_{r_0}(x_0);$$
then for any $0 < r < r_0$, f is Lipschitz on $S_r(x_0)$. Indeed, verify first that for any $x \in S_r(x_0)$ and $y \in S_{r_0 - r}(x)$ one has
$$|f(y) - f(x)| \leq \frac{M-m}{r_0 - r} \|y - x\|.$$
Then, for $x_1, x_2 \in S_r(x_0)$, divide the segment $[x_1, x_2]$ in n equidistant intervals, with $n \geq \dfrac{\|x_2 - x_1\|}{r_0 - r}$ and apply the above inequality on each interval; summing the n corresponding inequalities, the result follows.
(See, Corollary 2.4., Ch. I., Ekeland - Temam [1].

4. Let X be a Banach space and $f:X \to R$ twice G-differentiable on X; then

$$<f''(x)y, y> = \frac{d^2}{dt^2} f(x+ty)\Big|_{t=0}, \quad \forall x, y \in X.$$

Hint. For $\varphi(t) = f(x+ty)$, $t \in R$ one has: $<f'(x+ty), y> = \varphi'(t)$.

5. Let $f:X \to R$ be G-differentiable; there are equivalent:

i) f is convex
ii) f' is monotone, i.e. $<f'(x) - f', x - y> \geq 0, \quad \forall x, y \in X;$
iii) $f(y) - f(x) \geq <f'(x), y - x>, \quad \forall x, y \in X.$

Hint. See Prop. 15.4. and Prop. 5.5., Ch. I., Ekeland - Temam [1].

6. For $f:X \to \bar{R}$ we have $f^{**} \leq f$ and $f^{***} = f^*$.

7. i) Let $f:X \to \tilde{R}$ be convex; then $D(f^*) \neq \phi$ iff there exists a nonvertical hyperplane H in $X^* \times R$ such that Epi f is contained in one of the half-spaces determined by H.
ii) Epi f^* is convex and closed.

8. Let $f:X \to \tilde{R}$ be convex, $x \in \operatorname{Int} D(f)$ and $F(y) = f'_+(x,y), y \in X$; then $F^* = I_{\partial f(x)}$.

Hint. $F(y) = \inf_{t \downarrow 0} F_t(y)$, where $F_t(y) = \dfrac{f(x+ty) - f(x)}{t}$;
hence $F^* = \sup_{t \downarrow 0} F_t^*$ and it is an easy matter to verify that

$$F_t(x^*) = \frac{f^*(x^*) - <x^*, x> - f(x)}{t}, \quad \forall x^* \in X^*.$$

We now can apply the Proposition 2.14.

9. Let $f:X \to \tilde{R}$ be proper and convex; then $x^* \in \partial f(x)$ iff $x \in \partial f^*(x^*)$.

10. A function $f:X \to R$ is affine (i.e. the function $x \to f(x) - f(0)$ is linear) iff $\pm$ f are convex.

11. A function $f:X \to \tilde{R}$ is convex and l.s.c. iff it is the pointwise supremun of a family of continuos affine functions on X.

Hint: $f(x) = f^{**}(x) = \sup_{x^* \in X^*} (<x^*, x> - f^*(x^*)), x \in X.$

12. Let $\varphi:R \to R$ be convex, even and l.s.c. and $f(x) = \varphi(\|x\|)$, $x \in R$; then f is convex, l.s.c. and $f^*(x^*) = \varphi^*(\|x^*\|)$, $x^* \in X^*$. In particular if
$$f(x) = \frac{1}{p}\|x\|^p,\ p > 1,\ \text{then } f^*(x^*) = \frac{1}{q}\|x^*\|^q,\ q = \frac{p}{p-1}.$$

Hint. Consider first the case when φ is an affine continuous function and use then the exercise 10.

13. Verify that $\|\cdot\| = \|\cdot\|_1 + \|\cdot\|_2$ is an equivalent norm on $L^2[0,1]$ which is not G-differentiable.

14. Let $x, y \in X$; we say that x is orthogonal to y, denoted $x \perp y$, whenever $\|x\| \le \|x + \lambda y\|$, $\forall \lambda \in R$.
Prove that in a Hilbert space $x \perp y$ if and only if $<x,y>=0$.
Find examples showing that
$x \perp y$ need not imply $y \perp x$;
$x \perp y$ and $x \perp z$ need not imply that $x \perp (y + z)$.

15. Let $x^* \in X^*$; then $x \perp \ker x^*$ if and only if $|<x^*, x>| = \|x\|\,\|x^*\|$.

Hint. See Theorem 3, §1., Ch. II., Diestel [1].

16. A Banach space is smooth if and only if for any $0 \neq x \in X$ and $y \in X$, there exists a unique $\alpha \in R$ such that $x \perp \alpha x + y$.

Hint. See Theorem 4, §1., Ch. II., Diestel [1].

17. No selection of a set-valued duality mapping can be norm-to weak* continuous.

Hint. Let $\tilde{J}$ be a selection of J, $x^* \in Jx$ and $x^* \neq \tilde{J}x$; since J is monotone, then $<\tilde{J}(x + ty) - x^*, y> \ge 0$, $\forall y \in X$, $t \ge 0$.
Supposing $\tilde{J}$ has the above continuity property, we get $\tilde{J}x = x^*$ (See Gossez [1]).

18. The norm in ℓ^1 is not F-differentiable at any point.

Hint. See Example 1.14.a), Phelps [1].

19. If $f:X \to R$ is continuous and convex, then its subdifferential map is norm-to-weak* upper semi continuous.

Hint. For this generalization of Theorem 4.12 see Proposition 1.5., Phelps [1].

20. Let X, Y be two vector topological spaces and $D \subset Y$; we introduce the following notations:

$$\overset{-1}{T_+}(D) = \{x \in X; Tx \subset D\} \quad \overset{-1}{T_-}(D) = \{x \in X; Tx \cap D \neq \phi\}.$$

Prove that the following statements are equivalent:

a) T is upper-semicontinuous;

b) for any open set $V \subset Y, \overset{-1}{T_+}(V)$ is open in X;

c) for any closed set $W \subset Y, \overset{-1}{T_-}(W)$ is closed in X;

d) if $D \subset Y$, then $\overset{-1}{T_-}(\bar{D}) \supseteq \overline{\overset{-1}{T_-}(D)}$.

21. In the conditions of the exercise 20 we say that T is lower-semicontinuous at the point $x \in X$ if for any open set $V \subset Y$ such that $Tx \cap V \neq \phi$, there exists a neighborthood $U \in \mathcal{V}_x$ such that $Tx' \cap V \neq \Phi$ for each $x' \in U$. Prove that the following statement are equivalent:

a) T is lower-semicontinuous;

b) for any open set $V \subset Y, \overset{-1}{T(V)_-}$ is open in X;

c) for any closed set $W \subset Y, \overset{-1}{T_+}(W)$ is closed in X;

d) if $D \subset Y$, then $\overset{-1}{T_+}(\bar{D}) \supseteq \overline{\overset{-1}{T_+}(D)}$.

See Borisovich-Gel'man... [1].

BIBLIOGRAPHICAL COMMENTS

§1 and **§2**. Section 1 and 2 are devoted to a short introduction into convex analysis and the main properties of convex function in a Banach space are presented. For a study of convex functions on an interval of the real line one can consult Bourbaki [1]. The subdifferential map of a convex function was studied by Asplund [3], Asplund-Rockafellar [1], Brondsted-Rockafellar [1], Minty [1], [4], Moreau [1], [2], [3], Rockafellar [2], [4], [7]; the results on conjugate convex functions are to be found in Brondsted [1], Mosco [2], Rockafellar [1]. For a more detailed study we recommend the monographs of Barbu-Precupanu [1], Ekeland-Temam [1], Phelps [1], Giles [1], Holmes [1], Köning [1], Vainberg [1].
For the Gateaux and Fréchet differentiability one can consult the books of Dunford-Schwartz [1], Ladas-Laksmikantham [1], De Figueiredo. Many interesting applications and examples are to be found in Phelps [1].

§3. The study of the properties of Banach spaces connected with the differentiability of the norm was initiated by Mazur [1], Smulian [2] and developed by Cudia [1], [2], Day [4], Godini [1], Lindenstrauss [1]. Smooth Banach spaces were introduced by Krein [1] and studied later by Klee [2], among others.
The results in this section can be found in Beauzamy [1] and Diestel [1]. The importance of the modulus of smoothness will be emphasized in the second section of the next chapter where it is to be seen that they are the dual of the modulus of convexity and therefore usefull in the study of the uniform convexity; see also Ditzian-Totik [1].

§4. The concept of duality mapping is due to Beurling-Livingston [1] and was carried out by Asplund [2], Browder [4], [5], [15] and Browder-De Figueiredo [1]. General properties of the duality mappings can be found in De Figueiredo [1]. The characterization of the duality mappings as the subdifferential of a convex function in Theorem 4.4. is due to Asplund [2] and the general continuity property in Theorem 4.2. was obtained by Browder [15]. We also mention that the duality mappings in nonreflexive Banach spaces were considered by Gossez [3].

§5. Positive duality mappings were studied by Calvert [1], [2] in connection with the property of T-accretivity which will be considered in §1, Chapter VI; for the results presented here one can consult Picard [1]. For general properties of the Banach lattices we recommend the books of Cristescu [1] and Schaefer [1].

CHAPTER II

CHARACTERIZATIONS OF SOME CLASSES OF BANACH SPACES BY DUALITY MAPPINGS

In this chapter we shall characterize some classes of Banach spaces, among which strictly convex spaces, uniformly convex spaces and reflexive Banach spaces in terms of properties of the duality mapping such as continuity, injectivity or surjectivity. Some applications to L^p and l^p spaces are given.

In what follows we shall suppose, without restriction of the generality, that X is a real Banach space.

§.1. STRICTLY CONVEX BANACH SPACES

1.1. DEFINITION. We say that X is strictly convex if for all $x, y \in X, x \neq y,\ \|x\| = \|y\| = 1$, one has $\|yx + (1-\lambda)y\| < 1,\ \ \forall \lambda \in (0,1)$.

1.2. PROPOSITION: The following assertions are equivalent:
i) X is strictly convex.
ii) The boundary of the unit ball contains no line segments.
iii) If $x \neq y$ and $\|x\| = \|y\| = 1$, then $\|x + y\| < 2$.
iv) If for $x, y, z \in X$ we have $\|x - y\| = \|x - z\| + \|z - y\|$, then there exists $\lambda \in [0,1]$ so that $z = \lambda x + (1-\lambda)y$.
v) Any $x^* \in X^*$ assumes its sumpremum at most in one point of the unit ball.

PROOF. The implications i) $\Rightarrow$ ii), i) $\Rightarrow$ iii) and iii) $\Rightarrow$ ii) are straighforward.

ii) $\Rightarrow$ i) Let $x, y \in X, x \neq y, \|x\| = \|y\| = 1$ and $\lambda_o \in (0,1)$ such that $\lambda_o x + (1-\lambda_o)y = 1$; we shall prove that the segment line $[x,y]$ is on the unit ball, which is impossible.

Take $\lambda_o < \lambda < 1$; then from

$$\lambda_o x + (1-\lambda_o)y = \lambda_o/\lambda[\lambda x + (1-\lambda)y] + (1-\lambda_o/\lambda)y$$

we obtain

$$1 = \|\lambda_o x + (1-\lambda_o)y\| \leq \lambda_o/\lambda\|\lambda x + (1-\lambda)y\| + 1 - \lambda_o/\lambda.$$

This yields $\|\lambda x + (1-\lambda)y\| \geq 1$, hence: $\|\lambda x + (1-\lambda)y\| = 1$.

The situation $0 < \lambda < \lambda_o$ can be proved analogously.

i) $\Rightarrow$ iv) Let $x, y, z \in X$ so that $\|x-y\| = \|x-z\| + \|z-y\|$; we can suppose $\|x-z\| \neq 0, \|z-y\| \neq 0$ and $\|x-z\| \leq \|z-y\|$. We have

$$\left\|\frac{1}{2}\cdot\frac{x-z}{\|x-z\|} + \frac{1}{2}\cdot\frac{z-y}{\|z-y\|}\right\|$$

$$\geq \left\|\frac{1}{2}\cdot\frac{x-z}{\|x-z\|} + \frac{1}{2}\cdot\frac{z-y}{\|x-z\|}\right\| - \left\|\frac{1}{2}\cdot\frac{z-y}{\|x-z\|} - \frac{1}{2}\cdot\frac{z-y}{\|z-y\|}\right\|$$

$$= \frac{1}{2}\cdot\frac{\|x-y\|}{\|x-z\|} - \frac{1}{2}\cdot\frac{\|z-y\| - \|x-z\|}{\|x-z\|}$$

$$= \frac{1}{2}\cdot\frac{\|x-z\| + \|z-y\| - \|z-y\| + \|x-z\|}{\|x-z\|} = 1.$$

Hence $\left\|\frac{x-z}{\|x-z\|} + \frac{z-y}{\|z-y\|}\right\| = 2$. Then: $\frac{x-z}{\|x-z\|} = \frac{z-y}{\|z-y\|}$, and this yields

$$z = \frac{\|z-y\|}{\|x-z\| + \|z-y\|}.x + \frac{\|x-z\|}{\|x-z\| + \|z-y\|}.y.$$

iv) $\Rightarrow$ iii) Consider $x \neq y$ such that $\|x\| = \|y\| = \left\|\frac{x+y}{2}\right\| = 1$; then $\|x+y\| = \|x\| + \|y\|$. Consequently there is $\lambda \in (0,1)$ so that $z = 0 = \lambda x - (1-\lambda)y$, i.e. $x = \frac{1-\lambda}{\lambda}y$. Hence $\lambda = \frac{1}{2}$ so that x=y, which is a contradiction.

i) $\Rightarrow$ v) Suppose that for $x^* \in X^*$ there are two vectors $x_1 \neq x_2$, $\|x_1\| = \|x_2\| = 1$ with $<x^*, x_1> = <x^*, x_2> = \|x^*\|$. For $\lambda \in (0,1)$ we have

$$\|x^*\| \|\lambda x_1 + (1-\lambda)x_2\| \geq < x^*, \lambda x_1 + (1-\lambda)x_2 >$$
$$= \lambda < x^*, x_1 > + (1-\lambda) < x^*, x_2 > = \|x^*\|.$$

Then $1 \leq \|\lambda x_1 + (1-\lambda)x_2\| < 1$, which is absurd.

v) $\Rightarrow$ iii) Let be $x, y \in X$, so that $x \neq y, \|x\| = \|y\| = 1$ and $\|x+y\| = 2$. By the Hahn-Banach Theorem there exists an $x^* \in X^*$ such that $\|x^*\| = 1$ and $< x^*, \frac{x+y}{2} > = \left\|\frac{x+y}{2}\right\| = 1$, hence $< x^*, x > + < x^*, y > = 2$. As $< x^*, x > \leq 1$ and $< x^*, y > \leq 1$, it follows that

$$< x^*, x > = < x^*, y > = \|x^*\| = 1$$

contradicting iii). ■

1.3. THEOREM i) If X^* is smooth, then X is strictly convex.
ii) If X^* is strictly convex, then X is smooth.

PROOF. i) Supose that X^* is smooth and X is not strictly convex; then by Proposition 1.2. (v), there exists $x^* \in X^*$ and $x_1 \neq x_2, \|x_1\| = \|x_2\| = 1$ with $< x^*, x_1 > = < x^*, x_2 > = \|x^*\|$. But since $x_1, x_2 \in X^{**}$, X^* cannot be smooth.

ii) Let X^* be strictly convex and suppose X is not smooth; then there exists $x \in X$ and $x_1^*, x_2^* \in X^*, x_1^* \neq x_2^*$ with $\|x_1^*\| = \|x_2^*\| = 1$ so that $< x_1^*, x > = < x_2^*, x > = \|x\|$. This contradicts Proposition 1.2. (v) applied to X^*. Consequently X^* can not be strictly convex. ■

1.4. COROLLARY. Let X be reflexive; then X is strictly convex (respectively smooth) if and only if X^* is smooth (respectively strictly convex).

1.5. COROLLARY. Let X be a Banach space with X^* strictly convex; then any duality mapping is single-valued and norm-to-weak*-continuous.

PROOF. The statement is a consequence of the above theorem, Corollary 4.5 and theorem 4.12, Ch. I. ■

EXAMPLES. Each Hilbert space is strictly convex since by Remark 3.2, Ch. I., it is smooth.

We shall see later in §4 that the ℓ^p and L^p spaces with $1 < p < +\infty$ are strictly convex.

The spaces ℓ^1 and ℓ^∞ are not strictly convex.

Indeed, take $e_1=(1,0, \ldots)$ and $e_2=(0, 1, 0, \ldots)$, then $\|e_1\|_1 = \|e_2\|_1 = 1$, and $\|e_1 + e_2\|_1 = 2$, hence l^1 is not strictly convex.

Consider further $x=e_1+e_2$ and $y = e_1-e_2$; then $\|x\|_\infty = \|y\|_\infty = 1$ and $\|x+y\|_\infty = 2$, i.e. l^∞ is not stricly convex. (Here $\|\cdot\|_1$ and $\|\cdot\|_\infty$ are the l^1 and l^∞ norms, respectively).

Consequently R^N, $N>1$, with respect to the norms $\|x\|_1 = \sum_{i=1}^{N} |x_i|$ and $\|x\|_\infty = \max_{1 \le i \le N} |x_i|$ is not strictly convex.

1.6. PROPOSITION. The Banach space X is strictly convex if and only if the function $h(x) = \|x\|^2$ is strictly convex.

PROOF. Suppose X strictly convex and let be $x, y \in X$, $\lambda \in (0,1)$; we have

$$\begin{aligned}
&\|\lambda x + (1-\lambda)y\|^2 \le (\lambda\|x\| + (1-\lambda)\|y\|)^2 \\
&= \lambda^2\|x\|^2 + 2\lambda(1-\lambda)\|x\|.\|y\| + (1-\lambda)^2\|y\|^2 \\
&\le \lambda^2\|x\|^2 + \lambda(1-\lambda)\left(\|x\|^2 + \|y\|^2\right) + (1-\lambda)\|y\|^2 \\
&= \lambda\|x\|^2 + (1-\lambda)\|y\|^2. \qquad (1.1)
\end{aligned}$$

Hence the function h is convex.

Suppose now that there are $x, y \in X, x \neq y$ with

$$\|\lambda_o x + (1-\lambda_o)y\|^2 = \lambda_o\|x\|^2 + (1-\lambda_o)\|y\|^2 \text{ for some } \lambda_o \in (0,1).$$

Then, from (1.1) we obtain $2\|x\|\,\|y\| = \|x\|^2 + \|y\|^2$.

Hence $\|x\| = \|y\| = \|\lambda_o x + (1-\lambda_o)y\|$, which is impossible.

Conversely, let h be strictly convex and $x, y \in X$ so that $x \neq y$, $\|x\| = \|y\| = 1$ and with $\|\lambda x + (1-\lambda)y\| = 1$ for some $\lambda \in (0,1)$. Then

$$\|\lambda x + (1-\lambda)y\|^2 = 1 = \lambda\|x\|^2 + (1-\lambda)\|y\|^2$$

which is a contradiction. ■

1.7. LEMMA. Consider a duality mapping J of weight φ on X, a sequence $\{x_n\} \subseteq X$ such that $x_n \xrightarrow{n} x_o$ and $x_n^* \in Jx_n, n \in N$; then there exists a subsequence $\{x^*_{n'}\} \subseteq \{x^*_n\}$ and $x^*_o \in Jx_o$ so that $<x^*_{n'}, x_{n'}> \xrightarrow[n']{} <x^*_o, x_o>$ and $<x^*_{n'}, x_k> \xrightarrow[n']{} <x^*_o, x_k>$, $k \in N \cup \{0\}$.

PROOF. Let X_0 be the linear closed subspace generated by $x_0, x_1, x_2, \ldots$. Since X_0 is separable, the unit ball is compact and metrizable in the weak* topology on X^*_0. Consider the restrictions of x^*_n, $n \in N$, to X_0 and denote them again for simplicity by x^*_n. As the sequence $\{x^*_n\}_n$ is bounded in X^*_0, there exists a subsequence $\{x^*_{n'}\}_{n'} \subset \{x^*_n\}_n$ and $x^*_o \in X^*_o$ so that

$$<x^*_{n'}, x> \xrightarrow[n']{} <x^*_o, x>, \quad \forall x \in X_o.$$

But then

$$<x^*_{n'}, x_{n'}> - <x^*_o, x_o>$$

$$= <x^*_{n'}, x_{n'}> - <x^*_{n'}, x_o> + <x^*_{n'}, x_o> - <x^*_o, x_o> \xrightarrow[n']{} 0.$$

By the Hahn-Banach Theorem, we can extend x^*_o to all X and denote thus extention also by x^*_o. Then

$$\|x^*_o\| \le \lim_{n'} \|x^*_{n'}\| = \lim_{n'} \varphi(\|x_{n'}\|) = \varphi(\|x_o\|),$$

hence $\|x^*_o\| \le \varphi(\|x_o\|)$. Moreover

$$<x^*_o, x_o> = \lim_{n'} <x^*_{n'}, x_{n'}> = \lim_{n} \varphi(\|x_{n'}\|)\|x_{n'}\| = \varphi(\|x_o\|)\|x_o\|.$$

This yields $\|x^*_o\| = \varphi(\|x_o\|)$, so that $x^*_o \in Jx$. ■

1.8. THEOREM. (Petryshyn). Let J be a duality mapping of weight φ on the Banach space X; then X is strictly convex if and only if J is strictly monotone, i.e.

$$<x^* - y^*, x - y> > 0, \text{ whenever } x, y \in X, x \neq y \text{ and } x^* \in Jx, y^* \in Jy.$$

PROOF. Suppose X is strictly convex; as in the proof of the Proposition 4.7. b), Ch. I. we can prove that

$$<x^* - y^*, x - y> \ge (\varphi(\|x\|) - \varphi(\|y\|)(\|x\| - \|y\|), \; x^* \in Jx, y^* \in Jy. \quad (1.2)$$

Suppose that for some $x, y \in X$ such that $x \neq y$, $x^* \in Jx$ and $y^* \in Jy$, we have

$$<x^* - y^*, x - y> = 0 \quad . \quad (1.3)$$

Then by (1.2) we must have $\|x\| = \|y\|$. Since $\|x^*\| = <x^*, x_{/\|x\|}>$, it follows

$$<x^*, y_{/\|y\|}> < \|x^*\|, \quad \text{i.e.} \quad <x^*, y> < \|x^*\|.\|y\|.$$

Analogously we obtain $<y^*, x> < \|y^*\|.\|x\|$ and consequently

$$<x^* - y^*, x - y> = \|x^*\|\,\|x\| + \|y^*\|.\|y\| - <x^*, y> - <y^*, x>$$
$$> \|x^*\|\,.\|x\| + \|y^*\|.\|y\| - \|x^*\|.\|y\| - \|y^*\|.\|x\| = 0.$$

This contradicts (1.3).
Conversely, suppose that J is strictly monotone but X is not strictly convex. Select two elements $x, y \in X$, $x \neq y, \|x\| = \|y\| = 1$ with $\|\lambda x + (1-\lambda)y\| = 1$, $\forall \lambda \in (0,1)$; then, for $\lambda_n \xrightarrow[n]{} 0$, $\lambda_n \in (0,1)$; we have $\|y + \lambda_n(x-y)\| = 1$, $n \in N$.
Consider $y^* \in Jy$ and $x_n^* \in J(y + \lambda_n(x-y))$; then we may write

$$< y^*, \lambda_n(x-y) > \leq \psi(\|y + \lambda_n(x-y)\|) - \psi(\|y\|) = 0$$

and

$$< x_n^*, -\lambda_n(x-y) > \leq -\psi(\|y + \lambda_n(x-y)\|) + \psi(\|y\|) = 0.$$

Hence

$$< y^*, x - y > \leq 0 \text{ and } < x_n^*, x - y > \geq 0. \tag{1.4}$$

Let now $x_n = y + \lambda_n(x-y)$; then $x_n \xrightarrow[n]{} y$ and by Lemma 1.7 there is a subsequence $\{x_{n'}^*\}_{n'} \subseteq \{x_{n'}^*\}_n$ and $y_o^* \in Jy$ such that

$$< x_{n'}^* x - y > \xrightarrow[n']{} y_o^*, x - y.$$

Then (1.4) implies $< y_o^*, x - y > = 0$.
If we interchange the role of x and y in the above argument, we obtai the existence of an $x_o^* \in Jx$ so that $< x_o^*, y - x > = 0$.
Hence $< y_o^* - x_o^*, y - x > = 0$, i.e. J is not strictly montone. ■

1.9. COROLLARY. All duality mappings on a strictly convex Banach space are strictly monotone.

As a direct consequence of the above Theorem and Corollary 4.5, Ch. I, we have the following

1.10. THEOREM. Let X be a Banach space; then X is smooth and strictly convex if and only if there is a duality mapping on X which is single-valued and strictly monotone.

1.11. PROPOSITION. Let X be a Banach space with a strictly convex dual, J a duality mapping on X, $K \subseteq X$ a convex set and $x_o \in K$; then

$$\|x_o\| = \inf_{x \in K} \|x\| \text{ if and only if } < Jx_o, x_o > \leq < Jx_o, x >, \quad \forall x \in K.$$

PROOF. Let be $x_o \in K$ with $< Jx_o, x_o > \leq < Jx_o, x >$, $\forall x \in K$; then

$$\|x_o\| \cdot \|Jx_o\| \leq \|x\| \cdot \|Jx_o\|, \text{ i.e. } \|x_o\| \leq \|x\|, \quad \forall x \in K$$

and therefore $\|x_o\| = \inf_{x \in K} \|x\|$.
Conversely, suppose that $x_o \in K$ is so that $\|x_o\| = \inf_{x \in K} \|x\|$; then we have

$$\|x_o + \lambda(x - x_o)\| \geq \|x_o\|, \text{ for any } x \in K, \lambda \in [0,1].$$

If φ is the weight of J and $\psi(t) = \int_0^t \varphi(s)ds$, then

$$0 \geq \psi(\|x_0\|) - \psi(\|x_0 + \lambda(x - x_0)\|) \geq < J(x_0 + \lambda(x - x_0)), \lambda(x - x_0) >$$

Hence

$$< J(x_0 + \lambda(x - x_0)), x_0 - x > \leq 0, \quad \forall \lambda \in [0,1]$$

and letting now $\lambda \to 0$, we obtain

$$< Jx_0, x_0 > \leq < Jx_0, x >, \quad \forall x \in K.$$

■

1.12. PROPOSITION. Let X be a smooth Banach space; then for any duality mapping J of weight φ, we have

$$\psi(\|x + y\|) = \psi(\|x\|) + \int_0^1 < J(x + ty), y > dt, \quad x, y \in X.$$

PROOF. By Corollary 4.5, Ch. I, we have

$$< Jx, y > = \frac{d}{dt}\psi(\|x + ty\|)_{|t=0}.$$

Consequently

$$\frac{d}{dt}\psi(\|x + ty\|)_{|t=u} = \frac{d}{ds}\psi(\|x + uy + sy\|)_{|s=0} = < J(x + uy), y >, u \in R.$$

By the theorem 4.12, Ch. I., the function $t \to < J(x + ty), y >$ is continuous; hence

$$\int_0^1 < J(x + uy), y > du = \int_0^1 \frac{d}{dt}\psi(x + ty)_{|t=u} du = \psi(\|x + y\|) - \psi(\|x\|).$$

■

1.13. COROLLARY. Let X^* be strictly convex; then the statement of Proposition 1.12 is true.

§.2. UNIFORMLY CONVEX BANACH SPACES.

2.1. DEFINITION. A Banach space X is said to be:

i) uniformly convex if for every sequences $\{x_n\}, \{y_n\} \subseteq X$ with $\|x_n\| = \|y_n\| = 1, n \in N$ and $\|x_n + y_n\| \xrightarrow[n]{} 2$, we have $\|x_n - y_n\| \xrightarrow[n]{} 0$;

ii) uniformly convex at $x \in X, \|x\| = 1$, if for every sequence $\{x_n\} \subseteq X$ with $\|x_n\| = 1, n \in N$ and $\|x_n + x\| \xrightarrow[n]{} 2$, we have $\|x_n - x\| \xrightarrow[n]{} 0$;

locally uniformly convex if it is uniformly convex at any $x \in X, \|x\| = 1$;

iii) weakly uniformly convex at $x^* \in X^*, \|x^*\| = 1$, if for every sequences $\{x_n\}, \{y_n\} \subseteq X$, with $\|x_n\| = \|y_n\| = 1, n \in N$ and $< x^*, x_n + y_n > \xrightarrow[n]{} 2$, we have $\|x_n - y_n\| \xrightarrow[n]{} 0$;

weakly uniform convex if it is uniformly convex at any $x^* \in X^*, \|x^*\| = 1$.

2.2. REMARK. It is clear that the uniform convexity implies both the local and the weak uniform convexity. Moreover we have the following geometric interpretation:
Let X be uniformly convex; if the sequence of the middle points of the line segments $[x_n, y_n]$, $\|x_n\| = \|y_n\| = 1$ converges to a point of the boundary of the unit ball, then the lenghts of the segments converges to zero.
A simple example of uniformly convex spaces are the Hilbert spaces as the paralelogram's identity shows.

2.3. PROPOSITION. A Banach space X is
i) uniformly convex if and only if for each $\varepsilon > 0$, there is $\delta > 0$, so that $\|x\| = \|y\| = 1$ and $\|x - y\| \geq \varepsilon$ implies $\|x + y\| \leq 2(1 - \delta)$;
ii) uniformly convex at $x \in X, \|x\| = 1$ if and only if for each $\varepsilon > 0$, there is d= d(x)>0 so that $\|y\| = 1$ and $\|y - x\| \geq \varepsilon$ implies $\|x + y\| \leq 2(1 - \delta)$.
iii) weakly uniformly convex at $x^* \in X^*, \|x^*\| = 1$ if and only if for each $\varepsilon > 0$, there is $\delta = \delta(x^*) > 0$ so that $\|x\| = \|y\| = 1$ and $\|y - x\| \geq \varepsilon$ implies $< x^*, x + y > \leq 2(1 - \delta)$.

PROOF. i) Let be X uniformly convex and assume that the property in i) is not true; then there exists $\varepsilon > 0$, such that for every $\delta = \frac{1}{n}$, there are $\|x_n\| = \|y_n\| = 1$, with $\|x_n - y_n\| \geq \varepsilon$, but $\|x_n + y_n\| \geq 2\left(1 - \frac{1}{n}\right)$, $n \in N$. Then $\|x_n + y_n\| \xrightarrow[n]{} 2$, hence $\|x_n - y_n\| \xrightarrow[n]{} 0$ and this is impossible.
Conversely, suppose that X has the property i) but is not uniformly convex; then there are $\|x_n\| = \|y_n\| = 1$ with $\|x_n + y_n\| \xrightarrow[n]{} 2$ but $\|x_n - y_n\| \not\xrightarrow[n]{} 0$. Hence there is $\varepsilon > 0$, so that $\|x_n - y_n\| \geq \varepsilon, \forall n \in N$. By hypothesis there is $\delta > 0$ with $\|x_n + y_n\| \leq 2(1 - \delta)$ and this is a contradiction.
We can prove ii) and iii) analogously. ■

2.4. REMARK. One can replace in i) the condition $\|x\| = \|y\| = 1$ by $\|x\| \leq 1, \|y\| \leq 1$. (Exercise 8)

2.5 NOTATIONS. For $0 < \varepsilon \leq 2$ we define the following functions
i) the modulus of uniform convexity of the Banach space X

$$\Delta(\varepsilon) = \frac{1}{2} \inf_{\substack{\|x\|=\|y\|=1 \\ \|x-y\| \geq \varepsilon}} (2 - \|x + y\|);$$

ii) the modulus of uniform convexity at $x \in X, \|x\| = 1$

$$\Delta(\varepsilon,x) = \frac{1}{2} \inf_{\substack{\|y\|=1 \\ \|x-y\|\geq\varepsilon}} (2 - \|x+y\|), \quad 0<\varepsilon\leq 2;$$

iii) the modulus of weakly uniform convexity at $x^* \in X^*, \|x^*\| = 1$

$$\Delta(\varepsilon,x^*) = \frac{1}{2} \inf_{\substack{\|y\|= \|y\|=1 \\ \|x-y\|\geq\varepsilon}} (2 - \langle x^*, x+y \rangle), \quad 0<\varepsilon\leq 2.$$

Sometimes, in order to avoid possible confusions, we shall use the notation Δ_X for the modulus of uniform convexity of the Banach space X.

2.6. COROLLARY. X is uniformly convex, respectively locally uniformly convex, respectively weakly uniformly convex, if and only if for every $\varepsilon \in (0,2]$ we have: $\Delta(\varepsilon) > 0$, respectively $\Delta(\varepsilon,x) > 0 \ \forall \|x\| = 1$, respectively $\Delta(\varepsilon,x^*) > 0 \ \forall \|x^*\| = 1$.

2.7. PROPOSITION. A locally uniformly convex space is strictly convex.

PROOF. Let $x, y \in X, x \neq y$ with $\|x\| = \|y\| = 1$; then there is $\varepsilon > 0$ so that $\|y-x\| \geq \varepsilon$. Proposition 2.3. i) asserts the existence of a d such that $\|x+y\| \leq 2(1-\delta) < 2$. Then $\|x+y\| < 2$ and by Proposotion 1.2. (iii) X is strictly convex. ■

2.8. PROPOSITION. Let X be locally uniformly convex and $\{x_n\} \subseteq X$ a sequence such that

a) $\{x_n\}$ converges weakly to some $x \in X$ (we write $x_n \xrightarrow[n]{w} x$);

b) $\|x_n\| \xrightarrow[n]{} \|x\|$;

then $x_n \xrightarrow[n]{} x$.

PROOF. We can suppose without restriction that $x, x_1 \neq 0$. If we denote $y = x/\|x\|, y_n = x_n/\|x_n\|, n \in N$, then $\|y\| = \|y_n\| = 1, n \in N$ and $y_n \xrightarrow[n]{w} y$. This yields

$$2 = 2\|y\| \leq \underline{\lim_n}\|y_n + y\| \leq \overline{\lim_n}\|y_n + y\| \leq \|y\| + \lim_n\|y_n\| = 2.$$

Thus $\lim_n \|y_n + 2\| = 2$. As X is locally uniformly convex, it follows that $\lim_n \|y_n - y\| = 0$; it is now an easy matter to obtain that $\lim_n \|x_n - x\| = 0$.

■

2.9. THEOREM. (Milman-Pettis) A uniformly convex Banach space is reflexive.

PROOF. Let be $x^{**} \in X^{**}$ with $\|x^{**}\| = 1$. By Goldstine's theorem, the unit ball of X is weak*-dense in the ball of X^{**}, so that we can find a net $\{x_\iota\}_{\iota \in I}$ with

$$\|x_\iota\| \le 1, \forall\, \iota \in I \text{ and } < x^*, x_\iota > \xrightarrow[\iota]{} < x^*, x^{**} >, \forall\, x^* \in X^*. \tag{2.1}$$

As $1 = \|x^{**}\| \le \underline{\lim}_\iota \|x_\iota\| \le \overline{\lim}_\iota \|x_\iota\| \le 1$, then $\lim_\iota \|x_\iota\| = 1$. A simple computation shows now that we can suppose $\|x_\iota\| = 1, \forall\, \iota \in I$.
Fix $\varepsilon > 0$ and take $\delta > 0$ corresponding to e in the property of the uniform convexity. There exist $x_0^* \in X^*$ and $\iota_0 \in I$ so that $\|x_0^*\| = 1$, $< x_o^*, x^{**} > > 1 - \delta$ and $< x_o^*, x_\iota > > 1 - \delta$ for $\iota > \iota_0$.
Thus $< x_o^*, \frac{x_\iota + x_{\iota'}}{2} > > 1 - \delta$, $\iota' > \iota_0$, and this yields $\|x_\iota + x_{\iota'}\| > 2(1 - \delta)$. Then, by the choice of δ, $\|x_\iota + x_{\iota'}\| < \varepsilon$ for $\iota, \iota' > \iota_0$, such that the net $\{x_\iota\}_{\iota \in I}$ is (norm) Cauchy. Since X is complete, there is $x \in X$ so that $x_\iota \xrightarrow[\iota]{} x$. It follows from (2.1) that $x = x^{**}$, hence X is reflexive. ■

2.10. DEFINITION. A convex function $f : X \to R$ is called
i) uniformly strictly convex if for every $\varepsilon > 0$.

$$\inf_{\substack{\|x\|=1 \\ \|x-y\| \ge \varepsilon}} \left\{ f(x) - 2f\left(\tfrac{x+y}{2}\right) + f(y) \right\} > 0 \tag{2.2}$$

ii) Locally uniformly strictly convex if for any $x \in X$.

$$\inf_{\|x-y\| \ge \varepsilon} \left\{ f(x) - 2f\left(\tfrac{x+y}{2}\right) + f(y) \right\} > 0 \tag{2.3}$$

2.11. PROPOSITION. A Banach space is (locally) uniformly convex if and only if the function $f(x) = \frac{1}{2}\|x\|^2$ is (locally) strictly convex.

PROOF. Suppose that (2.3) is true; then for $x, y \in X$ with $\|x\| = \|y\| = 1$, we have $\inf_{\|x-y\| \ge \varepsilon} \left\{ 4 - \|x + y\|^2 \right\} > 0$, and this implies that X is locally uniformly convex.
Suppose now X locally uniformly convex and let $\varepsilon > 0$ and $x, y \in X$, with $\|x - y\| \ge \varepsilon$. Denote $u = x/_{\|x\|}$, $v = y/_{\|y\|}$ and $\varepsilon_0 = \varepsilon/_{\|x\|}$; then there exists $\delta > 0$ such that for $z \in X, \|z\| = 1$ and $\|u - z\| \ge \varepsilon_0$ we have $1 - \left\| \frac{u + z}{2} \right\| > \delta$.
We shall distinct three cases:

1) Suppose $\|x\| - \|y\| \geq \varepsilon/2$. We have

$$\frac{1}{2}\|x\|^2 - \left\|\frac{x+y}{2}\right\|^2 + \frac{1}{2}\|y\|^2 \geq \frac{1}{2}\|x\|^2 - \frac{1}{4}(\|x\| + \|y\|)^2 + \frac{1}{2}\|y\|^2 = \frac{1}{4}(\|x\| - \|y\|)^2 \geq \frac{\varepsilon^2}{16}$$

hence f satisfies (2.3).

II) Suppose $0 \leq \|x\| - \|y\| < \frac{\varepsilon}{2}$.

Then $\alpha = \|x\|/_{\|y\|} \geq \frac{\varepsilon}{2}\,1$ and $x + y = \|y\|(u + v + (\alpha - 1)u)$. We have

$$\zeta(x,y) = \frac{1}{2}\|x\|^2 - \left\|\frac{x+y}{2}\right\|^2 + \frac{1}{2}\|y\|^2$$

$$= \|y\|^2\left[\frac{1+\alpha^2}{2} - \left\|\frac{u+v+(\alpha-1)u}{2}\right\|^2\right] \geq \|y\|^2\left[\frac{1+\alpha^2}{2} - \left(\left\|\frac{u+v}{2}\right\| + \frac{\alpha-1}{2}\right)^2\right]$$

$$= \|y\|^2\left[\left(\frac{1+\alpha^2}{2}\right)^{1/2} + \left(\left\|\frac{u+v}{2}\right\| + \frac{\alpha-1}{2}\right)\right]\left[\left(\frac{1+\alpha^2}{2}\right)^{1/2} - \left(\left\|\frac{u+v}{2}\right\| + \frac{\alpha-1}{2}\right)\right]$$

$$\geq \|y\|^2\left(\alpha + \left\|\frac{u+v}{2}\right\|\right)\left(1 - \left\|\frac{u+v}{2}\right\|\right) \geq \|y\|^2\left(1 - \left\|\frac{u+v}{2}\right\|\right).$$

We used above the following inequalites:

$$\left(\frac{1+\alpha^2}{2}\right)^{1/2} + \frac{\alpha-1}{2} \geq \alpha \text{ and } \left(\frac{1+\alpha^2}{2}\right)^{1/2} + \frac{\alpha-1}{2} \geq 1.$$

Further we note that

$$\|u - v\| = \frac{1}{\|x\|}\left\|x - y - \frac{\|y\| - \|x\|}{y}.y\right\| \geq \frac{1}{\|x\|}\cdot[\|x - y\| + (\|x\|) - \|y\|] \geq \frac{\varepsilon}{\|x\|} = \varepsilon_0.$$

Hence $1 - \left\|\frac{u-v}{2}\right\| \geq \delta$. Therefore if $\|y\| \geq \frac{\varepsilon}{2}$, then $\zeta(x,y) \geq \frac{\varepsilon^2}{4}.\delta > 0$.

If $\|y\| \geq \frac{\varepsilon}{2}$, then the relation $\|x - y\| \geq \varepsilon$ implies that $\|x\| > \frac{\varepsilon}{2}$; hence $\|y\| > \|x\| - \frac{\varepsilon}{2} > 0$ such that $\zeta(x,y) \geq \left(\|x\| - \frac{\varepsilon}{2}\right)^2 \delta > 0$.

III) The case $0 \leq \|y\| - \|x\| < \frac{\varepsilon}{2}$ can be treated analogously.

Finally we remark that one can prove in a similar way the assertion relative to the uniform convexity. ■

In order to establish the connection between the uniform convexity and the differentiability of the norm on a Banach space, we give the

2.12. PROPOSITION. (Lindenstrauss' duality formulas). For each $\tau > 0, x \in X, \|x\| = 1$ and $x^* \in X, \|x^*\| = 1$ we have:

$$\rho_{X^*}(\tau) = \sup_{0<\varepsilon\le 2}\left[\tfrac{\tau\varepsilon}{2} - \Delta_X(\varepsilon)\right]; \quad \rho_X(\tau) == \sup_{0<\varepsilon\le 2}\left[\tfrac{\tau\varepsilon}{2} - \Delta_{X^*}(\varepsilon)\right] \quad ; \tag{2.9}$$

$$\rho_{X^*}(\tau,x^*) = \sup_{0<\varepsilon\le 2}\left[\tfrac{\tau\varepsilon}{2} - \Delta_X(\varepsilon,x^*)\right]; \quad \rho_X(\tau,x) == \sup_{0<\varepsilon\le 2}\left[\tfrac{\tau\varepsilon}{2} - \Delta_{X^*}(\varepsilon,x)\right] . \tag{2.10}$$

PROOF. We shall prove only the first formula, all other having similar proofs.
Let be $\tau > 0, 0 < \varepsilon \le 2, x, y \in X, \|x\| = \|y\| = 1$ and $x_0^*, y_0^* \in X^*$ with $\|x_0^*\| = \|y_0^*\| = 1$ and $< x_0^*, x + y > = \|x + y\|, y_0^* - x + y = \|x - y\|$. Then

$$\begin{aligned}
2\rho_{X^*}(\tau) = & \sup_{\|x^*\|=\|y^*\|=1} (\|x^* + \tau y^*\| + \|x^* - \tau y^*\| - 2) \\
& \ge \|x_0^* + \tau y_0^*\| + \|x_0^* - \tau y_0^*\| - 2 \\
& \ge < x_0^* + \tau y_0^*, x > + < x_0^* - \tau y_0^*, y > - 2 \\
& = < x_0^*, x + y > + \tau < y_0^*, x - y > - 2 \\
& = \|x + y\| + \tau\|x - y\| - 2 \ge \|x + y\| + \tau\varepsilon - 2.
\end{aligned}$$

Hence $2 - \|x + y\| \ge \tau\varepsilon - 2\rho_{X^{**}}(\tau)$ and therefore $\Delta_X(\varepsilon) \ge \frac{\tau\varepsilon}{2} - \rho_{X^*}(\tau)$.

As $0 < \varepsilon \le 2$ was arbitrarily, it follows that $2\rho_{X^*}(\tau) \ge \sup_{0<\varepsilon\le 2}\left(\frac{\tau\varepsilon}{2} - \Delta_X(\varepsilon)\right)$.
Conversely, let be $x^*, y^* \in X^*$ with $\|x^*\| = 1, \|y^*\| = 1$ and $\delta > 0$. There exist $x_0, y_0 \in X, \|x_0\| = \|y_0\| = 1$ so that $\|x^* + \tau y^*\| \le < x^* + \tau y^*, x_0 > + \delta$ and $\|x^* - \tau y^*\| \le < x^* + \tau y^*, y_0 > + \delta$.
Denote $\varepsilon_0 = |< y^*, x_0, y_0 >|$; then $0 < \varepsilon_0 \le 2$ and $\|x_0 - y_0\| \ge \varepsilon_0$. Hence

$$\begin{aligned}
& \|x^* + \tau y^*\| + \|x^* - \tau y^*\| - 2 \\
& \le < x^*, x_0 + y_0 > + \tau < y^*, x_0 - y_0 > - 2 + 2\delta \\
& \le \|x_0 + y_0\| - 2 + \tau|< y^*, x_0 - y_0 >| + 2\delta \\
& \le \tau\varepsilon - 2\Delta_X(\varepsilon_0) + 2\delta \le 2 \sup_{0<\varepsilon\le 2}\left[\frac{\tau\varepsilon}{2} - \Delta_X(\varepsilon)\right] + 2\delta.
\end{aligned}$$

As $\delta > 0$ was arbitrarily, we obtain $\rho_{X^*}(\tau) \le \sup_{0<\varepsilon\le 2}\left[\frac{\tau\varepsilon}{2} - \Delta_X(\varepsilon)\right]$. ■

2.13. THEOREM. Let X be a Banach space; the following statements are equivalent:
i) X is uniformly convex (respectively weakly uniformly convex).
ii) X^* is uniformly smooth (respectively locally uniformly smooth).

PROOF. i) $\Rightarrow$ ii) Suppose that X is uniformly convex but X^* is not uniformly smooth, that is $\Delta_X(\varepsilon) > 0$, $\forall\, \varepsilon \in (0,2]$ and $\lim\limits_{\tau \to 0} \frac{\rho_{X^*}(\tau)}{\tau} = \alpha > 0$. Let be $\tau_n \to 0$, $\tau_n < 1$; by the first formula in (2.4) there are $\varepsilon_n \in (0,2]$ so that $\frac{\tau_n \varepsilon_n}{2} - \Delta_X(\varepsilon_n) > \frac{\alpha}{2}.\tau_n$, $\forall\, n \in N$. But this implies

$$0 < \Delta_X(\varepsilon_n) < \tau_n\left(\frac{\varepsilon_n}{2} - \frac{\alpha}{2}\right). \tag{2.6}$$

Hence $\varepsilon_n - \alpha > 0$, $\forall\, n \in N$, and consequently $\varepsilon_n > \alpha$.
Then, as Δ_X is non-increasing, we have that $\Delta_X(\alpha) \le \Delta_X(\varepsilon_n)$. This implies $\Delta_X(\alpha) \le \lim\limits_n \Delta_X(\varepsilon_n) = 0$, i.e. X is not uniformly convex.
ii) $\Rightarrow$ i) Suppose X^* uniformly smooth; if there is an $\varepsilon_o \in (0,2]$ with $\Delta_X(\varepsilon_o) = 0$, then by the first formula in (2.4) we obtain

$$\frac{\rho_{X^*}(\tau)}{\tau} \ge \frac{\varepsilon_o}{2} - \frac{\Delta_X(\varepsilon_o)}{\tau} = \frac{\varepsilon_o}{2} < 0, \quad \forall\, \tau > 0$$

which is impossible.
Using the first formula in (2.5) we can prove analagously the connection between the weak uniform convexity and the local uniform smoothness. ■

The following result completes the duality between the uniform smoothness and the uniform convexity.

2.14. THEOREM. Let X be a Banach space; the following statements are equivalent:
i) X is uniformly smooth (respectively locally uniformly smooth);
ii) X^* is uniformly convex (respectively weakly uniform convex at any $x \in X$, $\|x\| = 1$.

PROOF. We can prove the equivalence i) $\Leftrightarrow$ ii) interchanging the role of X and X^* in the above proof and using the second formula of (2.4). Using the second formula of (2.5) we can prove in a similar way the statement concerning the equivalence of the local uniform smoothness of X and the weak uniform convexity of X^*. ■

2.15. COROLLARY. Uniformly smooth Banach spaces are reflexive.

PROOF. It is well-known that a Banach space is reflexive whenenver X^* is so. If X is uniformly smooth, then X^* is uniformly convex, hence reflexive. ■

2.16. THEOREM. Let X be a Banach space and J a duality mapping of weight φ on X; then

i) X^* is uniformly convex if and only if J is single-valued and norm-norm uniformly continuous on the unit sphere of X.

ii) X^* is weakly uniformly convex at any $x \in X$ with $\|x\| = 1$, if and only if J is single-valued and norm-norm continuous from X into X^*.

PROOF. By Proposition 4.7. (f), Ch. I. we can suppose without restiction of the generality that J is the normalized duality mapping. Moreover we recall that if $f(x) = \|x\|$ and X is smooth, then

$$f'(x) = Jx/\|x\| \tag{2.7}$$

i) Suppse X^* uniformly convex and let $\varepsilon > 0$; then there is $\delta > 0$ such that if $x^*, y^* \in X^*, \|x^*\| = \|y^*\| = 1$, $\|x^* + y^*\| > 2 - \delta$, then we have $\|x^* - y^*\| < \varepsilon$.

Let $x, y \in X$, $\|x\| = \|y\| = 1$ and suppose $\|x - y\| < \delta$. Then

$$\|Jx + Jy\| \geq \langle Jx + Jy , y \rangle$$
$$= \langle Jx, x \rangle + \langle Jy , y \rangle - \langle Jx, x - y \rangle \geq 2 - \|x - y\| > 2 - \delta .$$

Hence, $\|Jx - Jy\| < \varepsilon$ i.e. J is the norm-norm uniformly continuous on the unit ball of X.

Conversely suppose that J is single-valued and norm-norm uniformly continuous on the unit sphere of X; then X is smooth and by (2.7) it follows that f'(x) has the same continuity property. Applying now Lemma 1.17, Ch. I, to $F(t) = \|x + ty\|$, $x, y \in X$ fixed, we obtain

$$F'(0) \leq \frac{F(t) - F(0)}{t} \leq F'(t), \quad t > 0.$$

This yields

$$\langle f'(x), y \rangle \leq \frac{\|x + ty\| - \|x\|}{t} \leq f''(x + ty), y \rangle \tag{2.8}$$

and thus the norm is uniformly Fréchet differentiable on the unit sphere.

Hence, by the theorem 3.12. Ch. I, X is uniformly smoothand and therefore by theorem 2.14, X^* is uniformly convex.

ii) Suppose X^* weakly uniformly convex at any $\|x\| = 1$ and let $x_n \xrightarrow[n]{} x$; put $y_n = x_n/\|x_n\|$ and $y = x/\|x\|$; we have

$2 \geq \langle Jy_n + Jy , y \rangle = \langle Jy_n, y_n \rangle + \langle Jy , y \rangle - \langle Jy_n, y_n - y \rangle \geq 2 - \|y_n - y\|$.

Hence $\langle Jy_n + Jy , y \rangle \xrightarrow[n]{} 2$. Then: $Jy_n \xrightarrow[n]{} Jy$ and this yields $Jx_n \to Jx$.

The converse assertion is a consequence of the relation (2.8), Theorem 3.12 Ch. I and Theorem 2.14. ■

We end our considerations in this section with the following property for the duality mappings.

2.17. PROPOSITION. Let X be smooth and locally uniformly convex; then for every duality mapping J on X and sequence $\{x_n\}_n \subset X$, we have

$$\langle Jx_n - Jx, x_n - x\rangle \xrightarrow[n]{} 0 \text{ implies } x_n \xrightarrow[n]{} x.$$

PROOF. The following identity

$$\langle Jx_n - Jx, x_n - x = (\varphi(\|x_n\|) - \varphi(\|x\|)(\|x_n\| - \|x\|) + (\varphi(\|x\|)\|x\| - \langle Jx, x_n\rangle) + (\varphi(\|x_n\|)\|x\| - \langle Jx_n - x\rangle)$$

yields

$$\|x_n\| \xrightarrow[n]{} \|x\| \text{ and } \langle Jx, x_n\rangle \xrightarrow[n]{} \|Jx\|\,\|x\| = \langle Jx, x\rangle.$$

Then $\langle Jx, x_n + x\rangle = \langle Jx, x_n\rangle + \langle Jx, x\rangle \xrightarrow[n]{} \|Jx\|\,\|x\|$.

On the other hand we have

$$\langle Jx, x_n + x\rangle \le \|x_n + x\|\,\|Jx\| \le (\|x_n\| + \|x\|).\|Jx\|$$

so that

$$\lim_n \langle Jx, x_n + x\rangle = 2\langle Jx, x\rangle = 2\|Jx\|.\|x\|$$
$$\le \underline{\lim}_n \|x_n + x\|.\|Jx\| \le \overline{\lim}_n \|x_n + x\|.2\|Jx\|$$
$$\le \lim_n(\|x_n\| + \|x\|)\|Jx\| = 2\|x\|.\|Jx\|.$$

Consequently $\lim_n \|x_n + x\| = 2.\|x\|$ and this implies $\lim_n \left\| \frac{x_n}{\|x_n\|} + \frac{x}{\|x\|} \right\| = 2.$

By the local uniform convexity of X we must have $\|x_n - x\| \xrightarrow[n]{} 0$, so that the result is proved. ■

§.3. DUALITY MAPPINGS IN REFLEXIVE BANACH SPACE

We intend to present the following characterization of the reflexivity due to R. C. James:

3.1. THEOREM. A Banach space is reflexive if and only if each functional $x^* \in X^*$ attains its supremum on the unit ball of X.

We need some preparatory results and to this purpose we introduce the following notations.
Let $\{x_n^*\}_n \subset X^*$ be bounded; then

$L(x_n^*)=\left\{x^* \in X^*; \liminf_n < x_n^*, x > \leq < x^*, x > \leq \limsup_n < x_n^*, x >, \forall x \in X\right\}$

$co(x_n^*)(=co(x_1^*, x_2^* \ldots)) =$ the convex set generated by $\{x_n\}_n$.

3.2 LEMMA. Let X be a Banach space, $\theta \in (0,1) \lambda_i > 0$ with $\sum_{i=1}^{\infty} \lambda_i = 1$ and $\{x_n^*\}_n \subset X^*$ so that $\|x_n^*\| \leq 1, \forall n \in N$ and $\|x^* - y^*\| \geq \theta$ for all $x^* \in co(x_n^*)$, $y^* \in L(x_n^*)$. Then there exist $\{y_n^*\}_n \subset X^*, \|y_n^*\| \leq 1$ and $\alpha \in [\theta, 2]$ with the following properties

$$\left\|\sum_{i=1}^{\infty} \lambda_i (y_i^* - y^*)\right\| = \alpha \text{ and } \left\|\sum_{i=1}^{n} \lambda_i (y_i^* - y^*)\right\| < \alpha\left(1 - \theta \sum_{n+1}^{\infty} \lambda_i\right), \forall\ y^* \in L(y_i^*)\ n \in N.$$

PROOF. Select a sequence $\{\varepsilon_k\}_k$ of positive numbers such that

$$\sum_{k=1}^{\infty} \lambda_k . \varepsilon_k . \left(\sum_{k=1}^{\infty} \lambda_i\right)^{-1} . \left(\sum_{k}^{\infty} \lambda_i\right)^{-1} < 1 - \theta. \tag{3.1}$$

For each $n \in N \cup \{0\}$ we shall construct inductively $\alpha_n > 0, y_n^* \in X^*$ and sequences $\{f_i^n\}_i, \{g_i^n\}_i \subset X^*$ in the following way

Take $\{g_i^0\}_i = \{x_i^*\}$ and suppose that we already have contruct by the iterrative procedure described below $\alpha_k, y_k^*, \{f_i^k\}_i, \{g_i^k\}_i$ for all $k < n$. Suppose also that $\|y_k^*\| \leq 1, \|f_i^k\| \leq 1, \|g_i^k\| \leq 1,\quad i \in N, k < n$.
Denote further

$\alpha_n = \inf\left(\sup_{y^* \in L(f_i)} \left\|\sum_1^{n-1} \lambda_i y_i^* + \left(\sum_n^{\infty} \lambda_i\right) z^* - y^*\right\|\right)$ where the infimum is taken over $z^* \in co\left(g_n^{n-1}, g_{i+1}^{n-1}, \ldots\right)$ and $f_i \in co\left(g_i^{n-1}, g_{i+1}^{n-1}, \ldots\right)]$.
We can easily see that $\alpha_n \leq 2$.

Select $y_n^* \in co(g_n^{n-1}, g_{n+1}^{n-1}, \ldots)$ and $f_i^n \in co(g_i^{n-1}, g_{i+1}^{n-1}, \ldots)$, $i \in N$ so that

$$\alpha_n \leq \sup_{y^* \in L(f_i^n)} \left\| \sum_1^{n-1} \lambda_i y_i^* + \left(\sum_n^{\infty} \lambda_i\right) y_i^* - y^* \right\| < (1 + \varepsilon_n) \tag{3.2}$$

and then select $y_0^* \in L(f_i^n)$ so that

$$\alpha_n (1 - \varepsilon_n) < \sum_I^{n-1} \lambda_i y_i^* + \left(\sum_n^{\infty} \lambda_i\right) \|y_n^* - y_0^*\| < \alpha_n (1 + \varepsilon_n).$$

Let be $x_0 \in X$, $\|x_0\| = 1$ with

$$\alpha_n(1 + \varepsilon_n) < \sum_I^{n-1} \lambda_i < y_i^*, x_o > + \left(\sum_n^{\infty} \lambda_i\right) < y_n^*, x_o > - < y_n^*, x_o > \tag{3.3}$$

Since $\liminf_i < f_i^n, x_o > \leq < y_o^*, x_o >$, there is a subsequence $\{g_i^n\}$ of $\{f_i^n\}$ such that for each $y^* \in L\{g_i^n\}$ we have

$$\liminf_i < f_i^n, x_o > = \lim_i < g_i^n, x_o > = < y^*, x_o > \leq < y_i^*, x_o >.$$

i

This means that (3.3) is satisfied if we replace y_0^* by any $y^* \in L(g_i^n)$. Then (3.2) and (3.3) yield

$$\alpha_n(1-\varepsilon_n) < \left\| \sum_I^{n-1} \lambda_i y_i^* + \left(\sum_n^{\infty} \lambda_i\right) y_n^* - y^* \right\| < \alpha_n(1+\varepsilon_n) \text{ for all } y^* \in L(g_i^n). \tag{3.4}$$

This achieves our construction. One can see that (3.4) is true also for all $y^* \in L(y_i^n)$. Moreover the α_n constructed as above are non-decreasing; hence there exists $\alpha = \lim_n \alpha_n$.

Since $\|x^* - y^*\| \geq 0$ for all $x^* \in co(x_n^*)$ and $y^* \in L(x_n^*)$, it follows that $\theta \leq \alpha \leq 2$. It is also clear that (3.4) implies $\alpha = \left\| \sum_1^{\infty} \lambda_i (y_i^* - y^*) \right\|$.

We only have to estimate $\left\| \sum_1^{n} \lambda_i (y_i^* - y^*) \right\|$ for $y^* \in L(y_i^*)$. We may write:

$$\left\|\sum_1^n \lambda_i(y_i^*-y^*)\right\|$$

$$=\left\|\left(\lambda_n+\sum_{n+1}^{\infty}\lambda_i\right)\left(\sum_n^{\infty}\lambda_i\right)^{-1}\left[\sum_1^{n-1}\lambda_i(y_i^*-y)\right]+\lambda_n\left(\sum_n^{\infty}\lambda_i\right)\left(\sum_n^{\infty}\lambda_i\right)^{-1}(y_n^*-y)\right\|$$

$$\le\lambda_n\left(\sum_1^{\infty}\lambda_i\right)^{-1}\left\|\sum_{i=1}^{n-1}\lambda_i(y_i^*-y^*)+\left(\sum_n^{\infty}\lambda_i\right)(y_n^*-y)\right\|$$

$$+\left(\sum_{n+1}^{\infty}\lambda_i\right)\left(\sum_n^{\infty}\lambda_i\right)^{-1}\left\|\sum_1^{n-1}\lambda_i(y_i^*-y^*)\right\|.$$

Using now (3.4) in the last inequality, we obtain

$$\left\|\sum_1^n\lambda_i(y_i^*-y^*)\right\|$$

$$<\left(\sum_{n+1}^{\infty}\lambda_i\right)\left[\left(\sum_{n+1}^{\infty}\lambda_i\right)^{-1}\left(\sum_n^{\infty}\lambda_i\right)^{-1}\lambda_n\alpha_n(1+\varepsilon_n)+\left(\sum_n^{\infty}\lambda_i\right)^{-1}\left\|\left(\sum_1^{n-1}\lambda_i\right)(y_n^*-y^*)\right\|\right].$$

Apply the above inequality for n succesively replaced with n-1, n-2, a.s.o; we obtain

$$\left\|\sum_1^n\lambda_i(y_n^*-y^*)\right\|<\left(\sum_{n+1}^{\infty}\lambda_i\right)\left[\left(\sum_{n+1}^{\infty}\lambda_i\right)^{-1}\left(\sum_n^{\infty}\lambda_i\right)^{-1}\lambda_n\alpha_n(1+\varepsilon_n)\right.$$

$$\left.+\left(\sum_n^{\infty}\lambda_i\right)^{-1}\left(\sum_{n-1}^{\infty}\lambda_i\right)^{-1}\lambda_{n-1}\alpha_{n-1}(1+\varepsilon_{n-1})+\left(\sum_{n-1}^{\infty}\lambda_i\right)^{-1}\left\|\sum_1^{n-2}\lambda_i(y_i^*-y^*)\right\|\right]$$

....

$$<\left(\sum_{n-1}^{\infty}\lambda_i\right)\left[\sum_2^n\left(\sum_{k+1}^{\infty}\lambda_i\right)^{-1}\left(\sum_k^{\infty}\lambda_i\right)^{-1}\lambda_k\alpha_k(1+\varepsilon_k)+\left(\sum_2^{\infty}\lambda_i\right)\left\|\lambda_1(y_1^*-y^*)\right\|\right]$$

$$<\left(\sum_{n+1}^{\infty}\lambda_i\right)\left[\sum_1^n\left(\sum_{k+1}^{\infty}\lambda_i\right)^{-1}\left(\sum_k^{\infty}\lambda_i\right)^{-1}\lambda_k\alpha_k(1+\varepsilon_k)\right].$$

Finally using (3.1) inthe last inequality, we have

$$\left\|\sum_1^n \lambda_i(y_n^* - y^*)\right\| < \alpha\left(\sum_{n+1}^\infty \lambda_i\right)\left[\sum_1^n \left(\sum_{k+1}^\infty \lambda_i\right)^{-1}\left(\sum_k^\infty \lambda_i\right)^{-1}\lambda_k + (1-\theta)\right]$$

$$= \alpha\left(\sum_{n+1}^\infty \lambda_i\right)\left[\sum_1^n\left(\left(\sum_{k+1}^\infty \lambda_i\right)^{-1} - \left(\sum_k^\infty \lambda_i\right)^{-1}\right) + (1+\theta)\right]$$

$$= \alpha\left(\sum_{n-1}^\infty \lambda_i\right)\left[\left(\sum_{n+1}^\infty \lambda_i\right)^{-1} - \left(\sum_1^\infty \lambda_i\right)^{-1} + (1-\theta)\right]$$

$$= \alpha\left(\sum_{n+1}^\infty \lambda_i\right)\left[\left(\sum_{n+1}^\infty \lambda_i\right)^{-1} - \theta\right] = \alpha\left(1 - \theta\sum_{n+1}^\infty \lambda_i\right).$$

Thus our proof is complete. ■

3.3. PROPOSITION. Let X be a Banach space; there are equivalent:

i) X is not reflexive

ii) for any $\theta \in (0,1)$ there exist two sequences $\{x_n\}_n \subset X$, $\|x_n\| \le 1$ and $\{x_n^*\}_n \subset X^*$, $\|x_n^*\| \le 1$, so that

$$\lim_n < x_n^*, x_k > 0, \quad \forall\, k \in N \text{ and } \lim_k < x_n^*, x_k > = \theta, \forall\, n \in N. \tag{3.5}$$

iii) for any $\theta \in (0,1)$ there is a sequence $\{x_n^*\}_n \subset X$, $\|x_n^*\| \le 1$ such that $\|x^* - y^*\| \ge \theta$, $\forall\, x^* \in co(x_n^*)$ and $y^* \in L(x_n^*)$.

iv) for any $\theta \in (0,1)$ and $y_n \ge 0$ with $\sum_1^\infty y_n = 1$, there exist $\alpha \in [\theta, 2]$ and $\{y_n^*\}_n \subset X^*$, $\|y_n^*\| \le 1$ such that for any $y^* \in L(y_n^*)$ we have

$$\left\|\sum_1^\infty \lambda_i(y_i^* - y^*)\right\| = \alpha \text{ and } \left\|\sum_1^\infty \lambda_i(y_i^* - y^*)\right\| < \alpha\left(1 - \theta\sum_{n+1}^\infty \lambda_i\right), \quad \forall\, n \in N. \tag{3.6}$$

v) there exists $y^* \in X^*$ which does not achieve its norm.

PROOF. i) $\Rightarrow$ ii) If X is non-reflexive, then by the Eberlein-Smulian Theorem there exists a closed separable non-reflexive subspace Y of X. As $Y \ne Y^{**}$, for a fixed $\theta \in (0,1)$, there is $y^{**} \in Y^{**}$ with $\|y^{**}\| = 1$ and $d(y^{**}, Y) > \theta$. Let $\{y_n\}_n$ be dense in Y. We shall make use of the following theorem of Helly which we state for completness.

Let Z be a Banach space, $z_1^*, \ldots, z_n^* \in Z^*$, $c_1, \ldots, c_n \in C$; for each $\varepsilon > 0$ there is $z \in Z$ such that $\|z\| \le M + \varepsilon$ and $< z_i^*, z > = c_i, 1 \le i \le n$, if and only if for any $\alpha_1, \ldots, \alpha_n \in C$

$$\left|\sum_1^n \alpha_i c_i\right| \le M \left\|\sum_1^n a_i z_i^*\right\|.$$

Consider for $n \in N$ fixed, $Z = Y^*$, $z_1^* = y_1, \ldots, z_{n-1}^* = y_{n-1}, z_n^* = y^{**}$, $c_1 = c_2 = \ldots = c_{n-1} = 0, c_n = \theta$; then for any $\alpha_1, \ldots, \alpha_n \in C$ we have

$$\left|\sum_1^n \alpha_i c_1\right| = |\alpha_n|\theta = |\alpha_n|.\theta.d(y^{**}, Y)^{-1}.d(y^{**}, Y)$$

$$\le |\alpha_n|.\theta.d(y^{**}, Y)^{-1}.\left\|y^{**} + \sum_1^{n-1} \alpha_n^{-1}.\alpha_i x_i\right\|$$

$$= \theta.d(y^{**}, Y)^{-1}\left\|\sum_1^{n-1} \alpha_i x_i + \alpha_n y^{**}\right\|.$$

Let be $\varepsilon > 0$ with $\theta.d(y^*, Y)^{-1} + \varepsilon \le 1$ and $M = 1 - \varepsilon$; then by Helly"s theorem there is $y_n^* \in Y^*$ with $\|y_n^*\| \le 1, < y_n^*, y_j > = 0$, $1 \le j \le n-1$ and $< y^{**}, y_n^* > = \theta$. Since $\{y_n\}_n$ is dense in Y and $\sup_n \|y_n^*\| \le 1$, we obtain

$$\lim_{n\to\infty} < y_n^*, y > = 0, \quad \forall y \in Y. \tag{3.7}$$

Moreover, as the unit ball of Y is weak*-dense in the unit ball of Y^{**} (Goldstine's theorem), for any $k \in N$, there is $x_k \in Y$ with $\|x_k\| \le 1$ and $|< y_n^*, x_k > - \theta| = |< y_n^*, x_k - y^{**} >| \le \frac{1}{k}$, for $n \le k$.

It follows that $\lim_k < y_n^*, x_k > = < y_n^*, y^{**} > = \theta$, $\forall n \in N$. By (3.7) we also have $\lim_n < y_n, x_k > = 0$, $\forall k \in N$.

Now we may use the Hahn-Banach Theorem to obtain the extensions $x_n^* \in X^*$ of $y_n^*, n \in N$, with $\|x_n^*\| = \|y_n^*\| \le 1$, and satisfying (3.5).

ii) $\Rightarrow$ iii) Let be $\theta \in (0,1)$ and $\{x_k\} \subset X, \{x_n^*\} \subset X^*$ the sequences given by ii). If $x^* \in co(x_n^*)$, then obviously $\lim_k < x^*, x_k > = \theta$ and if $y^* \in L(x_n^*)$ then $< y^*, x_k > = 0$, $\forall k \in N$. Hence

$$\|x^* - y^*\| \geq \lim_k < x^* - y^*, x_k > = \theta.$$

iii) $\Rightarrow$ iv) is exactly the Lemma 3.2.

iv) $\Rightarrow$ v) Let $\lambda_n \geq 0, \sum_1^\infty \lambda_n = 1$ be such that $\lambda_{n+1} < \delta \lambda_n$, $\forall n \in N$ for some δ with $0 < \delta < \frac{\theta^2}{2}$. Then

$$\alpha\theta - 2\delta \geq \theta^2 - 2\delta > 0 \text{ and } \sum_{n+2}^\infty \lambda_k < \delta \sum_{n+1}^\infty \lambda_k . \tag{3.8}$$

We shall prove that the functional $\sum_1^\infty \lambda_k(y_k^* - y^*)$, with y_i^* and y^* given by iv), does not attain its norm on the unit ball.
In fact, let be $x \in X, \|x\| \leq 1$; since $\liminf_k < y_k^*, x > \leq < y^*, x >$ and $\alpha \geq \theta$, there exists n such that

$$< y_{n+1}^* - y^*, x > < \theta^2 - 2\delta \leq \alpha\theta - 2\delta.$$

Using now (3.6) and (3.8), we have

$$\sum_1^\infty \lambda_k < y_k^* - y^*, x > < \sum_1^\infty \lambda_k < y_k^* - y^*, x >$$
$$+ (\alpha\theta - 2\delta)\lambda_{n+1} + \sum_{n+2}^\infty \lambda_k < y_k^* - y^*, x >$$
$$\leq \left\| \sum_1^\infty \lambda_k(y_k^* - y^*) \right\| + (\alpha\theta - 2\delta)\lambda_{n+1} + 2 \sum_{n+2}^\infty \lambda_k$$
$$< \alpha\left(1 - \theta. \sum_{n+1}^\infty \lambda_i\right) + (\alpha\theta - 2\delta)\lambda_{n+1} + 2\delta \sum_{n+1}^\infty \lambda_i$$
$$= \alpha - (\alpha\theta - 2\delta)\left(\sum_{n+1}^\infty \lambda_i\right) + (\alpha\theta - 2\delta)\lambda_{n+1}$$
$$= \alpha - (\alpha\theta - 2\delta)\left(\sum_{n+2}^\infty \lambda_i\right) < \alpha .$$

This proves that the norm of the functional $\sum_1^\infty \lambda_k(y_k^* - y^*)$ is not achieved.

v) $\Rightarrow$ i) Consider $x^* \in X^*, \|x^*\| = 1$ which does not achieve its norm and suppose X reflexive. Then by Theorem of Hahn-Banach, there exists $x^{**} \in X^{**} = X$ with $\|x^{**}\| = 1$ and $< x^*, x^{**} > = 1$ and this is impossible. ■

PROOF OF THEOREM 3.1. It is a direct consequence of the equivalence i) $\Leftrightarrow$ v). ■

3.4. THEOREM. Let X be a Banach space and J a duality mapping of weight φ; then X is reflexive if and only if $\bigcup_{x\in X} Jx = X^*$.

PROOF. Suppose X reflexive and let $x_0^* \in X^*$; by the Hahn-Banach Theorem, there is $x_0 \in X$ with $\|x_0\| = 1$ and $< x_0^*, x_0 > = \|x_0\|$.
Since φ has the property of Darboux, there is $t_0 \geq 0$ with

$$\varphi(\|t_0 x_0\|) = \varphi(t_0) = \|x_0^*\| .$$

Since we also have $< x_0^*, t_0 x_0 > = \|x_0^*\|.\|t_0 x_0\|$, it follows that $x_0^* \in J(t_0 x_0)$.
Conversely, suppose that for each $x^* \in X$ there is $x \in X$ such that $x^* \in Jx$, and consider $y = x/_{\|x\|}$. Then obviously $\|y\| = 1$ and $< x^*, y > = \|x^*\|$. Hence each continuous functional attains its supremum on the unit ball and then by James' theorem 3.1., X is reflexive. ■

3.5. COROLLARY. Let X be a reflexive Banach space and J a duality mapping of weight φ; then J^{-1} is the duality mapping on X^* of weight φ^{-1}.

PROOF. By the above theorem we that that

$$J^{-1}(x^*) = \{x \in X; x^* \in Jx\} \neq \phi, \qquad \forall x^* \in X^* .$$

Let J_* be the duality maping of weight φ^{-1} on X^*; then it is clear that $x \in J^{-1}(x^*)$ if and only if $< x^*, x > = \|x\|.\|x^*\|$ and $\|x\| = \varphi^{-1}(\|x^*\|)$ or equivalently if and only if $x \in J_*(x^*)$. ■

3.6 PROPOSITION. Let X be a real Banach space and J a duality mapping of weight φ; then X is reflexive, strictly convex and with a strictly convex dual if and only if J is a bijection of X onto X^*.
In this case, if for a sequence $\{x_n\}_n \subseteq X$,

$$< Jx_n - Jx, x_n - x \xrightarrow[n]{} 0 \tag{3.9}$$

then $x_n \xrightarrow[n]{w} x$.

PROOF. Suppose X refelexive, strictly convex and X^* strictly convex too; then by Corollary 4.5 Ch. I., J is single-valued, by Corollary.1.9, J is injective and by Theorem 3.4 it is also surjective.
Conversely, suppose J bijective from X onto X^*; then by Theorem .3.4 X is reflexive and by Corollary .4.5 Ch. I. and Corollary 1.4, X^* is strictly convex.

Moreover, by the Corollary 3.5., J^{-1} is the duality mapping of weight φ^{-1} on X^*; since J^{-1} is single valued, X^* is smooth, hence X is strictly convex.
In order to prove the last part of the Proposition we recall the identity

$$\begin{aligned} &< Jx_n - Jx, x_n - x > \\ &= < Jx_n, x_n > - < Jx_n, x > - < Jx, x_n > + < Jx, x > \\ &= [\varphi(\|x_n\|) - \varphi(\|x\|)](\|x_n\| - \|x\|) + [\varphi(\|x\|)\, \|x_n\| - < Jx, x_n >] \\ &\qquad + [\varphi(\|x_n\|)\, \|x\| - < Jx_n, x >] \end{aligned}$$

Since each paranthesis in the last equality is positive, then by (3.9) each one converges to zero; hence

$$\|x_n\| \xrightarrow[n]{} \|x\| \text{ and } < Jx, x_n \xrightarrow[n]{} \varphi(\|x\|).\|x\| = < Jx, x > .$$

Since X is reflexive, it suffices to prove that if for a subsequence $\{x_{n'}\} \subseteq \{x_n\}, x_{n'} \xrightarrow{w} y$, then $y = x$.
As $< Jx, x_{n'} > \xrightarrow[n']{} < Jx, y > = < Jx, x >$, we obtain that $\|x\| \le \|y\|$. But $\|y\| \le \|x\|$ so that $\|x\| = \|y\|$. Hence $< Jx, y/_{\|x\|} > = < Jx, x/_{\|x\|} > = \|Jx\|$.
As X is strictly convex, the functional Jx attains its supremum in at most one point and consequently x = y. ■

In order to obtain more properties of duality mappings on reflexive Banach spaces, we need some results due to Bishop-Phelps. We start with the

3.7. DEFINITION. Let X be a real Banach space;
i) a convex set $K \subseteq X$ is a convex cone if it is closed under positive scalar multiplication;
ii) let $C \subseteq X$ and $x_0 \in C$; the convex cone K supports C at x_0 if $(K + x_0) \cap C = \{x_0\}$.

For $x^* \in X^*$, $\|x^*\| = 1$ and $k \ge 0$ we define

$$K(x^*, k) = \{x \in X, \|x\| \le k < x^*, x >\}.$$

3.8. REMARK. It is clear that $K(x^*, k)$ is a closed convex set. Moreover, if k > 1, then $\operatorname{Int} K(x^*, k) \ne Æ$; indeed, there exists $x_0 \in X$ so that $\|x_0\| = 1$ and $< x^*, x_0 > > \frac{1}{k}$; then the continuity of the norm and of x^* yield the desired result.

3.9. LEMMA. Let $C \subseteq X$ be closed and convex, $x^* \in X^*$ with $\|x^*\| = 1$ and bounded on C and $k > 0$; then for every $z \in C$, there exists $x_0 \in X$ so that $x_0 \in K(x^*, k) + z$ and $K(x^*, k)$ supports C at x_0.

PROOF. Consider the following partial order relation on C

$$x > y \text{ if and only if } x - y \in K(x^*, k).$$

Fix $z \in C$ and denote $Z = \{x \in C; x > z\}$; then $Z = C \cap [K(x^*, k) + z]$ and as $K(x^*, k)$ is closed, Z is closed too.
In order to prove the result we only have to show that Z has a maximal element x_0; indeed, then $x_0 \in K(x^*, k) + z$ and $K(x^*, k)$ suports C at x_0 because if $y \in [K(x^*, k) + x_0] \cap C$, then $y - x_0 \in K(x^*, k)$ hence $y = x_0$.
Let W be a chain in Z. Then $\{<x^*, x>\}_{x \in W}$ is a montone and bounded net in R; indeed if $x > y$, then $x - y \in K(x^*, k)$. Consequently $\|x - y\| \leq k < x^*, x - y >$, so that $<x^*, x> - <x, y> \geq 0$.
It is now clear that the above net converges to its least upper bound, in particular it is a Cauchy net. Then the relation

$$\|x - y\| \leq k(<x^*, x> - <x^*, y>), x, y \in W, \ x > y, \tag{3.10}$$

shows that W itself is a Cauchy net (in the norm topology) and hence converges to some $y_0 \in Z$.
Letting $x \to y_0$ in (3.10) we finally see that $y_0 > y$, for any $y \in W$.
Thus every chain has an upper bound and by Zorn's Lemma Z has a maximal element x_0. ■

3.10. LEMMA. Let $x^*, y^* \in X^*$, $\|x^*\| = \|y^*\| = 1$ and $\varepsilon > 0$; if for any $x \in X$, with $\|x\| \leq 1$, $<x^*, x> = 0$ implies $|<y^*, x>| \leq \frac{\varepsilon}{2}$, then either $\|x^* + y^*\| \leq \varepsilon$ or $\|x^* - y^*\| \leq \varepsilon$.

PROOF. By the Hahn-Banach Theorem there is $z^* \in X^*$ such that $z^* = y^*$ on $\ker x^*$ and $\|z^*\| \leq \frac{\varepsilon}{2}$. Since $y^* - z^*$ is zero on $\ker x^*$, then $y^* - z^* = \alpha x^*$, for some $\alpha \in R$. We have

$$\big|1 - |\alpha|\,\big| = \big|\|y^*\| - \|z^* - y^*\|\big| \leq \|z^*\| \leq \frac{\varepsilon}{2}.$$

Therefore, if $\alpha \geq 0$, then

$$\|x^* - y^*\| = \|(1 - \alpha)x^* - z^*\| \leq |1 - \alpha| + \|z^*\| \leq \varepsilon.$$

For $\alpha < 0$ we have

$$\|x^* + y^*\| = \|(1 + \alpha)x^* + z^*\| \leq |1 + \alpha| + \|z^*\| \leq \varepsilon. \quad ■$$

3.11. LEMMA. Let $x^*, y^* \in X^*$, $\|x^*\| = \|y^*\| = 1$, $\varepsilon > 0$ and $k > 1 + \frac{2}{\varepsilon}$; if y^* is non-negative on $K(x^*, k)$, then $\|x^* - y^*\| \le \varepsilon$.

PROOF. Let $x \in X$, $\|x\| = 1$ and $< x^*, x > > \frac{1}{k}\left(1 + \frac{2}{\varepsilon}\right)$. If $y \in X$ is such that $\|y\| \le \frac{2}{\varepsilon}$ and $< x^*, y > = 0$, then

$$\|x \pm y\| \le 1 + \frac{2}{\varepsilon} < k < x^*, x > = k < x^*, x \pm y >. \; x^* \in Jx$$

Hence $x \pm y \in K(x^*, k)$ and thus $< y^*, x \pm y > \ge 0$.
Then we have $|< y^*, y >| \le < y^*, x > \le \|x\| = 1$.
Consequently if $\|z\| \le 1$ and $< x^*, z > = 0$, then we must have $|< y^*, z >| \le \frac{\varepsilon}{2}$. Then by the Lemma 3.10, we have that

either $\|x^* + y^*\| \le \varepsilon$ or $\|x^* - y^*\| \ge \varepsilon$.

Select $x_0 \in X$, so that $\|x_0\| = 1$ and $< x^*, x_0 > > \max\left(\varepsilon, \frac{1}{k}\right)$; then $< x^*, x_0 > > \frac{1}{k}\|x_0\|$, i.e. $x_0 \in K(x^*, k)$. Hence $< y^*, x_0 > \ge 0 >$ and consequently $\|x^* + y^*\| \ge <, x^* + y^*, x_0 > > \varepsilon$.Then we must have $\|x^* - y^*\| \le \varepsilon$.

■

3.12. THEOREM. (Bishop-Phelps). Let C be a closed bounded convex set in a Banach space X; then the set of all functionals in X^* that achieve their maximum on C is dense in X^*.

PROOF. We can suppose $0 \in C$; observe also that it suffies to approximate only vectors $x^* \in X$, with $\|x^*\| = 1$.
Let $0 < \varepsilon < 1$ and $k > 1 + \frac{2}{\varepsilon} > 1$; then by the Remark 3.8, $K(x^*, k)$ is a closed convex cone with non empty interior. We apply Lemma 3.9 to $z=0$; then there exists $x_0 \in C$ with $x_0 \in K(x^*, k)$ and $[K(x^*, k) + x_0] \cap C = \{x_0\}$. By a well-known separation theorem there is $y^* \in X^*$, $y^* \ne 0$, $\|y^*\| = 1$ such that

$$\sup_{x \in C} < y^*, x > \le < y^*, x_0 >$$
$$\le \inf_{x \in K(x^*, k) + x_0} < y^*, x > = \inf_{x' \in K(x^*, k)} < y^*, x' > + < y^*, x_0 > .$$

It follows that $< y^*, x' > \ge 0$ for all $x' \in K(x^*, k)$ and then by Lemma 3.11, $\|x^* - y^*\| \le \varepsilon$.

3.13. COROLLARY. Let J be a duality mapping of weight φ on the Banach space X; then the set $\{x^*; x^* \in Jx, x \in X\}$ is dense in X.

PROOF. Since we have $\varphi_2(\|x\|)J_1(x) = \varphi_1(\|x\|)J_2x$ (see Proposition 4.7. (f), Ch. I), we only need to prove the assertion for the normalized duality mapping. We first observe that the set $\{x^*; x^* \in Jx, x \in X\}$ coincides with the set of the functionals which achieve their supremum on the unit ball. Indeed, if $x^* \in Jx$, then x^* achieves its supremum in $x/_{\|x\|}$; if $x^* \in X^*$ is such that $\|x^*\| = \langle x^*, x_o \rangle$ for some x_0 with $\|x_0\| = 1$, then a simple computation shows that $x^* \in J(x_0\|x^*\|)$. Hence the assertion is a consequence of Theorem 3.12. ■

3.14. PROPOSITION. Let X be a Banach space; if the norms on X and X^* are Fréchet differentiable, then X is reflexive.

PROOF. Consider J and J_* the normalized duality mappings on X and X^*; they are single-valued and by theorem 2.16 also norm-norm continuous. We clearly have $J_*(Jx) = x, \ \forall x \in X$.
By Corollary 3.13. the set $\{Jx, x \in X\}$ is dense in X^*; then the continuity of J^* yields that the set $\{J_*Jx, x \in X\}$ is dense in X^{**} in the norm topology; hence $X^{**} = X$. ■

3.15 COROLLARY. If the norms on X and X^* are Fréchet differentiable then any duality mapping J is a homeomorphism of X onto X^*.

PROOF. Indeed, then X is reflexive and both X and X^* are locally uniformly smooth; hence by Proposition 3.6 J is bijective and by Theorem 2.16, J and $J_* = J^{-1}$ are norm-to-norm continuous. ■

3.16. PROPOSITION. Let X be a Banach space; if the norms on X and X^* are twice Fréchet differentiable, when on X there is an equivalent hilbertian norm.

PROOF. It follows that X is refelxive and that for the normalized duality mappings on X and X^* we have

$$J_*J = I \quad \text{and} \quad JJ_* = I^* \text{ (identity maps on X and } X^*). \tag{3.11}$$

By hypothesis J and J_* are differentiable (see the formula (2.7)); then using the chain rule (Theorem 1.14. Ch. I) to differentiate the first equality in (3.11) at $x_0 \in X$ and the second at Jx_0, we obtain

$$J'_*(Jx_0)J'(x_0) = I \quad \text{and} \quad J'[J_*(Jx_0)]J'_*(Jx_0) = J'(x_0)J'_*(Jx_0) = I^*.$$

Denote $A = J'(x_o) \in L(X; X^*)$ and $B = J'_*(Jx_o) \in L(X^*; X)$; since AB=I and BA= 1^*, A is a bijection.
Define the following bilinear form on $X \times X$

$$[x, y] = < Ax, y >, x, y, \in X.$$

Observe first that $< Ax, x > \geq 0$, $\forall x \in X$; indeed this is a consequence of the formula

$$< J'(x_o)x, x > = \frac{d^2}{dt^2} \psi(\|x_o + ty\|)|_{t=0}$$

where $y \in X$ and $\psi(x) = \|x\|^2/2$. Therefore the bilinear form $[x, y]$ is positive definite. It is also symmetric; in effect for $x_o, x, y \in X$, let $\phi(t, s) = \psi(\|x_o + tx + sy\|)$; then

$$< J'(x_o)x, y > = \lim_{t \to 0} \frac{< J(x_o + tx), y > - < Jx_o, y >}{t}$$

$$= \lim_{t \to 0} \frac{\frac{\partial \phi}{\partial s}(t, 0) - \frac{\partial \phi}{\partial s}(0, 0)}{t} = \frac{\partial^2 \phi}{\partial t \partial s}(0, 0)$$

and

$$< J'(x_o) y, x > = \lim_{s \to 0} \frac{< J(x_o + sy, x > - < Jx_o, y >}{s}$$

$$= \lim_{s \to 0} \frac{\frac{\partial \phi}{\partial t}(0, s) - \frac{\partial \phi}{\partial t}(0, 0)}{s} = \frac{\partial^2 \phi}{\partial s \partial t}(0, 0).$$

Since $\frac{\partial^2 \phi}{\partial t \partial s}(0, 0) = \frac{\partial^2 \phi}{\partial s \partial t}(0, 0)$, we get

$$< J'(x_o)x, y > = < J'(x_o)y, x >, \text{ i.e. } < Ax, y > = < Ay, x >.$$

Denote now $p(x) = [x, x]^{1/2}$; then p is a semi-norm on X and we have

$$|[x, y]| \leq p(x). p(x), \quad \forall x, y \in X. \tag{3.12}$$

If $p(x) = 0$, then $< Ax, y > = 0$, $\forall y \in X$, hence Ax=0. Since A is a bijection, then x=0, so that p is a norm on X. Moreover p is equivalent to the initial norm. Indeed, $p^2(x) = < Ax, x > \leq \|A\| \, \|x\|^2$, hence $p(x) \leq \|A\|^{1/2}.\|x\|$.
Let be $X_1 = \{x \in X; p(x) \leq 1\}$; the inequality (3.12) yields $|< Ay, x >| \leq p(y)$, for $x \in X_1, y \in X$. Since A is surjective, it follows

that X_1 is weakly bounded, hence norm bounded; then there is M>0 with $\|x\| \leq M.p(x)$, $\forall\, x \in X$ and thus p and $\|\cdot\|$ are equivalent. ■

§.4. DUALITY MAPPINGS L^p SPACES

Consider the Banach space $L^p(\Omega, \Sigma, \mu)$, $p > 1$ (μ being a measure on (Ω, Σ)) with the norm $\|f\|_p = \left(\int_\Omega |f|^p d\mu\right)^{1/p}$; for any $x = (x_k)_{K \in N} \in \ell^p$ we denote $\|x\|_p = \left(\sum_{k=1}^{\infty} |x_k|^p\right)^{1/p}$. In order to establish Clarkson's result on the uniform convexity of $L^p(\Omega, \Sigma, \mu)$, we need some preliminary lemmas.

4.1. LEMMA. For $p \geq 2$ and a,b,$\in$ C we have

$$|a + b|^p + |a - b|^p \leq 2^{p-1}\left(|a|^p + |b|^p\right).$$

PROOF. We note that $\left(\alpha^p + \beta^p\right) \leq \left(\alpha^2 + \beta^2\right)^{p/2}$ for all $\alpha, \beta > 0$.

In effect, denote $\alpha^2 + \beta^2 = m^2$, $m > 0$; then $\left(\frac{\alpha}{m}\right)^2 + \left(\frac{\beta}{m}\right)^2 = 1$. Hence $\left(\frac{\alpha}{m}\right)^p + \left(\frac{\beta}{m}\right)^p \leq \left(\frac{\alpha}{m}\right)^2 + \left(\frac{\beta}{m}\right)^2 = 1$ i.e. $\alpha^p + \beta^p \leq m^p = \left(\alpha^2 + \beta^2\right)^{p/2}$.

Thus we may write

$$|a + b|^p + |a - b|^p \leq \left(|a + b|^2 + |a - b|^2\right)^{p/2} = 2^{p/2}\left(|a|^2 + |b|^2\right)^{p/2}.$$

By Hölder's inequality for $\frac{2}{p} + \frac{p-2}{p} = 1$ we have

$$|a|^2 + |b|^2 \leq \left(|a|^p + |b|^p\right)^{2/p}.(1 + 1)^{\frac{p-2}{p}} = \left(|a|^p + |b|^p\right)^{2/p}.2^{\frac{p-2}{p}}$$

This yields the result. ■

As a direct consequence we obtain the

4.2 COROLLARY. For $f, g \in L^p(\Omega, \Sigma, \mu)$, $p \geq 2$, we have

$$\|f+g\|_p^p + \|f-g\|_p^p \le 2^{p-1}\left(\|f\|_p^p + \|g\|_p^p\right). \tag{4.1}$$

4.3 LEMMA. For $1 < p < 2$, $\frac{1}{p} + \frac{1}{q} = 1$ and $0 \le a \le 1$, we have

$$\left(1+a^q\right)^{p-1} \le \frac{1}{2}\left((1+a)^p + (1-a)^p\right). \tag{4.2}$$

PROOF. For $a = 1$ the result is true. Let $a < 1$; we shall prove that

$$f(a) = \frac{1}{2}(1+a)^p + \frac{1}{2}(1-a)^p - \left(1+a^q\right)^{p-1} \ge 0.$$

The binomial development yields

$$\frac{1}{2}(1+a)^p + \frac{1}{2}(1-a)^p = \sum_{n=0}^{\infty}\binom{p}{2n}a^{2n} = \sum_{n=0}^{\infty}\frac{p(p-1)(2-p)\dots(2m-p-1)2n}{2n!}a^{2n},$$

$$\left(1+a^q\right)^{p-1} = \sum_{n=0}^{\infty}\binom{p-1}{n}a^{nq}$$

$$= \sum_{n=0}^{\infty}\left(\frac{(p-1)(2-p)\dots(2n-1-p)}{(2n-1)!}a^{(2n-1)q} - \frac{(p-1)(2-p)\dots(2n-p)}{(2n)!}a^{2nq}\right)$$

It follows that

$$f(a) = \sum_{n=0}^{\infty}\frac{(2-p)\dots(2n-p)}{(2n-1)!}a^{2n}\left[\frac{p(p-1)}{(2n-p)2n} + \frac{p-1}{2n}a^{2nq-2n} - \frac{p-1}{2n-p}a^{2nq-q-2n}\right]$$

$$= \sum_{n=1}^{\infty}\frac{(2-p)\dots(2n-p)}{(2n-1)!}a^{2n}\left[\frac{1-a^{\frac{2n-p}{p-1}}}{\frac{2n-p}{p-1}} - \frac{1-a^{\frac{2n}{p-1}}}{\frac{2n}{p-1}}\right].$$

(note that $2nq - 2n = \frac{2n}{p-1}$ and $2nq - q - 2n = \frac{2n-p}{p-1}$).
Since $p < 2$, we only need to prove that in the above sum each parenthesis is ≥ 0. But this is a consequence of the fact that the function $t \to \frac{1}{t}(1-a^t)$ is non-increasing $0 \le a < 1$. ■

4.4. LEMMA. For any $a, b \in R$ and $1 < p < 2$ we have

$$|a+b|^q + |a-b|^q \le 2\left(|a|^p + |b|^p\right)^{q-1}, \frac{1}{p}+\frac{1}{q}=1. \tag{4.3}$$

PROOF. Let a + b = 2α and a - b = 2β ; then a = α + β , b = α - β and (4.3) is equivalent with

$$2^q\left(|\alpha|^q + |\beta|^q\right) \le 2\left(|\alpha+\beta|^p + |\alpha-\beta|^p\right)^{q-1}.$$

We can suppose without loss of generality that α β >0. Let m = β/α; then $0 \le m \le 1$ and the inequality (4.3) becames

$$2^q(1+m^q) \le 2\left((1+m)^p + (1-m)^p\right)^{q-1}.$$

But this one is equivalent with

$$(1+m^q)^{1/q-1} = (1+m^q)^{p-1} \le 2^{-1}\left((1+m)^p + (1-m)^p\right),$$

which is exactly (4.2). ■

4.5. LEMMA. Let k > 1 and $f, g \in L^1(\Omega, \Sigma, \mu)$; we have

$$\left[\left(\int_\Omega |f| d\mu\right)^k + \left(\int_\Omega |g| d\mu\right)^k\right]^{1/k} \le \int_\Omega \left(|f|^k + |g|^k\right)^{1/k} d\mu \tag{4.4}$$

PROOF. The existence of the integral on the right side of (4.4) is a consequence of the inequality

$$\left(|f|^k + |g|^k\right)^{1/k} \le (|f| + |g|).$$

On the other hand, for $\frac{1}{k}+\frac{1}{k'}=1$, we have

$$\left(|a_1|^k + |a_2|^k\right)^{1/k} = \sup\left\{|a_1 b_1 + a_2 b_2| ; \left(|b_1|^{k'} + |b_2|^{k'}\right)^{1/k'} = 1\right\} \tag{4.5}$$

Indeed, we consider on R^2 the ^{k} norm; then its dual is R^2 endowed with $^{k'}$ norm and we can compute the norm of an element $(a_1, a_2) \in R^2$ using the following formula

$$\|(a_1, a_2)\|_k = \sup\left\{|x^*(a_1, a_2)|, x^* \in R^2, \|x^*\|_{k'} = 1\right\}.$$

Then taking in (4.5) $a_1 = \int_\Omega |f| d\mu$ and $a_2 = \int_\Omega |g| d\mu$, we obtain

$$\left[\left(\int_\Omega |f|d\mu\right)^k + \left(\int_\Omega |g|d\mu\right)^k\right]^{1/k}$$

$$= \sup\left\{\left|b_1 \cdot \int_\Omega |f|d\mu + b_2 \cdot \int_\Omega |g|d\mu\right|; \left(|b_1|^{k'} + |b_2|^{k'}\right)^{1/k'} = 1\right\}$$

$$\leq \sup\left\{\int_\Omega (|f|\,|b_1| + |g|\,|b_2|)d\mu; \left(|b_1|^{k'} + |b_2|^{k'}\right)^{1/k'} = 1\right\}$$

$$\leq \int_\Omega \left(|f|^k + |g|^k\right)^{1/k}\left(|b|^{k'} + |b_2|^{k'}\right)^{1/k'} d\mu = \int_\Omega \left(|f|^k + |g|^k\right)^{1/k'} d\mu .$$

4.6. COROLLARY. Let $1 < p < 2$ and $f, g \in L^p(\Omega, \Sigma, \mu)$; we have

$$\|f + g\|_p^q + \|f - g\|_p^q \leq 2\left(\|f\|_p^p + \|g\|_p^p\right)^{q-1} \quad \text{where} \quad \frac{1}{p} + \frac{1}{q} = 1. \qquad (4.6)$$

PROOF. It is clear that $q > 2$; hence $k = \frac{q}{p} > 1$. We apply the above Lemma to $|f + g|^p$ and $|f - g|^p$, to obtain

$$\left(\int_\Omega |f + g|^p d\mu\right)^{q/p} + \left(\int_\Omega |f - g|^p d\mu\right)^{q/p} \leq [\int_\Omega \left(|f + g|^q + |f - g|^q\right)^{p/q} d\mu]^{q/p} .$$

We use now the inequality (4.3) to obtain

$$\|f + g\|_p^q + \|f - g\|_p^p \leq 2[\int_\Omega \left(|f|^p + |g|^p\right)^{(q-1)q/p} d\mu]^{q/p} = 2\left(\|f\|_p^p + \|g\|_p^p\right)^{q-1} . \blacksquare$$

4.7. THEOREM. (Clarkson) The space $L^p(\Omega, \Sigma, \mu)$ is uniformly convex for $1 < p < 1$.

PROOF. Consider two sequences $\{f_n\}, \{g_n\} \subset L^p(\Omega, \Sigma, \mu)$ so that $\|f_n\|_p = \|g_n\|_p = 1$ and $\|f_n + g_n\|_p \xrightarrow{n} 2$.

If $p \geq 2$, then by (4.1) we have $\|f_n + g_n\|_q^p + \|f_n - g_n\|_p^p \leq 2^p$.

If $0 < p < 1$, by (4..6) we have $\|f_n + g_n\|_p^q + \|f_n - g_n\|_p^q \leq 2^q$.

Hence, in both cases $\|f_n - g_n\|_p \xrightarrow{n} 0$. $\blacksquare$

We present further some properties of duality mappings in L^p-spaces and we effectively calculate them.

4.8. PROPOSITION. Each duality mapping on $L^p(\Omega, \Sigma, \mu)$, $1 < p < +\infty$, is an homeomorphism of $L^p(\Omega, \Sigma, \mu)$ onto $L^q(\Omega, \Sigma, \mu)$.

PROOF. The result is a consequence of the reflexivity and uniform convexity (see Theorem 2.16 and Proposition 3.6.) of the space $L^p(\Omega, \Sigma, \mu)$. ■

4.9. PROPOSITION. The duality mapping on $L^p(\Omega, \Sigma, \mu)$ corresponding to the weight $\varphi(t) = t^{p-1}$ is given by

$$Jf = |f|^{p-1} \text{ sign } f, f \in L^p(\Omega, \Sigma, \mu).$$

PROOF. We apply Asplund's Theorem 4.4. Ch. I to $\psi(t) = \int_0^t t^{p-1} dt = \frac{t^p}{p}$ and $f, g \in L^p(\Omega, \Sigma, \mu)$ to calculate

$$< Jf, g > = \frac{d}{dt}\psi(\|f + tg\|)\Big|_{t=0}$$

$$= \frac{1}{p}\frac{d}{dt}\left(\|f + tg\|^p\right)\Big|_{t=0} = \frac{1}{p}\frac{d}{dt}\int_\Omega |f + tg|^p d\mu\Big|_{t=0}$$

$$= \int_\Omega |f + tg|^{p-1} . g . \text{sign}(f + tg)\Big|_{t=0} d\mu$$

$$= \int_\Omega |f|^{p-1} . \text{sign } f . g \, d\mu = < |f|^{p-1} . \text{sign } f, g >. \quad ■$$

Using now Proposition 4.7.(f), Ch. I, we obtain

4.10. COROLLARY. The normalized duality mapping on $L^p(\Omega, \Sigma, \mu)$ has the form

$$Jf = |f|^{p-1} . \text{sign } f / \|f\|_p^{p-1}, \ f \in L^p(\Omega, \Sigma, \mu).$$

4.11. COROLLARY. The duality mapping on ℓ^p, $1 < p < +\infty$, corresponding to the weight $\varphi(t) = t^{p-1}$, has the form

$$Jx = \left(|x_k|^{p-1} . \text{sign } x_k\right)_{k \in N}, \quad x = (x_k)_k \in \ell^p. \tag{4.5}$$

4.12. PROPOSITION. Let be $\Omega \subset R^n$ bounded and consider the Sobolev space $H_o^{1,p}(\Omega)$ endowed with the norm $\|f\|_{1,p} = \left(\sum_{i=1}^{n} \left\| \frac{\partial f}{\partial x_i} \right\|_p^p \right)^{1/p}$, $1 < p < +$; the duality mapping on $H_o^{1,p}$ of weight $\varphi(t) = t^{p-1}$ is given by

$$Jf = -\sum_{i=1}^{n} \frac{\partial}{\partial x_i} \left(\left| \frac{\partial}{\partial x_i} \right|^{p-1} .\text{sign} \frac{\partial f}{\partial x_i} \right),\ f \in H_o^{1,p}(\Omega).$$

PROOF. Observe that the norm on $H_o^{1,p}(\Omega)$ is obtained by the composition of a linear differential operator on L^p with the L^p-norm; hence the norm on $H_o^{1,p}(\Omega)$ is Fréchet differentiable and for $f, g \in H_o^{1,p}(\Omega)$ we may write

$$< Jf, g > = \frac{1}{p} \frac{d}{dt} \left(\|f + tg\|_{1,p}^p \right) \Big|_{t=o} = \frac{1}{p} \frac{d}{dt} \sum_{i=1}^{n} \int_\Omega \left| \frac{\partial f}{\partial x_i} + \frac{\partial g}{\partial x_i} \right|^p d\mu \Big|_{t=o}$$

$$= \sum_{i=1}^{n} \int_\Omega \left| \frac{\partial f}{\partial x_i} \right|^{p-1} .\text{sign} \frac{\partial f}{\partial x_i} . \frac{\partial g}{\partial x_i} d\mu = -< \sum_{i=1}^{n} \frac{\partial}{\partial x_i} \left(\left| \frac{\partial f}{\partial x_i} \right|^{p-1} .\text{sign} \frac{\partial f}{\partial x_i} \right), g >. \blacksquare$$

4.13. COROLLARY. The normalized duality mapping on $H_o^{1,2}(\Omega)$ is $J = -\Delta$.

4.14. PROPOSITION. In ℓ^p spaces, $1 < p < +\infty$, every duality mapping is sequentially weak-weak continuous.

PROOF. We recall that in ℓ^p a sequence is weakly convergent if and only it is norm bounded and its components are convergent.
Let $x_n \xrightarrow{w} x$; then $\sup_n \|x_n\|_p < +\infty$ and $x_{n,k} \xrightarrow[n]{} x_k$, $\forall k \in N$. Denote $(Jx_n)_k$ the k-th component of Jx_n; from (4.5) we obtain

$$(Jx_n)_k = |x_{n,k}|^{p-1} .\text{sign}\ x_{n,k} \xrightarrow[n]{} |x_k|^{p-1} .\text{sign}\ x_k = (Jx)_k .$$

Since $\sup_n \|Jx_n\| = \sup_n \varphi(\|x_n\|) < +\infty$, it follows that $Jx_n \xrightarrow[n]{w} Jx$. $\blacksquare$

4.15. REMARK. The above result is not true in L^p-spaces. Indeed, we can construct

$$\{f_n\}_n \subseteq L^p([0,1]) \text{ so that } f_n \xrightarrow[n]{w} 0 \text{ but } Jf_n \not\xrightarrow[n]{w} 0.$$

To this purpose select $f_0 \in L^p([0,1])$ such that $f_0(t) \neq 0$ for every $t \in [0,1]$ and

$$\int_0^1 f_0(t)dt = 0, \int_0^1 |f_0(t)|^p dt = 1, \int_0^1 |f_0(t)|^{p-2}.f_0(t)dt \neq 0.$$

Extend f_0 by periodicity to all R_+ and define $f_n(t) = f_0(nt)$. Then we have

$$\|f_n\|^p = \int_0^1 |f_0(nt)|^p dt = \frac{1}{n}\int_0^n |f_0(t)|^p dt = \frac{n}{n} = 1, n \in N.$$

Moreover, for each $a \in [0,1]$, we have

$$\int_0^a f_n(t)dt = \int_0^a |f_0(nt)|dt = \frac{1}{n}.\int_0^{a.n} f_0(t)dt$$

$$= \frac{[an]}{n}.\int_0^1 f_0(t)dt + \frac{1}{n}\int_{[an]}^{an} f_0(t)dt = \int_{[an]}^{an} f_0(t)dt \xrightarrow[n]{} 0.$$

(here $[an]$ denotes the greatest integer $\leq an$).

Now it is an easy matter to see that $f_n \xrightarrow[n]{w} 0$.

On the other hand, we may write:

$$\int_0^1 (Jf_n)(t)dt = \int_0^1 |f_n(t)|^{p-1}.\text{sign } f_n(t)\,dt = \int_0^1 |f_0(nt)|^{p-1}.\text{sign } f_0(nt)\,dt$$

$$= \frac{1}{n}\int_0^n |f_0(nt)|^{p-2}.f_0(t)dt = \int_0^1 |f_0(t)|^{p-2}.f_0(t)dt \neq 0.$$

Hence Jf_n can not converge weakly to zero.

4.16. THEOREM. (Beurling-Livingston). Let X be a Banach space, J a duality mapping of weight φ on X and X_0 a reflexive subspace of X; then

i) for every $x_0 \in X$ and $x_0^* \in X^*$ we have

$$J(X_0 + x_0) \cap (X_0^\perp + x_0^*) \neq$$

where $X_o^\perp = \{x^* \in X^*; < x^*, x > = 0, \quad \forall x \in X_0\}$.

ii) If X and X^* are strictly convex, then the above intersection contains exactly one element.

PROOF. i) Denote $f(x) = \psi(\|x + x_o\|) - < x_o^*, x >$; then f is convex and weakly lower semi-continuous. As φ is strictly monotone and continuous, there is a positive constant $M \geq \|x_0\|$ so that

$$\varphi\left(\frac{\|x\|-\|x_0\|}{2}\right) \geq 2(\|x_0^*\|+1), \text{ for } \|x\| > M .$$

Then for $x \in X, \|x\| > M$ and $a = \|x\| - \|x_0\|$, we have

$$f(x) \geq \psi(\|x + x_0\|) - \|x_0^*\| \, \|x\|$$
$$\geq \psi(\|x\| - \|x_0\|) - \|x_0^*\| \, \|x\| = \int_0^a \varphi(t)dt - \|x_0^*\| \, \|x\|$$
$$\geq \int_{a/2}^a \phi(t)dt - \|x_0^*\| \, \|x\| \geq \varphi\left(\frac{\|x\|-\|x_0\|}{2}\right)\frac{\|x\|-\|x_0\|}{2} - \|x_0^*\| \, \|x\|$$
$$\geq (2\|x_0^*\|+1).\left(\frac{\|x\|-\|x_0\|}{2}\right) - \|x_0^*\| \, \|x\| = \|x\| - (\|x_0^*\|+1)\|x_0\| .$$

Thus $\lim_{\|x\| \mapsto \infty} f(x) = + \infty$.

Fix $z_0 \in X_0$; then there exists $C > 0$ such that $f(y) > f(z_0)$ for $\|y\| > C$. Denote $K = \{y \in X_0; \|y\| \leq C\}$. As Y is reflexive, K is weakly compact in Y and thus f attains its infimum minimum on K at a point $y_0 \in K$; consequently we have

$$f(y_0) \leq f(z_0) \leq f(y), \forall\, y \in X_0.$$

We shall prove now that there exists a subgradient of f at y_0 belonging to $X_0^{\perp}$. Let $K_0 = \{(x, t) \in X \times R; f(x) - f(y_0) < t\}$ and $K_1 = X_0^x\{0\}$. Then K_0 is open and $K_0 \cap K_1 =$; hence there exists $y_0^* \in X^*$ and $\alpha \in R$ so that the vertical hyperplane

$$H = \{(x,t) \in X \times R < y_0^*, x > + \alpha = t\}$$

strictly separes K_0 and K_1. It follows that

$$< y_0^*, x > + \alpha < t \text{ for } (x, t) \in K_0 \tag{4.6}$$

and

$$< y_0^*, y > + \alpha \geq 0 \quad \text{for} \quad y \in Y . \tag{4.7}$$

Then (4.7) implies $\alpha \geq 0$ and $< y_0^*, y > = 0, \ \forall\, y \in X_0$, i.e. $y_0^* \in X_0^{\perp}$. On the other hand, as $(x, f(x) - f(y_0) + \varepsilon) \in K_0, \ \forall\, x \in X$ and $\varepsilon > 0$, (4.6) yields

$$< y_0^*, x > + \alpha < f(x) - f(y_0) + \varepsilon, \quad \forall\, x \in X, \ \varepsilon > 0. \tag{4.8}$$

In particular, for $x = y_0$, we obtain $<y_o^*, y_o> + \alpha = \alpha < \varepsilon, \forall\, \varepsilon > 0$, i.e. $\alpha \leq 0$. It follows that $\alpha=0$; then (4.8) yields

$$f(x) - f(y_0) + \varepsilon > < y_0^*, x - y_0 >, \quad \forall\, \varepsilon > 0.$$

Hence

$$f(x) - f(y_0) \geq < y_0^*, x - y_0 >, \quad \forall\, x \in X. \tag{4.9}$$

Consequently $y_0^* \in \partial f(y_0)$.

Finally we remark that (4.9) can be written

$$\psi(\|x + x_0\|) - \psi(\|y_0 + x_0\|) \geq <x_0^* + y_0^*, x - y_0 >, \quad \forall x \in X;$$

hence

$$\psi(\|z\|) - \psi(\|y_0 + x_0\|) \geq <x_0^* + y_0^*, z - (x_0 + y_0) >, \quad \forall z \in X.$$

It follows that $x_0^* + y_0^* \in J(x_0 + y_0) = \partial\psi(\|x_0 + y_0\|)$ and obviously $x_0^* + y_0^* \in J(Y + x_0) \cap (Y^\perp + x_0^*)$.

ii) As X^* is strictly convex, J is single-valued. Suppose that there are two elements y_0 and $y_1 \in Y$ such that $J(y_0 + x_0) - x_0^* \in X_0^\perp$ and $J(y_1 + x_0) - x_0^* \in X_0^\perp$. Then $J(y_0 + x_0) - J(y_1 + x_0) \in X_0^\perp$ such that

$$<J(y_0 + x_0) - J(y_1 + x_0), y > = 0, \quad \forall y \in X_0.$$

In particular

$$<J(y_0 + x_0) - J(y_1 + x_0) >, (y_0 + x_0) - (y_1 + x_0) > = 0.$$

This contradicts the fact that J is strictly monotone. (Theorem 1.8.)
∎

Consider the space $L^p([0, 2\Pi]), 1 < p < +\infty$ and for $f \in L^p([0, 2\Pi])$ denote by $c_n(f), n \in Z$, the Fourier coefficients of f. Then we can present the following generalization of the classical result of Riesz-Fischer on Fourier series.

4.17. THEOREM. (Beurling-Livingston). Let Z_1, Z_2 be two non-void disjoint subsets of Z with $Z_1 \cup Z_2 = Z$, and $\{a_n\}_{n \in Z} \subset C$ so that there exist two functions $g_0 \in L^p([0, 2\Pi])$ and $h_0 \in L^q([0, 2\Pi]), \frac{1}{p} + \frac{1}{q} = 1$, with $c_n(g_0) = a_n$ for $n \in Z_1$ and $c_n(h_0) = a_n$ for $n \in Z_2$. Let J be a duality mapping of weight φ on $L^p([0, 2\Pi])$; then there exists a unique function $f_0 \in L^p([0, 2\Pi])$ so that $c_n(f_0) = a_n, n \in Z_1$ and $c_n(Jf_0) = a_n, n \in Z_2$.

P R O O F . Let $V = \{g \in L^p([0, 2\Pi]); c_n(g) = 0, n \in Z_1\}$. Since $|c_n(f) - c_n(g)| \leq M\|f - g\|_p$ for all $f, g \in V$, it follows that V is closed, hence relexive. Moreover

$$V^\perp = \{h \in L^q([0, 2\Pi]), c_n(h) = 0, \quad \forall n \in Z_2\}$$

Indeed, it is clear that $\{e^{int}\}_{n \in Z_2} \subset V$ and thus for each $h \in V^\perp$ we have $c_n(h) = \frac{1}{2\Pi}\overline{<h, e^{int}>} = 0, \quad \forall n \in Z_2.$

Converseley, let $h \in L^q([0,2\Pi])$ be such that $c_n(h) = 0, \ \forall n \in Z_2$. For each $f \in V$ we have $\lim_n \left\| f - \sum_{n \in Z_2} e^{int} c_n(f) \right\|_p = 0$. Therefore

$$\int_0^{2\Pi} f(t)h(t)dt = \lim_n \sum_{n\in Z_2} c_n(f). \int_0^{2\Pi} e^{int} h(t)dt = \lim_n \sum_{n \in Z_2} c_n(f). \overline{c_n(h)} = 0 .$$

This shows that $h \in V^{\perp}$. We can now easily see that
$g_0 + V = \left\{ g \in L^p([0,2\Pi]);\ c_n(g) = a_n,\ \ n \in Z_1 \right\}$ and
$h_0 + V^{\perp} = \left\{ h \in L^q([0,2\Pi]);\ c_n(h) = a_n,\ \ n \in Z_2 \right\}$
We can use now Theorem 4.16. ii), to obtain a unique $F_0 \in J(V + g_0) \cap (V^{\perp} + h_0)$; then $f_0 = J^{1}F_0$ has obviously the required properties. ■

§.5. DUALITY MAPPINGS IN BANACH SPACES WITH THE PROPERTY (h) AND $(\Pi)_1$.

In this section we shall be dealing with some classes of Banach spaces which can be characterized by means of duality mappings.

5.1. DEFINITION. We say that the Banach space X has the property (h) if $x_n \xrightarrow[n]{w} x$ and $\|x_n\| \xrightarrow[n]{} \|x\|$ implies $x_n \xrightarrow[n]{} x$.
It is to be observed that by Proposition 2.8. every locally uniformly convex space has the property (h).

5.2. PROPOSITION. Let X be reflexive and so that X^* is strictly convex and with the property (h); then every duality mapping on X is norm-norm continuous.

PROOF. Let $x_n \xrightarrow[n]{} x$; then since X is reflexive, we know that $Jx_n \xrightarrow[n]{w} Jx$. But is

$$\|Jx_n\| = \varphi(\|x_n\|) \xrightarrow[n]{} \varphi(\|x_n\|) = \|Jx\|;$$

the property (h) of X^* implies $Jx_n \xrightarrow[n]{} Jx$, i.e. the norm-norm continuity of J. ■

The next Corollary is now a consequence of Proposition 2.7.

5.3. COROLLARY. Let X be reflexive with X^* locally uniformly convex; then every duality maping on X is norm-norm continuous.

5.4. DEFINITION. We say that the operator $T:X \to X^*$ has
i) the property (S) if
$x_n \xrightarrow{w} x$ and $< Tx_n - Tx, x_n - x > \xrightarrow[n]{} 0$ implies $x_n \xrightarrow[n]{} x$.
ii) the property (Q) if
$x_n \xrightarrow[n]{w} x$ and $\|Tx_n\| \xrightarrow[n]{} \|Tx\|$ implies $< Tx_n, x > \xrightarrow[n]{} < Tx, x) >$.

5.5. PROPOSITION. Let X be a relfexive Banach space such that X and X^* are strictly convex; then every duality mapping J on X is norm-norm continuous and has the property (S) if J^{-1} has the same properties.

PROOF. Suppose that the duality maping J of a weight φ is norm-norm continuous and has the property (S); by Corollary 3.5., J^{-1} is a duality mapping on X^*.
We shall first prove that J^{-1} has the property (S).
Let be $x_n^* \xrightarrow[n]{w} x^*$ and $< J^{-1}x_n^* - J^{-1}x^*, x_n^* - x^* > \xrightarrow[n]{} 0$.
Denote $x_n = J^{-1}x_n^*$ and $x = J^{-1}x$. Then

$$< J^{-1}x_n^* - J^{-1}x^*, x_n^* - x^* > = < Jx_n - Jx, x_n - x >$$
$$= (\varphi(\|x_n\|) - \varphi(\|x\|))(\|x_n\| - \|x\|) + (\varphi(\|x\|).\|x_n\| - < Jx, x_n >)$$
$$+ (\varphi(\|x_n\|)\cdot\|x\| - < Jx_n, x >) \xrightarrow[n]{} 0 .$$

It follows

$$\|x_n\| \xrightarrow[n]{} \|x\|, < Jx_n, x > \xrightarrow[n]{} < Jx, x >, < Jx, x_n > \xrightarrow[n]{} < Jx, x > .$$

We infer that $x_n \xrightarrow[n]{w} x$; as X is reflexive, it is sufficient to prove that if for a subsequence $\{x_{n'}\}_{n'} \subset \{x_n\}_n$ we have $x_{n'} \xrightarrow[n]{w} y$, then x = y.

But if $x_{n'} \xrightarrow[n]{w} y$, then $< Jx, x_{n'} > \xrightarrow[n']{} < Jx, y > = < Jx, x >$.
This yields $\|x\| \le \|y\|$. Since $\|y\| \le \lim_{n'} \|x_{n'}\| = \|x\|$, we obtain $\|x\| = \|y\|$.
Hence $< Jx, y > = \varphi(\|y\|)\cdot\|y\|$ and $\|Jx\| = \varphi(\|y\|)$. Therefore Jx=Jy and then, by the injectivity of J, x = y.
Since J has the property (S), $x_n \xrightarrow[n]{} x$. As J is norm-norm continuous, $x_n^* = Jx_n \xrightarrow[n]{} Jx = x^*$, i.e. J^{-1} has the property (S).
Let us prove that J^{-1} is also norm-norm continuous. If $x_n^* \xrightarrow[n]{} x^*$, then $x_n = J^{-1}x_n^* \xrightarrow[n]{w} J^{-1}x^* = x$. This yields

$$< Jx_n = Jx, x_n - x > = < x_n^* - x, J^{-1}x_n^* - J^{-1}x^* > \xrightarrow[n]{} 0.$$

The property (S) of J gives us $J^{-1}x_n^* = x_n \xrightarrow[n]{} x = J^{-1}x^*$.
The converse statement is a consequence of the reflexivity of X and the relation $J = (J^{-1})^{-1}$. ■

5.6. PROPOSITION. Let X be a smooth Banach space; then X has the property (h) if and only if there is a duality mapping J on X with the properties (S) and (Q).

PROOF. Assume X has the property (h) and consider a duality mapping J of weigh φ. Let be $\{x_n\}_n \subset X$ such that $x_n \xrightarrow[n]{w} x$ and $< Jx_n - Jx, x_n - x \xrightarrow[n]{} 0$. Then the identity

$$< Jx_n - Jx, x_n - x > = [(\varphi(\|x_n\|) - \varphi(\|x\|))](\|x_n\| - \|x\|)$$
$$+ (\varphi(\|x\|).\|x_n\| - < Jx, x_n >) + [\varphi(\|x_n\|)\|x\| - < Jx_n, x >]$$

yields $x_n \xrightarrow[n]{} \|x\|$. Hence $x_n \xrightarrow[n]{} x$, i.e. J has the property (S).
Let be now $\{x_n\}_n \subset X$ such that $x_n \xrightarrow[n]{w} x$ and $\|Jx_n\| \xrightarrow[n]{} \|Jx\|$, i.e. $\varphi(\|x_n\|) \longrightarrow \varphi(\|x\|)$. Then $\|x_n\| \xrightarrow[n]{} \|x\|$ and again the property (h) yields $x_n \xrightarrow[n]{} x$.
Since J is norm-to- w*-continuous, we have $< Jx_n, x > \xrightarrow[n]{} < Jx, x >$ i.e. J has also the property (Q).
Conversely, suppose that some J has the properties (S) and (Q) and consider a sequence $\{x_n\}_n \subset X$ with $x_n \xrightarrow[n]{w} x$ and $x_n \longrightarrow \|x\|$.
Then

$$\|Jx_n\| = \varphi(\|x_n\|) \xrightarrow[n]{} \varphi(\|x\|) = \|Jx\|$$

so that by the property (Q), $< Jx_n, x > \xrightarrow[n]{} < Jx, x >$. Therefore
$< Jx_n - Jx, x_n - x = = \varphi(\|x_n\|)\|x_n\| - < Jx_n, x > < Jx, x_n > + < Jx, x > \xrightarrow[n]{} 0$.
The property (S) yields $x_n \xrightarrow[n]{} x$, i.e. X has the property (h). ■

5.7. REMARK. We implicitely obtained that in a smooth Banach space with the property (h) every duality mapping has the properties (S) and (Q).

From the Propositions 5.5. and 5.6. we directly obtain the

5.8. COROLLARY. Let X be relfexive, strictly convex and smooth and J a duality mapping on X; then X has the property (h), J is norm-norm continuous and J^{-1} has the property (Q) if and only if X^* has the property (h), J^{-1} is norm-norm continuous and J has the property (Q).

A class of spaces with the property (h) is the following

5.9. DEFINITION. A Banach space X has the property (I) if it is smooth and there is a duality mapping on X sequentially weak-to-weak*-continuous.

5.10. REMARK. By Proposition 4.14., and Remark 4.15. the ^{p} spaces, $1 < p < +\infty$ have the property (I) but the L^p spaces not.

5.11 PROPOSITION. A Banach space with the property (I) has the property (h).

PROOF. Let be $\{x_n\}_n \subset X$ such that $x_n \xrightarrow[n]{w} x$ and $\|x_n\| \xrightarrow[n]{} \|x\|$; then, by the Proposition 1.12.

$$\psi(\|x_n - x\|) = \psi(\|x_n\|) + \int_0^1 < J(x_n - tx), -x > dt.$$

Letting $n \longrightarrow \infty$, we have

$$\lim_n \psi(\|x_n - x\|) = \psi(\|x\|) + \int_0^1 < J(x - tx), - x > dt$$
$$= \psi(\|x\|) + \psi(\|x - x\|) - \psi(\|x\|) = 0$$

Hence $\|x_n - x\| \xrightarrow[n]{} 0$. ■

5.12. PROPOSITION. Let be reflexive, strictly convex and with the property (I); then X is locally uniformly convex.

PROOF. Let be $x, x_n \in X, n \in N$ such that $\|x\| = \|x_n\| = 1$, $n \in N$ and $\|x + x_n\| \longrightarrow 2$; we shall prove that $x_n \xrightarrow[n]{w} x$; then the above proposition yields the desired result.
Since X is reflexive, it is sufficient to show that for every subsequence $\{x_{n'}\}_{n'} \subseteq \{x_n\}_n$ with $x_{n'} \xrightarrow[n']{w} y$ we have $x = y$.

Proposition 1.12. shows that

$$\psi(\|x_{n'} + x\|) = \psi(\|x_{n'}\|) + \int_0^1 < J(x_{n'} + tx), x > dt.$$

Letting $n' \longrightarrow \infty$ we obtain

$$\psi(2) - \psi(1) = \int_0^1 < J(y + tx), x > dt.$$

It is an easy matter to see that we also have

$$\psi(2) - \psi(1) = \int_0^1 < J(x + tx), x > dt = \int_0^1 \varphi(\|(x + tx)\|).\|x\| dt.$$

It follows that

$$\int_0^1 [< J(y + tx), x > - \varphi(1 + t)]\, dt = 0 \tag{5.1}$$

We note now that

$$< J(y + tx), x > \leq \varphi(\|y + tx\|) \leq \varphi\left(\underline{\lim}_{n'} \|x_n\| + t\|x\|\right) = \varphi(1 + t) .$$

Consequently by (5.1)

$$< J(y + tx), x > = \varphi(1 + t) = \varphi\|(y + tx)\| = \|J(y + tx)\|, \ t \in [0, 1] \tag{5.2}$$

Hence the functional $J(y + tx)$ attains its norm both in X and in $\frac{y + tx}{\|y + tx\|}$. Since X is strictly convex, then $x = \frac{y + tx}{\|y + tx\|}$, $t \in [0, 1]$. Thus $x = y/_{\|y\|}$. We shall prove that $\|y\| = 1$. In fact, by (5.2) we have $\|J(x\|y\| + tx)\| = \varphi(1 + t)$, $\forall\, t \in [0, 1]$; in particular, $\|J(x\|y\|)\| = \varphi(1)$. This yields $\varphi(\|y\|) = \varphi(\|x\|\ \|y\|) = \|J(x\|y\|)\| = \varphi(1)$, so that $\|y\| = 1$. Hence $x = y$ and this achieves the proof. ■

The last class of Banach spaces we discuss is given by the

5.13. DEFINITION. A Banach space has the property $(\Pi)_1$ if there exists a directed set $\{X_\alpha\}_{\alpha \in A}$ of finite dimensional subspaces of X so that $\bigcup_{\alpha \in A} X_\alpha = X$, and linear continuous projections $P_\alpha : X \longrightarrow X_\alpha$, with $\|P_\alpha\| = 1, \forall\, \alpha \in A$.

If the above set A is countable, $X_n \subseteq X_{n+1}$, $\forall\, n \in N$ and $P_n x \xrightarrow[X]{} x$ then we say that X is a Banach space with a projection scheme $\{X_n, P_n\}$ of type $(\Pi)_1$.

Examples. All Hilbert spaces have the property $(\Pi)_1$: we can take $\{X_\alpha\}_{\alpha \in A}$ the set of all finite dimensional subspaces of X and the respective orthogonal projections. If the space is separable than it has a projection scheme of type $(\Pi)_1$.

It is not difficult to see that the spaces ^{p}, $1 \leq p < +\infty$ are with projection scheme of type $(\Pi)_1$; we only have to consider the subspaces X_n generated by $e_1, e_2, \ldots$, where $\{e_k\}_{k \in N}$ is the canonical basis in ℓ^p.

More general, all Banach spaces with a monotone Schauder basis are with projections scheme of type $(\Pi)_1$; in particular so are the spaces c_o and C(K), where K is a metric compact space.
Moreover we can prove that:

5.14. PROPOSITION. The space $L^p(\Omega, \Sigma, \mu), 1 \le p < \infty$, with μ a σ-finite measure on (Ω, Σ), has the property $(\Pi)_1$.

PROOF. Consider $(\Omega_1 \dots, \Omega_n)$ a finite family of disjoint subsets of Ω with $\mu(\Omega_k) < +\infty, 1 \le k \le n$ and denote $X_{\Omega_1 \dots, \Omega_n}$, the finite dimensional subspace of $L^p(\Omega, \Sigma, \mu)$ generated by the characteristic functions $\chi_{\Omega_1}, \dots, \chi_{\Omega_n}$; define P by

$$Pf = \sum_{k=1}^{n} \mu(\Omega_k)^{-1}\left(\int_{\Omega_k} f\, d\mu\right)\chi_{\Omega_k}, f \in L^p(\Omega, \Sigma, \mu).$$

We shall prove that $\|P\| = 1$. Indeed, for $f \in L^p(\Omega, \Sigma, \mu)$ we may write:

$$\|Pf\|_p^p = \int_\Omega \left|\sum_{k=1}^{n} \mu(\Omega_k)^{-1}\left(\int_{\Omega_k} f\, d\mu\right).\chi_{\Omega_k}\right|^p d\mu$$

$$\le \int_\Omega \sum_{k=1}^{n} \mu(\Omega_k)^{-1}\left|\int_{\Omega_k} f d\mu\right|^p \chi_{\Omega_k} d\mu$$

$$= \sum_{k=1}^{n} \mu(\Omega_k)^{1-p} \cdot \left|\int_{\Omega_k} f \cdot \chi_{\Omega_k} d\mu\right|^p$$

$$\le \sum_{k=1}^{n} \mu(\Omega_k)^{1-p} . \mu(\Omega_k)^{p/q} . \left(\int_\Omega |f|^p d\mu\right)$$

$$= \sum_{k=1}^{n} \int_{\Omega_k} |f|^p d\mu = \int_\Omega |f|^p d\mu = \|f\|_p^p .$$

Hence $\|P\| \le 1$ and this yields $\|P\| = 1$.

We finally note that as each $f \in L^p(\Omega, \Sigma, \mu)$ is a limit of a sequence of linear combinations of characteristic functions, the union of all possible $X_{\Omega_1 \dots, \Omega_n}$ is dense in X. ■

5.15. COROLLARY. The space $L^p(\Omega, \Sigma, \mu), 1 \le p < \infty$ has a projection scheme of type $(\Pi)_1$.

PROOF. For each $n \in N$, consider $\Omega_k = \left[\frac{k-1}{n}, \frac{k}{n}\right], 1 \le k \le n$ and the corresponding $X_n = X_{\Omega_1, \dots, \Omega_n}$ and P_n; then it is an easy matter to see that $(X_n, P_n)_{n \in N}$ has the required properties. ■

5.16. PROPOSITION. Let X be a reflexive Banach space with a projection scheme $\{X_n, P_n\}$ of type $(\Pi)_1$ so that $P_m P_n = P_{\min(m,n)}$; then $\{P_n^* X^*, P_n^*\}$ is a projection scheme of type $(\Pi)_1$ for X^* satisfying $\dim P_n^* X^* = \dim X_n$.

PROOF. It is obvious that P_n^*, $n \in N$, is a projection on X^* and since $P_n^* : X_n^* \longrightarrow P_n^* X^*$ is an homeormorphism, it follows that $\dim P_n^* X^* = \dim X_n^* = \dim X_n$.

We note farther that $P_n P_m = P_n$ for $m \ge n$ implies $P_m^* P_n^* = P_n^*$, hence $P_n^* X^* \subset P_m^* X^*$.

We also have $X^* = \overline{\bigcup_{n \in 1} P_n^* X^*}$. Indeed, otherwise we would find $x \in X \setminus \{0\}$ such that $< y^*, x > = 0$ for all $y^* \in \bigcup_{n \in 1} P_n^* X^*$. Hence $< x^*, P_n x > = 0$ for all n and $x^* \in X^*$ and therefore $x = 0$, which is a contradiction.

Finally we show that $P_n^* x^* \xrightarrow[n]{} x^*$ for every $x^* \in X^*$. Given $\varepsilon > 0$ we may select n and $y^* \in P_n^* X^*$ so that $\|x^* - y^*\| < \varepsilon$ to obtain

$$\|P_m^* x^* - x^*\| \le \|P_m^*(x^* - y^*)\| + \|y^* - x^*\| \le 2\varepsilon .$$

This ends the proof. ■

5.17. PROPOSITION. Let X be a Banach space with the property $(\Pi)_1$ and J a duality mapping of weight φ; then:

i) $P_\alpha^*(Jx) \subseteq Jx, \quad \forall x \in X_\alpha, \alpha \in A$.

In particular, if X is smooth, we have

$$P_\alpha^*(Jx) \subseteq Jx, \quad \forall x \in X_\alpha.$$

ii) If X is reflexive, then

$$\operatorname{Im} P_\alpha^* \subseteq Jx(X_\alpha), \forall \alpha \in A.$$

PROOF. i) Let be $x^* \in Jx^*$; then $P_\alpha^* x^* \in Jx, \forall \alpha \in A$. Indeed, we have

$$< P_\alpha^* x^*, x > = < x^*, P_\alpha x > = < x^*, x > = \|x^*\| . \|x\| = \varphi(\|x\|)\|x\|.$$

Hence

$$\|x^*\| = < P_\alpha^* x^*, \frac{x}{\|x\|} > \le \|P_\alpha^* x^*\| \le \|x^*\|,$$

i.e. $\|P_\alpha^* x^*\| = \|x^*\| = \varphi(\|x\|), \quad \forall x \in X_\alpha$.

ii) Let be $x^* \in \operatorname{Im} P_\alpha^*$; by the Theorem 4.16. of Beurling-Livingston we have

$$J(X_\alpha) \cap \left(X_\alpha^\perp + x^*\right) \ne \phi, \ \forall \alpha \in A.$$

But $X_\alpha^\perp = (\operatorname{Im} P_\alpha) = \operatorname{Ker} P_\alpha^*$ and therefore

$$J(X_\alpha) \cap (\operatorname{Ker} P_\alpha^* + x^*) \ne \phi, \ \forall \alpha \in A.$$

Select $x \in X_\alpha$ and $y^* \in \operatorname{Ker} P_\alpha^*$ such that $y^* + x^* \in Jx$; then by i) we obtain:

$$x^* = P_\alpha^*(y^* + x^*) \in Jx$$ ■

5.18. COROLLARY. Let X be reflexive, smooth and with the property $(\Pi)_1$; then $J(X_\alpha) = \operatorname{Im} P_\alpha^* = X_\alpha^*$.

EXERCISES

1. A Banach space is strictly convex if and only if for any $0 \ne x \in X$ and $y \in X$, there exists a unique $\alpha \in R$ with $(\alpha x + y) \perp x$.

   Hint: For the definition of the orthogonality, see Exercise 14, Ch. I; for the proof of the result see Theorem 5, §1, Ch. II, Diestel [1].

2. X is strictly convex if and only if for every $1 < p < +\infty$ and all $x, y \in X, x \ne y$

$$\left\|\frac{x+y}{2}\right\|^p \le \frac{1}{2}\left(\|x\|^p + \|y\|^p\right)$$

   Hint: The condition is necessary: suppose x and y are not colinear and prove using iv) in Proposition 1.2., that $\|x + y\| \le \|x\| + \|y\|$; if $x = \alpha y$, use the inequality

$$\left(\frac{1+\alpha}{2}\right)^p < \frac{1}{2}(1 + \alpha^p), \alpha \in R_+ \setminus \{1\}.$$

   (see Proposition 1., §1., Ch. I, Part 3., Beauzamy [1].

3. Let X be reflexive and strictly convex and $x, y, z_n \in X, n \in N$ such that the sequences $\{\|x - z_n\|\}_{n\in N}, \{\|y - z_n\|\}_{n\in N}$ are convergent and $\lim_n \|x - z_n\| + \lim_n \|y - z_n\| = \|x - y\|$;
then there is $\lambda \in [0,1]$ with $z_n \xrightarrow{w} z = \lambda x + (1-\lambda)y$.

<u>Hint</u>: Observe that if $\{z_{n'}\}_{n'}$ is a subsequence of $\{z_n\}_n$ with $z_{n'} \xrightarrow[n']{w} z$, then $\|x - z\| + \|y - z\| = \|x - y\|$; apply Proposition 1.2. (iv) to find $\lambda \in (0,1)$ with $z = \lambda x + (1-\lambda)y$; then $\lambda = \lim_n \frac{\|y - z_n\|}{\|y - x\|}$ i.e., z is independent of the subsequence $\{z_{n'}\}_{n'}$.

4. Let X be reflexive and $K \subseteq X$ convex and closed; then there is $x_o \in K$ such that $\|x_o\| = \min_{x\in K} \|x\|$.
If X is strictly convex, then x_o is unique.

<u>Hint</u>: There is $\{x_n\}_n \subseteq K$ with $\|x_n\| \xrightarrow[n]{} m = \inf_{x\in K} \|x\|$ and $x_n \xrightarrow[n]{} x_o$; then $x_o \in K$ and $\|x_o\| = m$.
If X is strictly convex and $x_o \neq y_o$ are distinct elements of K with $\|x_o\| = \|y_o\| = m$, then for any $\lambda \in [0,1]$ we get $\|\lambda x_o + (1-\lambda)y_o\| = m$ which is impossible.

5. Prove that X is uniformly convex if and only if for every $\varepsilon > 0$, there is $\delta > 0$ such that $x, y \in X, \|x\| \le 1, \|y\|$ and $\|x - y\| \ge \varepsilon$ then $\|x + y\| \le 2(1-\delta)$.

(see Proposition 1, §1, Ch. II, Part 3, Beauzamy [1]).

6. Every quotient of a uniformly convex space is also uniformly convex.

(see Proposition 4, §1, Ch. II. Part 3, Beauzamy [1].

7. In a Hilbert space we have:

$$\Delta_X(\varepsilon) = 1 - \sqrt{1 - \frac{\varepsilon^2}{4}} \text{ and } \rho_X(\tau) = \sqrt{1+\tau^2} - 1.$$

8. Prove that for $p \ge 2, \Delta_{L^p[0,1]}(\varepsilon) = 1 - \left[1 - \left(\frac{\varepsilon}{2}\right)^p\right]^{1/p}$.

<u>Hint</u>: The inequality $\|f+g\|_p^p + \|f-g\|_p^p \le 2^{p-1}\left(\|f\|_p^p + \|g\|_p^p\right)$ yields

$\Delta_{L^p[0,1]}(\varepsilon) \ge 1 - \left[1 - \left(\frac{\varepsilon}{2}\right)^p\right]^{1/p}$. Use the functions $f(t) \equiv 1$, $t \in [0,1]$

and $g(t) = \begin{cases} 1 & \text{for} \quad 0 \le t \le 1 - \left(\frac{\varepsilon}{2}\right)^p \\ -1 & \text{for} \quad 1 - \left(\frac{\varepsilon}{2}\right)^p < t \le 1 \end{cases}$

to prove that we have the equality.

9. Prove that:
for $1 < p < 2$ there exists $C_p > 0$ with $\varepsilon^2 . C_p \le \Delta_{L^p[0,1]}(\varepsilon)$;

for $p \ge 2$ there exists $C_p > 0$ with $\varepsilon^p . C_p \le \Delta_{L^p[0,1]}(\varepsilon)$.

(see Theorem 1., §5, Ch. II, Diestel [2]).

10. Let X be a real Banach space; then $\Delta_X(\varepsilon) \le 1 - \sqrt{1 - \frac{\varepsilon^2}{4}}$ (Day-Nördlander's theorem)
(see Theorem 1, §3, Ch. III, Diestel [2] or Nördlander's [1]).

11. For any Banach space
$\sqrt{1+\tau^2} - 1 \le \rho_X(\tau)$, $\tau \ge 0$.

<u>Hint</u>: Use Lindestrauss' formula and exercise 10.

12. A reflexive locally uniformly convex Banach space is weakly uniformly convex.

13. Let X be separable with X^* uniformly convex and let $\{x_n\}$ be a bounded sequence in X. There exist a subsequence $\{x_{n_k}\}$ and a point $x_0 \in X$ such that $J\{x_{n_k} - x\} \xrightarrow{w} 0$.

(for the solution, see S. Reich [1] and J. R. Webb [6]).

14. Find the analogue of Theorem 4.17 for the 2Π-periodic Sobolev space $W_\Pi^{m,p}$, $p > 1$, $m \in Z$.

BIBLIOGRAPHICAL COMMENTS

§1. The strictly convex Banach spaces were introduced by Clarkson [1]; he proved that on every separable Banach space there is an equivalent strictly convex norm. Theorem 1.3. can be found in Day [4], Klee [2] or in Smulian [1], [2]. The characterization of the strict convexity by means of duality mappings in Theorem 1.8. belongs to Petryshyn [8]. The usefull formula in Proposition 1.4. is due to Asplund [2] and is basic in our discussions.

§2. The (locally) uniformly convex Banach spaces were considered by Clarkson [1] (respectively Lovaglia [1]). The weak local uniform convexity was introduced by Lindenstrauss [3]. Day [1] gave an example of a reflexive Banach space which is not isomorphic to a locally uniformly convex Banach space and Lovaliglia [1] constructed a locally uniformly convex space not isomorphic to an uniformly convex space.

Theorem 2.9. is due independently to Milman [1] and Pettis [1]; the proof given here can be found in Lindenstrauss-Tzafriri [1]. A counterexample for the converse of Theorem 2.9. was given by Day [1] who also studied other aspects of the uniform convexity [2], [3].

The Lindenstrauss duality formulas are proved in Linderstrauss-Tzafriri [1]. The results in Theorems 2.10. and 2.14. were obtained by Smulian [2] and Cudia [2]; the proofs presented here are due to Lindenstrauss [1].

As related papers we mention those of Figiel [1], Fan-Gilcksberg [1], Godini [1], Milman V. D. [1], Zizler [1]; usefull monographs to be consulted are Holmes [1], [2], Krasnoselski-Zabreiko [1], Singer [1].

§3. The proof of the reflexivity criterion in Theorem 3.1. follows James [3] (see also James [1], [2] and Klee [3]).

Properties of the duality mappings in reflexive Banach spaces were investigated by Petryshyn [10]. Our presentation of Theorem 3.12. follows Bishop-Phelps [1].

Proposition 3.14. is due to Restrepo [1], [2] and Proposition 3.16. was obtained independtly by Bonic-Reis [1] and Sundaresan [1], [2]; for related topics see also Istratescu [1].

§4. The uniform convexity of L^p-spaces was established by Clarkson [1]; the proof presented here is taken from the book of Hirzebruch-Scharlau [1]. Proposition 4.14. and the remark which follows are taken from Browder-De Figueiredo [1].

Duality mappings in Sobolev spaces are presented in Lions [1]; for Sobolev space we one can consult Adams' book [1].

Theorem 4.16. generalizes a result of Beurling-Livingston [1] and was proved by Browder [7]; the proof given here can be found in Asplund [2] or in De Figueiredo [1]. The aplication to the Fourier sereis is due to Beurling-Livingston [1].

§5. The results on the characterizations of the Banach spaces with the property (h) by means of the duality mappings are due to Petryshyn [10]. The property (I) was considered by Gossez-Lami Dozo [1]. Banach spaces with the property $(\Pi)_1$ were introduced independently by De Figueiredo and Lindenstrauss; the results exposed here are taken from De Figueiredo [1]. The importance of this class of spaces will be underlined in §2, Chapter IV through the "the Galerkin approximations". More properties of duality mappings are to be found in J. R. L. Webb [1], Bru-Heinich [1] and Kolomy [1].

CHAPTER III

RENORMING OF BANACH SPACES

This chapter is mainly concerned with the basic renorming theorems: the renorminy of $c_0(\Gamma)$ and the theorems of Lindenstrauss and Trojanski. Moreover Aspund's averaging techinque is presented.

§.1. CLASSICAL RENORMING RESULTS.

In the problem of renorming Banach spaces to improve the initial norm the following function space $c_0(\Gamma)$ plays a basic role.

1.1. DEFINITION. Let Γ be an arbitrary set; we introduce the space $c_0(\Gamma)=\{f:\Gamma\to R;\ \forall\ \varepsilon>0,\ |f(\gamma)|>\varepsilon$ only for a finite number of $\gamma\in\Gamma\}$. We endowe $c_0(\Gamma)$ with the norm $\|f\|_\infty=\sup_{\gamma\in\Gamma}|f(\gamma)|$.

1.2. REMARK. Let $f\in c_0(\Gamma)$ and $E_1=\{\gamma;\ |f(\gamma)|=\|f\|\}$; then $E_1\neq\varnothing$ and consists of a finite number of elements. Denote by $f_1=\chi_{\Gamma\setminus E_1}\cdot f$ where $\chi_{\Gamma\setminus E_1}$ is the characteristic function of the set $\Gamma\setminus E_1$; it is clear that $f_1\in c_0(\Gamma)$. Let $E_2=\{\gamma;|f_1(\gamma)|=\|f_1\|\}$; then E_2 is a finite and $E_1\cap E_2=\varnothing$. Denote $f_2=\chi_{\Gamma\setminus E_2}\cdot f_1$ and continue by induction the construction; we get a sequence $\{E_j\}_{j\in N}$ of finite disjoint sets. Denote $E(f)=\bigcup_{j=1}^{\infty}E_j$; then $f=0$ on $\Gamma\setminus E(f)$. In effect, if for a $\gamma_0, f(\gamma_0)\neq 0$ then for ε small enough the set $\{\gamma;\ |f(\gamma)|>|f(\gamma_0)|-\varepsilon\}$, is finite and contains γ_0. But each of its points, in particular γ_0, is contained in some E_j. Hence the support of f is E(f) and we can arrange its elements in a sequence $\{\gamma_n\}_{n\in N}$ so that if $\gamma_n \in E_j$ and $\gamma_m \in E_{j+1}$, then n < m; this also provides

$$|f(\gamma_{n+1})|\le|f(\gamma_n)|,\ \forall n\in N.$$

We shall need the following

1.3 LEMMA. Let $\{a_n\}_n$ and $\{b_n\}_n$ be two non increasing sequences of real numbers and β a permutation of N; then

$$\sum_n a_n b_{\beta(n)} \le \sum_n a_n b_n \tag{1.1}$$

if the both involved series are convergent. Moreover for each $m\in N$ for which β does not permutes the set {1, 2, ..., m} onto its self, we have

$$(a_m - a_{m+1})(b_m - b_{m+1}) \le \sum_n a_n b_n - \sum_n a_n b_{\beta(n)}. \tag{1.2}$$

PROOF. It is an easy matter to verify the following identity

$$\sum_n a_n b_n - \sum_n a_n b_{\beta(n)} = \sum_n (a_n - a_{n+1})(b_1 + \ldots b_n - b_{\beta(1)} - \ldots - b_{\beta(n)}). \tag{1.3}$$

As $\{a_n\}_n, \{b_n\}_n$ are non-increasing, we have

$$a_{n+1} \le a_n \quad \text{and} \quad \sum_{k=1}^{n} b_{\beta(k)} \le \sum_{k=1}^{n} b_k,\ \forall n\in N$$

and thus (1.3) yields (1.1).

Suppose now that for $m\in N$, $\{\beta(1), \ldots, \beta(m)\}$ does not coincide with $\{1, \ldots, m\}$; then there is a k_m with $\beta(k_m)>m$. Therefore we have

$$\sum_{k=1}^{m} b_{\beta(k)} \le b_1 + b_2 + \ldots + b_{m-1} + b_{m+1}$$

Consequently

$$b_m - b_{m+1} = b_1 + b_2 + \ldots + b_{m-1} + b_m - b_1 - b_2 - \ldots - b_{m-1} - b_{m+1}$$

$$\leq \sum_{k=1}^{m} b_k - \sum_{k=1}^{m} b_{\beta(k)}$$

Then using (1.3) we obtain

$$(a_m - a_{m+1})(b_m - b_{m+1}) \leq (a_m - a_{m+1})\left(\sum_{k=1}^{m} b_k - \sum_{k=1}^{m} b_{\beta(k)}\right)$$

$$\leq \sum_{n=1}^{\infty} (a_n - a_{n+1})\left(\sum_{k=1}^{n} b_k - \sum_{k=1}^{n} b_{\beta(k)}\right) = \sum_{n=1}^{\infty} a_n b_n - \sum_{n=1}^{\infty} a_n \cdot b_{\beta(n)}. \quad \blacksquare$$

We can now give the

1.4. THEOREM. On $c_0(\Gamma)$ there is an equivalent locally uniformly convex norm.

PROOF. (Rainwater). Define the mapping $T: c_0(\Gamma) \longrightarrow \ell^2(\Gamma)$ by

$$(Tf)(\gamma) = \begin{cases} 2^{-n} f(\gamma_n) & \text{if } \gamma = \gamma_n \in E(f) \\ 0 & \text{if } \gamma \notin E(f) \end{cases}.$$

It is clear that

$$\sum_{\gamma \in \Gamma} |(Tf)(\gamma)|^2 = \sum_{\gamma \in E(f)} |(Tf)(\gamma)|^2 = \sum_{n=1}^{\infty} 2^{-2n} |f(\gamma_n)|^2 \leq \|f\|^2$$

hence

$$2^{-1}\|f\|_\infty = 2^{-1} f(\gamma_1) \leq \|Tf\|_{\ell^2(\Gamma)} \leq \|f\|_\infty, \forall f \in c_0(\Gamma) \tag{1.4}$$

Define $p(f) = \| Tf \|_{\ell^2(\Gamma)}$; we shall prove that p(f) is a locally uniformly convex norm on $c_0(\Gamma)$. By (1.4), this norm, which is called Day's norm, will be equivalent to the initial norm on $c_0(\Gamma)$.

We first remark the trivial fact that $p(af) = |a| p(f), \forall f \in c_0(\Gamma),\ a \in R$.

We prove further that p is subadditive.

Let $f, g \in c_0(\Gamma)$ and $\{\gamma_n\}_n, \{\mu_n\}_n, \{\lambda_n\}_n \subseteq \Gamma$ be the supports of the functions f, g and f + g, respectively, constructed as in the Remark 1.2.

Then we have

$$p(f+g)=\left(\sum_n 2^{-2n}|f(\lambda_n)+g(\lambda_n)|^2\right)^{\frac{1}{2}}$$

$$\le\left(\sum_n 2^{-2n} f^2(\lambda_n)\right)^{\frac{1}{2}}+\left(\sum_n 2^{-2n}g^2(\lambda_n)\right)^{\frac{1}{2}}$$

$$\le\left(\sum_n 2^{-2n} f^2(\gamma_n)\right)^{\frac{1}{2}}+\left(\sum_n 2^{-2n}g^2(\mu_n)\right)^{\frac{1}{2}}$$

$$= p(x) + p(y)$$

where we used the Minkovski's inequality and the Lemma 1.3 with $a_n=2^{-2n}$ and b_n equal to $f^2(\gamma_n)$ and $g^2(\mu_n)$ respectively.

Let us prove that p is locally uniformly convex on $c_o(\Gamma)$. We have to show that if f, $\{f_n\}_n \subset c_o(\Gamma)$ are so that $p(f)=p(f_n)=1$, $n\in N$ and $\lim_{n\to\infty} p(f+f_n)=2$, then $\lim_{n\to\infty} p(f-f_n)=0$.

Let be $E(f)=\{\gamma_k\}_k$, $E(f_n)=\{\gamma_k^n\}_k$ and $E(f+f_n)=\{\lambda_k^n\}_k$. Then using again (1.1) for $a_n=2^{-2n}$ and b_n equal to $|f(\alpha_n)|^2$ and $|f_n(\gamma_k^n)|^2$, respectively, we obtain

$$0\le\sum_k 2^{-2k}\left[f(\lambda_k^n)-f_n(\lambda_k^n)\right]^2$$

$$=\sum_k 2^{-2k}\left[2f^2(\lambda_k^n)+2f_n^2(\lambda_k^n)-\left((f+f_n)(\lambda_k^n)\right)^2\right]$$

$$=\sum_k 2^{-2k}\left[2f^2(\gamma_k)+2f_n^2(\gamma_k^n)-\left((f+f_n)(\lambda_k^n)\right)^2\right]$$

$$=2p^2(f)+2p^2(f_n)-p^2(f+f_n)\xrightarrow[n]{}0\cdot$$

It follows that

$$\lim_{n\to\infty}\left[f(\lambda_k^n)-f_n(\lambda_k^n)\right]=0 \text{ uniformly for } k\in N. \tag{1.5}$$

Suppose that $p(f-f_n)\not\to 0$; then $\|f-f_n\|_\infty \not\to 0$.

Then for some $\varepsilon>0$ and a subsequence of $\{f_n\}_n$, which we denote for the safe of the simplicity again $\{f_n\}_n$ we have

$$\left\| f - f_n \right\|_\infty \geq \varepsilon,\ n \in N. \tag{1.6}$$

Let k_o be the largest integer for which $|f(\gamma_k)| \geq 2^{-3}\varepsilon$; then

$$|f(\gamma)| < 2^{-3} \cdot \varepsilon \leq \left|f(\gamma_{k_o})\right| \quad \text{for} \quad \alpha \notin \left\{\gamma_1, \ldots, \gamma_{k_o}\right\} \tag{1.7}$$

and therefore

$$\left(2^{-2k_o} - 2^{-2(k_o+1)}\right)\left(f^2(\gamma_{k_o}) - f^2(\gamma_{k_o+1})\right) = \delta > 0.$$

We also have, by (1.5)

$$\sum_k 2^{-2k}\left(f^2(\gamma_k) - f^2(\lambda_k^n)\right)$$

$$= \sum_k 2^{-2k}\left(f^2(\gamma_k) - f_n^2(\lambda_k^n)\right) + \sum_k 2^{-2k}\left(f_n^2(\lambda_k^n) - f^2(\lambda_k^n)\right)$$

$$= p^2(f) - p^2(f_n) + \sum_k 2^{-2k}\left(f_n^2(\lambda_k^n) - f^2(\lambda_k^n)\right)$$

$$= \sum_k 2^{-2k}\left(f_n^2(\lambda_k^n) - f^2(\lambda_k^n)\right) \xrightarrow[n]{} 0.$$

Thus there is $n_o \in N$ so that for $n \geq n_o$

$$\sum_k 2^{-2k} \cdot f^2(\gamma_k) - \sum_k 2^{-2k} \cdot f^2(\lambda_k^n)$$
$$< \delta = \left(2^{-2k_o} - 2^{-2(k_o+1)}\right)\left(f^2(\gamma_{k_o}) - f^2(\gamma_{k_o} + 1)\right)$$

Then by (1.2) it follows that for any $n \geq n_o$

$$\left\{\gamma_1, \gamma_2, \ldots, \gamma_{k_o}\right\} = \left\{\lambda_1^n, \lambda_2^n, \ldots, \lambda_{k_o}^n\right\}.$$

Passing to a subquence we may assume that for all n and k = 1, 2, ..., k_o we have $\lambda_k^n = \gamma_{\beta(k)}$, where $\left\{\gamma_{\beta(1)}, \gamma_{\beta(2)} \ldots \gamma_{\beta(k_o)}\right\}$ and $\left\{\gamma_1, \gamma_2, \ldots, \gamma_{k_o}\right\}$ coincide.

Then by (1.5) it follows that $f_n(\gamma) \xrightarrow[n]{} f(\gamma)$, $\forall \gamma \in \left\{\gamma_1, \gamma_2, \ldots, \gamma_{k_o}\right\}$.

Hence there exists $n_1 \in N$ so that for $n \geq n_1$ we have

$$|(f - f_n)(\gamma)| < 2^{-5} \cdot \varepsilon^2 \cdot (3 \cdot 2^{2k_o})^{-1} \text{ for } \gamma \in \left\{\gamma_1, \ldots, \gamma_{k_o}\right\}. \tag{1.8}$$

By (1.6) we can now pick for each $n \in N, \alpha_n \in \Gamma$ so that

$$|(f - f_n)(\alpha_n)| = \| f - f_n \|_\infty \geq \varepsilon. \tag{1.9}$$

Suppose $\alpha_n \notin \left\{\gamma_1, \ldots, \gamma_{k_o}\right\}$, $n \geq n_1$; then replacing $E(f_n) = \left\{\lambda_k^n\right\}_k$ by a sequence which starts with $\gamma_1, \ldots, \gamma_{k_o}, \alpha_n$, we get by Lemma 1.3.

$$\sum_{k=1}^{k_o} 2^{-2k} f_n^2(\gamma_k) + 2^{-2(k_o+1)} f_n^2(\alpha_n) \leq p^2(f_n).$$

Hence, using (1.7) and (1.8) we may write

$$2^{-2(k_0+1)} f_n^2(\alpha_n) \le p^2(f_n) - \sum_{k=1}^{k_0} 2^{-2k} f_n^2(\gamma_k)$$

$$= p^2(f) - \sum_{k=1}^{k_0} 2^{-2k} f_n^2(\gamma_n)$$

$$= \sum_{k=1}^{k_0} 2^{-2k} \left(f^2(\gamma_k) - f_n^2(\gamma_k)\right) + \sum_{k_0+1}^{\infty} 2^{-2k} \cdot f^2(\gamma_k)$$

$$\le \sum_{k=1}^{k_0} 2^{-2k} |(f - f_n)(\gamma_k)| (|f(\gamma_n)| + |f_n(\gamma_k)|) + 2^{-6} \cdot \varepsilon^2 \sum_{k_0+1}^{\infty} 2^{-2k}$$

$$\le 2 \cdot 2^{-5} \cdot \varepsilon^2 \cdot (3 \cdot 2^{2k_0})^{-1} + 2^{-6} \varepsilon^2 \cdot (3 \cdot 2^{2k_0})^{-1} \le 2^{-2k_0-4} \cdot \varepsilon^2 .$$

This yields $|f_n(\alpha_n)| \le 2^{-1} \cdot \varepsilon$. Then

$$|(f - f_n)(\alpha_n)| \le |f(\alpha_n)| + |f_n(\alpha_n)| \le 2^{-3} \cdot \varepsilon + 2^{-1} \cdot \varepsilon = \frac{5}{8} \varepsilon < \varepsilon .$$

If $\alpha_n \in \{\gamma_1, \gamma_2, \ldots, \gamma_{k_0}\}, n \ge n_1$, then by (1.8) $|(f - f_n)(\alpha_n)| \le 2^{-5} \cdot \varepsilon^2 < \varepsilon$.
In both cases, we obtain a contradiction of the relation (1.9). ■

1.5 THEOREM. (Klee). A Banach space X has an equivalent strictly convex norm if and only if there exists a strictly convex Banach space Y and an injective continuous linear operator $T: X \to Y$.

PROOF. If X admits an equivalent strictly convex norm, then we can take as Y the space X with that norm and T = I (the identity operator).

Conversely, suppose Y is a strictly convex Banach space and $T: X \to Y$ is an injective linear and continuous operator. We define

$$\| x \| = \left(\| x \|_X^2 + \| Tx \|_Y^2 \right)^{\frac{1}{2}}$$

where $\| \cdot \|_X$ and $\| \cdot \|_Y$ are the norms on X and Y, respectiverly. It is clear that $\| \cdot \|$ is an equivalent norm on X. As Y is strictly convex, then by Proposition 1.6. Ch. II, the function $Y \ni y \to \| y \|_Y^2$ is strictly convex. Then the injectivity of T implies that also the function $X \ni x \to \| Tx \|_Y^2$ is strictly convex and this yields that $X \ni x \to \| x \|^2$ is a strictly convex equivalent norm on X. ■

1.6. COROLLARY. If X is a Banach space such that for some Γ there is a continuous injective linear operator $T: X \to c_0(\Gamma)$, then admits an equivalent strictly convex norm.

We present further Asplund's method of averaging norms.

1.7. THEOREM. Let X be a Banach space so that both on X and X* there is an equivalent strictly convex norm; then there is an equivalent strictly convex norm on X so that the dual norm is also strictly convex.

PROOF. Let be $\|\cdot\|_1$ and $\|\cdot\|_2^*$ equivalent strictly convex norms on X and X* respectively. In general $\|\cdot\|_1$ and $\|\cdot\|_2^*$ are not dual norms.
Let $\|\cdot\|_1$ be the norm on X, dual to $\|\cdot\|_2^*$; it is an easy matter to prove that $\|\cdot\|_1$ and $\|\cdot\|_2$ are equivalent.
Denote $f_o(x) = \frac{1}{2}\| x \|_1^2$ and $g_o(x) = \frac{1}{2}\| x \|_2^2$; then we can suppose that we have

$$g_o \le f_o \le (1+c)\, g_o, \text{ for some } c > o.$$

Let be $f_1(x) = \dfrac{f_o(x) + g_o(x)}{2}$ and $g_1(x) = \inf\limits_{y \in X} \dfrac{f_o(x+y) + g_o(x-y)}{2}$.
Then f_1 and g_1 are convex functions on X and

$$g_o \le g_1 \le f_1 \le f_o \text{ and } f_1 \le \left(1 + \frac{c}{2}\right) g_o.$$

Moreover it is not difficult to verity that $x \to \sqrt{2 f_1(x)}$ and $x \to \sqrt{2 g_1(x)}$ are two norms on X, equivalent with $\|\cdot\|_1$ and $\|\cdot\|_2$ (this is left as Exercise 1). Now iterate this procedure to obtain

$$f_{n+1}(x) = \frac{f_n(x) + g_n(x)}{2} \text{ and } g_{n+1}(x) = \inf_{y \in X} \frac{f_n(x+y) + g_n(x-y)}{2}.$$

Thus we get two sequences $\{f_n\}_n$, $\{g_n\}_n$ of convex functions, homogeneous of degree two, satisfying the estimates

$$g_n \le g_{n+1} \le f_{n+1} \le f_n \text{ and } f_n \le (1 + 2^{-n} c) g_n, \; n \in N$$

and so that

$$x \to \sqrt{2 f_n(x)} \quad \text{and} \quad x \to \sqrt{2 g_n(x)}$$

are norms on X, equivalent with $\|\cdot\|_1$ and $\|\cdot\|_2$.
It follows that.

$$0 \le f_n - g_n \le 2^{-n} c\, g_n, \quad n \in N.$$

Hence the two sequences converge to a common limit h which is a convex function satisfying:

$$(1+2^{-n}c)^{-1}h \le g_n \le h \le f_n \le (1+2^{-n}c)h. \tag{1.10}$$

Consequently $x \to \sqrt{2h(x)}$ is a norm on X too, equivalent with $\|\cdot\|_1$ and $\|\cdot\|_2$.

We shall prove that the function h is strictly convex; then, by Proposition 1.6. Ch.II, the norm $\sqrt{2}$ is strictly convex.
To this purpose we prove by induction that the more exact estimate holds:

$$g_n \le f_n \le \left(1+4^{-n}c\right)g_n, \quad n \in N. \tag{1.11}$$

The statement is true for n = 0; suppose that (1.11) is true for $k \le n$ let $a = 1+4^{-n} \cdot c/2$. Then

$$f_{n+1} = \frac{f_n + g_n}{2} \le \frac{1}{2}\left(1+4^{-n} \cdot c + 1\right)g_n = a \cdot g_n \text{ with } a = \frac{1}{2}\left(1+4^{-n}c+1\right).$$

Consequently using the convexity of f and its property to be homogeneous of degree two, we obtain

$$\frac{1}{2}(f_n(x+y)+g_n(x-y)) \ge \frac{1}{2a^2} f_{n+1}(ax+ay) + \frac{1}{2a} f_{n+1}(x-y)$$

$$= \frac{1}{2} \cdot \frac{1+a}{a^2}\left[\frac{1}{1+a} f_{n+1}(ax+ay) + \frac{a}{1+a} f_{n+1}(x-y)\right]$$

$$\ge \frac{1+a}{2a^2} f_{n+1}\left(\frac{2ax}{1+a}\right) = \frac{2}{1+a} f_{n+1}(x).$$

Passing now to the inferior about $y \in X$ we get $\frac{2}{1+a} f_{n+1}(x) \le g_{n+1}(x)$.
Thus, for each $x \in X$

$$f_{n+1}(x) \le \frac{1+a}{2} g_{n+1}(x) = \frac{1}{2}\left(1+4^{-n}\cdot c/2\right) g_{n+1}(x) = \left(1+4^{-(n+1)}c\right) g_{n+1}(x).$$

From (1.11) we obtain

$$0 \le f_n - h \le f_n - g_n \le c\cdot 4^{-n} g_n \le c\cdot 4^{-n}\cdot f_0. \tag{1.12}$$

Define $f_n = 2^{-n} f_0 + h_n$; we can prove by induction that $h_n = \sum_{k=0}^{n-1} \frac{g_k}{2^{n-k}}$ and hence h_n are convex, $n \in N$. Moreover, from (1.12) we obtain

$$\left(2^{-n} - c\cdot 4^{-n}\right) f_0 + h_n \le h \le 2^{-n} f_0 + h_n, \qquad n \in N.$$

For $x, y \in X, x \ne y$ we way write

$$h(x) - 2\,h\left(\frac{x+y}{2}\right) + h(y)$$

$$\ge \left(2^{-n} - c\cdot 4^{-n}\right) f_0(x) + h_n(x) - 2^{-n} f_0\left(\frac{x+y}{2}\right) + \left(2^{-n} - c\cdot 4^{-n}\right) f_0(y) + h_n(y)$$

$$= 2^{-n}\left[f_0(x) - c\cdot 2^{-n} f_0(x) - f_0\left(\frac{x+y}{2}\right) + f_0(y) - c\cdot 2^{-n} f_0(y)\right]$$

$$+ h_n(x) - 2h_n\left(\frac{x+y}{2}\right) + h_n(y) \ge$$

$$\ge 2^{-n}\left\{ f_0(x) + f_0(y) - 2f_0\left(\frac{x+y}{2}\right) - c\cdot 2^{-n}\left[f_0(x) + f_0(y)\right]\right\}.$$

If we choose n large enough in order that $c\cdot 2^{-n}[f_0(x)+f_0(y)]$ be smaller then the positive quantity $f_0(x) + f_0(y) - 2f_0\left(\frac{x+y}{2}\right)$ (f_0 is strictly convex!) then we obtain

$$h(x) - 2h\left(\frac{x+y}{2}\right) + h(y) > 0.$$

Hence h is strictly convex. To end the proof we have only to show that the dual norm $x^* \to \sqrt{2h^*(x^*)}$ is also strictly convex, or equivalently that the conjugate function h* is strictly convex. To this purpose we first show that we have:

$$f^*_{n+1}(x^*) = \inf_{y^* \in X^*} \frac{1}{2}\left[f^*_n(x^*+y^*) + g^*_n(x^*-y^*)\right] \tag{1.13}$$

$$g^*_{n+1}(x^*) = \frac{1}{2}\left[f^*_n(x^*) + g^*_n(x^*)\right]. \tag{1.14}$$

Indeed, by Proposition 2.16., Ch. I, we have

$$f^*_{n+1}(x^*) = \frac{1}{2}[f_n + g_n]^*(2x^*)$$

$$\frac{1}{2}\inf_{z^* \in X^*}\left[f^*_n(2x^* - x^*) + g^*_n(z^*)\right]$$

$$\frac{1}{2}\inf_{y^* \in X^*}\left[f^*_n(x^* + y^*) + g^*_n(x^* - y^*)\right]$$

i.e. (1.13) is true. We use again Proposition 2.16, Ch. I, to estimate

$$g_{n+1}(x) = \inf_{y \in X} \frac{f_n(x+y) + g_n(x-y)}{2} = \inf_{z \in X} \frac{f_n(2x-z) + g_n(z)}{2} =$$

$$\frac{1}{2}\inf_{z \in X}\left[(f^*_n)^*(2x-z) + (g^*_n)^*(z)\right] = \frac{1}{2}(f^*_n + g^*_n)^*(2x) = \left(\frac{f^*_n + g^*_n}{2}\right)^*(x).$$

Then by theorem 2.17. Ch. I, we obtain: $g^*_{n+1}(x^*) = \frac{f^*_n(x^*) + g^*_n(x^*)}{2}$,
i.e. also (1.14) holds.
Then the estimate (1.10) yields

$$(1+2^{-n}c)^{-1}h^* \le f^*_n \le h^* \le g^*_n \le (1+2^{-n}c)h^*.$$

It is now clear that $h^* = \lim_n f^*_n = \lim_n g^*_n$. Finally, using the fact that g^*_0 is strictly convex, we can prove with the same argument as for the function h that h* is strictly convex. ■

1.8. THEOREM. Let X be a Banach space so that both on X an X* there is an equivalent (locally) uniformly convex norm; then there is an equivalent (locally) uniformly convex norm on X so that the dual norm is also (locally) uniformly convex; in particular there is an equivalent norm on X which is Fréchet differentiable together with the dual norm.

PROOF. We apply Asplund's method to construct the dual norms $\sqrt{2h}$ and $\sqrt{2h^*}$ on X and X* respectively and then we use Proposition 2.8. Ch. II, to prove that h and h* are (locally) uniformly strictly convex functions. (Exercise 9). ■

§.2. LINDENSTRAUSS' AND TROJANSKI'S THEOREMS.

This paragraph is concearned with the presentation of two fundamental renorming results for reflexive Banach spaces.

We start with the

2.1. THEOREM. (Lindenstrauss). Let X be a reflexive Banach space; then there is a set Γ and an injective continuous linear operator $T: X \to c_0(\Gamma)$.

We need a serie of preparatives.

2.2. LEMMA. Let X be a Banach space, X_0 a finite dimensional subspace, $n \in N$ and $\varepsilon > 0$; there exists a separable linear subspace $Z \supseteq X_0$ so that for any subspace $Y \supseteq X_0$ with $\dim Y/X_0 = n$, there is a linear continuous operator $T: Y \to Z$ with $\|T\| \le 1+\varepsilon$ and $Tx = x$, for each $x \in X_0$.

PROOF. Let P be a linear continuous projection of X on X_0 (it can be obtained in the following way: take $e_1,\ldots,e_{n_0}$ a basis in X_0 ; then $x = \sum_{k=1}^{n_0} \alpha_k(x) e_k, \forall x \in X_0$; observe that for each $1 \le k \le n_0, X_0 \ni x \to \alpha_k(x)$ is a linear continuous functional on X_0 which can be extended continuously by the Hahn-Banach theorem to all of X; then $Px = \sum_{k=1}^{n_0} \alpha_k(x) \cdot e_k,\ x \in X$ has the desired properties).

Let q be a positive integer. We can choose $x_k = x_k(q) \in X_0, 1 \le k \le m_q$ so that the balls $\{x \in X_0; \|x - x_k\| < q^{-1}\}_{1 \le k \le m_q}$ cover the ball $\{x \in X_0;\ \|x\| \le q\}$. Consider on R^n the norm $\|\lambda\| = \sum_{i=1}^{n} |\lambda_i|$ for $\lambda = (\lambda_i)_{1 \le i \le n}$ and denote $S_n = \{\lambda \in R^n; \|\lambda\| \le 1\}$; choose then $\lambda^j = \lambda^j(q) \in S_n, 1 \le j \le p_q$, so that the balls centred in λ^j and of radius q^{-1} cover S_n. Denote W = (I - P)X and define $f = f(q); W^n = \underbrace{Wx....xW}_{n} \to R^{m_q p_q}$ by

$$f_{k,j}(w_1,\dots,w_n) = \left\| \sum_{i=1}^{n} \lambda_i^j w_i + x_k \right\|, 1 \le k \le m_q, 1 \le j \le p_q.$$

Denote $\overline{S}_w$ the closed unit hall in W; we can find elements $(w_1^\ell,\dots,w_n^\ell) \in \overline{S}_w^n$, $w_i^\ell = w_i^\ell(q), 1 \le i \le n, 1 \le \ell \le r_q$, in such a way that for any $(w_1,\dots,w_n) \in S_w^n$, there exists ℓ with

$$\left| f_{k,j}(w_1,\dots,w_n) - f_{k,j}(w_1^\ell,\dots,w_n^\ell) \right| \le q^{-1}, 1 \le k \le m_q, 1 \le j \le p_q \quad (2.1)$$

We shall prove that the subspace Z for which we are looking for is the subspace spanned by $X_o \cup \{w_i^\ell(q);\ 1 \le i \le n;\ 1 \le \ell \le r_q, q \in N\}$.
Indeed, let be $Y \supset X_o$ with $\dim Y_{/X_o} = n$ and $(w_1,\dots,w_n) \in \overline{S}_w^n$ a basis in $(I-P)Y$; there is a positive constant α satisfying

$$\alpha \cdot \left\| \sum_{i=1}^{n} \lambda_i w_i \right\| \ge \sum_{i=1}^{n} |\lambda_i|, \text{ for any } \lambda = (\lambda_i)_{1 \le i \le n}. \quad (2.2)$$

Select now $q \in N$ so large that

$$2 \le \varepsilon(q-1) \quad \text{and} \quad q \ge 5 \cdot \varepsilon^{-1} \alpha \|I - P\|. \quad (2.3)$$

Then we can find an $\ell \in N, 1 \le \ell \le r_q$ such that (2.1) is satisfied for $(w_1^\ell,\dots,w_n^\ell)$ and $(w_1,\dots,w_n)$.
Let us finally define $T: Y \to Z$ by

$$T\left(\sum_{i=1}^{n} \lambda_i w_i + x \right) = \sum_{i=1}^{n} \lambda_i w_i^\ell + x, \ x \in X_o, \ (\lambda_i)_{1 \le i \le n} \in R^n.$$

We shall prove that $\| T \| \le 1 + \varepsilon$, or equivalently that

$$\left\| \sum_{i=1}^{n} \lambda_i w_i^\ell + x \right\| \le (1+\varepsilon) \left\| \sum_{i=1}^{n} \lambda_i w_i + x \right\|, \ x \in X_o, \ \sum_{i=1}^{n} |\lambda_i| = 1. \quad (2.4)$$

Suppose first that $\| x \| \geq q$; then we have

$$\left\| \sum_{i=1}^{n} \lambda_i w_i + x \right\| \geq q - 1.$$

This yields

$$\left\| \sum_{i=1}^{n} \lambda_i w_i + x \right\| \leq \left\| \sum_{i-1}^{n} \lambda_i w_i + x \right\| + \left\| \sum_{i=1}^{n} \lambda_i w_i^{\ell} \right\| + \left\| \sum_{i=1}^{n} \lambda_i w_i \right\|$$

$$\leq \left\| \sum_{i=1}^{n} \lambda_i w_i + x \right\| + 2 \leq \left\| \sum_{i=1}^{n} \lambda_i w_i + x \right\| + \varepsilon(q-1)$$

$$\leq \left\| \sum_{i=1}^{n} \lambda_i w_i + x \right\| + \varepsilon \left\| \sum_{i=1}^{n} \lambda_i w_i + x \right\| = (1+\varepsilon) \left\| \sum_{i=1}^{n} \lambda_i w_i + x \right\|.$$

Assume now $\| x \| < q$ and select $1 \leq k \leq m_q$ and $1 \leq j \leq p_q$ so that $\| x - x_k \| < q^{-1}$ and $\| \lambda - \lambda^j \| < q^{-1}$ where $\lambda = (\lambda_i)_{1 \leq i \leq n}$ is fixed of norm 1. We have

$$= \left\| \sum_{i=1}^{n} \lambda_i^j w_i^{\ell} + x_k + (x - x_k) + \sum_{i=1}^{n} \left(\lambda_i - \lambda_i^j\right) w_i^{\ell} \right\|$$

$$\leq f_{k,j}\left(w_1^{\ell}, \dots, w_n^{\ell}\right) + \| x - x_k \| + \sum_{i=1}^{n} \left| \lambda_i - \lambda_i^j \right| \leq f_{k,j}\left(w_1^{\ell}, \dots, w_n^{\ell}\right) + 2q^{-1}.$$

Analogously, we obtain

$$\left\| \sum_{i=1}^{n} \lambda_i w_i + x \right\| \geq f_{k,j}(w_1, \dots, w_n) - 2q^{-1}.$$

Use now (2.1), (2.2) and (2.3) to estimate

$$\left\| \sum_{i=1}^{n} \lambda_i w_i^{\ell} + x \right\| \leq q^{-1} + f_{k,j}(w_1, \dots, w_n) + 2q^{-1}$$

$$\leq \left\| \sum_{i=1}^{n} \lambda_i w_i + x \right\| + 5q^{-1}$$

$$\leq \left\| \sum_{i=1}^{n} \lambda_i w_i + x \right\| + \varepsilon \| I - P \|^{-1} \alpha^{-1} \sum_{i=1}^{n} | \lambda_i |$$

$$\leq \left\| \sum_{i=1}^{n} \lambda_i w_i + x \right\| + \varepsilon \cdot \| I - P \|^{-1} \cdot \left\| \sum_{i=1}^{n} \lambda_i w_i \right\|$$

$$= \left\| \sum_{i=1}^{n} \lambda_i w_i + x \right\| + \varepsilon \cdot \| I - P \|^{-1} \cdot \left\| (I-P)\left(\sum_{i=1}^{n} \lambda_i w_i + x \right) \right\|$$

$$\leq (1+\varepsilon) \left\| \sum_{i=1}^{n} \lambda_i w_i + x \right\| \qquad \blacksquare$$

2.3 LEMMA. Let X be a reflexive Banach space and $X_o \subseteq X$ a finite dimensional subspace; there exists a linear operator T: $X \to X$ with a separable range so that $\| T \| = 1$ and Tx = x, for every $x \in X_o$.

PROOF. We construct as in the above Lemma separable subspaces $X_o \subseteq Z_n \subseteq X$ corresponding to every $n \in N$ and $\varepsilon = n^{-1}$; let Z be the closed linear subspace spanned by $\bigcup_{n=1}^{\infty} Z_n$; then for every finite dimensional subspace $Y \supseteq X_o$, with $\dim Y / X_o = n$, there is an linear operator $T_Y: Y \to Z$ so that $T_Y x = x$ for any $x \in X_o$ and $\| T_Y \| \leq 1 + n^{-1}$. Define now

$$\tilde{T}_Y x = \begin{cases} T_Y x & x \in Y \\ 0 & x \notin Y \end{cases}$$

Denote by L(X; Z) the space of all mappings of X in Z and endow it with the topology of the pointwise convergence, considering the weak topology on Z; let us call it the weak-operator topology on L(X; Z). (Here we recall that the strong-operator topology on L(X; Z) is the topology of the point-wise convergence, considering the norm topology on Z). Since $L(X; Z) \approx \prod_{x \in X} Z_x (= Z^X)$ (where $Z_x = Z$, $\forall x \in X$) and Z is reflexive, Tihonov's compacity Theorem implies that the set $\prod_{x \in X} S_x$ (where S_x = S = unit ball in Z, $\forall x \in X$) is compact in L(X; Z). Thus there exists a subnet $\{\tilde{T}_{Y'}\}_{\dim Y' < +\infty}$ of the net $\{T_Y\}_{\dim Y < +\infty}$ which is convergent in the weak-operator topology on L(X; Z) to a mapping T: $X \to Z$ (the net $\{T_Y\}_{\dim Y < +\infty}$ has a cluster point in the above topology! this argument is known as Amir-Lindenstrauss' compacity argument).

It is clear that Tx = x, for any $x \in X_o$ and since

$$\| Tx \| \leq \underset{Y'}{\lim} \| T_{Y'}\, x \| \leq \| x \|,$$

it follows that $\| T \| = 1$. Moreover, the range of T is contained in Z which is separable.
We finally note that T is linear, since $T_{Y'}$ is linear on Y'. ■

2.4. LEMMA. Let X be a reflexive Banach space, $\{x_i\}_{1 \leq i \leq n} \subseteq X$, $\{x_j^*\}_{1 \leq j \leq m} \subseteq X^*$ and $\varepsilon > 0$; there exists a continuous linear operator $T: X \to X$ with a separable range and so that

$$\| T \| = 1,\ Tx_i = x_i,\ 1 \leq i \leq n \ \text{ and } \left\| T^* x_j^* - x_j^* \right\| < \varepsilon, 1 \leq j\, m.$$

PRROF. For every finite dimensional subspace subspace $Y \subseteq X$ containning the elements $\{x_i\}_{1 \leq i \leq n}$, consider the operator $T_Y: X \to X$ given by Lemma 2.3; then $\| T_Y \| = 1, T_Y X$ is separable and $T_Y x = x,\ \forall x \in Y$; in particular $T_Y x_i = x_i,\ 1 \leq i \leq n$. We use again Amir Lindenstrauss' compacity argument to get a subnet $\{T_{Y'}\}_{Y'}$ of the net $\{T_Y\}_{Y \supseteq \{x_i\}_{1 \leq i \leq n}}$ convergent in the weak-operator topology on L(X; X) to the identity operator I on X. But then $\{T_{Y'}^*\}_{Y'}$ converges to the identity operator I^* on X^*, in the weak-operator topology on $L(X^*, X^*)$. Consequently there exists a net $\{T_\alpha^*\}_\alpha \subset L(X{*}X{*})$ convergent in the strong-operator topology to I^*, consisting of finite convex combinations of elements of the net $\{T_{Y'}^*\}_{Y'}$. It is now clear why there exits

$T^* = \sum_{k-1}^{k_0} \lambda_k T_{Y_K'}^*$, with $\sum_{k=1}^{k_0} \lambda_k = 1,\ \lambda_k \geq 0$, so that

$$\left\| T^* x_j^* - x_j^* \right\| < \varepsilon,\ 1, \leq j \leq n.$$

Finally we note that $T = \sum_{k=1}^{k_0} \lambda_k T_{Y_k'}$ has all the required properties. ■

2.5. PROPOSITION. Let X be a reflexive Banach space and Y, Y_* two closed separable subspaces of X and X^* resepectively; there exists a separable subspace $W \supseteq Y$ and a linear projection P of X on W with:

$$\| P \| = 1 \text{ and } P^* x^* = x^* \text{ for any } x^* \in Y_*.$$

PROOF. Let $\{x_j^*\}_{j \in N}$ be a dense sequence in Y_*. With Lemma 2.4. we can construct inductively a sequence of separable subspaces $\{Y_n\}_{n \in N}$ and linear operators T_n: $X \to Y_n$ with $\|T_n\| = 1$, $\|T_n^* x_j^* - x_j^*\| \le \frac{1}{n}$, $1 \le j \le n$, $n \in N$ and so that $T_n x_i^k = x_i^k, 1 \le i \le n, 0 \le k \le n-1, n \in N$, where $\{x_i^k\}_{i \in N}$ is a dense subset of Y_k.

Indeed, take $Y_0 = Y$ and $\{x_i^\circ\}_{i \in N}$ a dense subset of Y; there exists T_1: $X \to X$ with $\|T_1\| = 1$, T_1X separable such that

$$T_1 x_1^\circ = x_1^\circ \text{ and } \|T_1^* x_1^* - x_1^*\| \le 1.$$

Let $Y_1 = T_1X$; there exists T_2: $X \to X$ with $\|T_2\| = 1, T_2X$ separable and such that

$T_2x = x$ for $x \in \{x_1^\circ, x_2^\circ, x_1^1, x_2^1\}$ and $\|T_2^* x_1^* - x_1^*\| \le \frac{1}{2}, \|T_2^* x_2^* - x_2^*\| \le \frac{1}{2}$.

The iterative procedure is by now clear.

Let W be the linear closed subspace spanned by $\bigcup_{n=1}^{\infty} Y_n$; then $W \supseteq Y$ and W is separable. On the other hand, by Amir Lindenstrauss' compacity argument, $\{T_n\}_{n \in N}$ has a cluster point P in the weak-operator topology on L(X; X). Then P has the required properties.

Indeed, it is clear that $PX \subseteq W$. From the construction we have that Pw = w, w ∈ W, hence $\|P\| = 1$. Since $\lim_n \|T_n^* x_j^* - x_j^*\| = 0, \forall j \in N$ we easily see that $P^*x^* = x^*$ for any $x^* \in Y_*$. ■

2.6. DEFINITION. We call density character of the Banach space X the smaller cardinal number λ so that there is a dense subset of X of cardinal λ.

2.7. PROPOSITION. Let X be a reflexive Banach space and λ an infinite cardinal number; consider Y and Y_* two subspaces of X, respectively X^*, with density characters $\le \lambda$; then there is a closed subspace $W \supseteq Y$ with the density character $\le \lambda$ and a linear projection P of X on W so that $\|P\| = 1$ and $P^* x^* = x^*$, for any $x^* \in Y$.

PROOF. We prove the statement by the method of the transfinite induction on the cardinal number λ.

By Proposition 2.5., the statement is true for $\lambda = \chi_0$. Suppose that the Proposition has been established for all cardinal numbers $< \lambda$.

Consider $Y \subseteq X$ and $Y_* \subseteq X^*$ with density characters λ; let μ be the first ordinal of cardinality λ; then since the ordinal of the set $\{\alpha;\alpha$ ordinal, $\alpha < \mu\}$ is μ, it follows that its cardinality is λ; therefore there are sequences $\{y_\alpha\}_{\alpha<\mu} \subset Y$ and $\{y_\alpha^*\}_{\alpha<\mu} \subset Y_*$ which are dense in Y, respectively in Y_*.

We shall construct by transfinite induction a family of linear projections $\{P_\alpha\}_{\omega \le \alpha < \mu}$ of X (here ω is the ordinal of N) satisfying:

$\|P_\alpha\| = 1$, $W_\alpha = P_\alpha X \supseteq \bigcup_{\omega \le \beta < \alpha} [W_\beta \cup \{y_\beta\}]$, $P_\alpha P_\beta = P_\beta$ and $P_\alpha^* y_\beta^* = y_\beta^*$ for $\beta<\alpha$

and so that the density character of $W_\alpha \le$ the cardinality of α.

By Propositon 2.5. P_ω exists.

Suppose all the P_β's have been constructed for $\omega \le \beta < \alpha$; let Y_α be the closed subspace spanned by $\bigcup_{\omega \le \beta < \alpha} [W_\beta \cup \{y_\beta\}]$ and Y_α^* be the closed subspace spanned by $\{y_\beta^*\}_{\beta<\alpha}$. As the density characters of Y_α and Y_α^* are $\le$ cardinality of $\alpha (< \lambda)$, we can apply the inductive hypothesis to obtain P_α with the required properties.

Finally we note that the net $\{P_\alpha\}_{\omega \le \alpha < \mu}$ has a cluster point P in the weak-operator topology on L(X; X) and it is routine now to check that P is a projection on $W = \overline{\bigcup_{\omega \le \alpha < \mu} W_\alpha}$ and has all the required properties. ■

2.8. LEMMA. Let X be a reflexive Banach space, γ an infinite ordinal number and $\{P_\alpha\}_{\omega \le \alpha \le \gamma}$ linear projections of X into X satisfying

$$\|P_\alpha\| = 1 \text{ and } P_\alpha P_\beta = P_\beta P_\alpha = P_\beta \text{ for } \omega \le p < \alpha\ \gamma.$$

Then for any $x \in X$ and $\varepsilon > 0$ the set $\{\alpha; \|P_{\alpha+1}x - P_\alpha x\| \ge \varepsilon\}$ is finite.

PROOF. Suppose that there is $\varepsilon > 0$, $x_0 \in X$ and an infinite sequence of ordinals $\omega \le \alpha_1 < \alpha_2 < \ldots < \gamma$ so that $\|P_{\alpha_{k+1}} x_0 - P_{\alpha_k} x_0\| \ge \varepsilon$, $k \in N$. Let $P_{2k-1} = P_{\alpha_k}$ and $P_{2k} = P_{\alpha_k+1}$. Then $\|P_n\| = 1$ and $P_n P_m = P_m P_n = P_m$ if m<n. Let now P be a cluster point of the sequence $\{P_n\}_{n \in N}$ in the weak operatorial topology on L(X); then P is a linear projection on X, which satisfies:

$$\| P \| = 1, P_n P = P_n, \ \forall n \in N \text{ and } PX = \overline{\bigcup_n P_n X}.$$

Since $P_n x - x = 0, \ \forall x \in P_m X, n > m$, then $\lim_n \| P_n x - x \| = 0$, for any $x \in Px$. Hence

$$0 = \lim_n \| P_n(Px) - Px \| = \lim_n \| P_n x - Px \|, \ \forall x \in X.$$

This yields $\lim_n \| P_{n+1} x_0 - P_n x_0 \| = 0$, contradicting the choice of x_0.

■

We can now give the

PROOF OF THEOREM 2.1. The proof will proceed by transfinite induction on the density character λ of X.
If $\lambda = \chi_0$, then the set Γ = N works. In fact, since X^* is separable we can select $\{x^*\}_{n \in N} \subseteq X^*$ dense in the unit ball of X^* and define $T: X \to c_0 = c_0(N)$ by $Tx = \{n^{-1} < x_n^*, x >\}_{n \in N}$.
It is clear that $\| T \| \leq 1$ and that $Tx = 0$ implies $< x_n^*, x >= 0, \ \forall n \in N$; this yields $<x^*, x> = 0, \ \forall x^* \in X^*$, so that $x = 0$.
Suppose that the theorem is true for all cardinals smaller then λ and that the density character of X is λ; denote by μ to the first ordinal of cardinality λ.
For any ordinal $\omega \leq \alpha < \mu$ we can construct inductively a subspace $W_\alpha \subseteq X$ and a proyection P_α of X on W_α with the properties

i) $\| P_\alpha \| = 1$.

ii) The density character of $W_\alpha \leq$ cardinality of α.

iii) $P_\alpha P_\beta = P_\beta P_\alpha$ for $\beta < \alpha$.

iv) $X = \overline{\bigcup_{\omega \leq \alpha < \mu} W_\alpha}$.

v) if α is a limit ordinal $> \omega$, then P_α is the limit in the weak operatorial topology of a subnet of $\{P_\beta\}_{\beta < \alpha}$.

Indeed, let $\{x_\iota\}_{\iota < \mu}$ be a dense subset in X. For $\alpha = \omega$ we use Propositon 2.5 with Y generated by $\{x_\iota\}_{\iota < \omega}$ to get a linear proyection P_ω so that $\| P_\omega \| = 1$, $P_\omega x_\iota = x_\iota$, $\iota < \omega$ and with $W_\omega = P_\omega X$ separable.
Assume all proyection $\{P_\beta\}_{\omega \leq \beta < \alpha}$ have been defined and satisfy the conditions (i) - (v).

It α is not a limit ordinal, i.e. if α has an immediate predecessor $\alpha - 1$, take then $x_o \in X \setminus W_{\alpha-1}$; by Proposition 2.7. for $Y_\alpha = W_{\alpha-1} \cup \{x_o\}$ and $Y_\alpha^* = P_{\alpha-1}^* X^*$, there is $W_\alpha \supseteq W_{\alpha-1} \cup \{x_o\}$ with the density character $\leq$ cardinality of α and a linear proyection P_α of X on W_α so that $\| P_\alpha \| = 1$ and $P_\alpha^* x^* = x^*$, $\forall x^* \in P_{\alpha-1}^* X^*$. Then $P_\alpha^* P_{\alpha-1}^* = P_{\alpha-1}^*$, i.e. $P_{\alpha-1} \cdot P_\alpha = P_{\alpha-1}$. Since $W_{\alpha-1} \subseteq W_\alpha$, then $P_\alpha \cdot P_{\alpha-1} = P_{\alpha-1}$. Consequently we have: $P_\alpha P_{\alpha-1} = P_{\alpha-1} P_\alpha = P_{\alpha-1}$, This yields for $\beta < \alpha$

$$P_\alpha P_\beta = P_\alpha \left(P_{\alpha-1} P_\beta \right) = \left(P_\alpha P_{\alpha-1} \right) P_\beta = P_{\alpha-1} P_\beta = P_\beta$$

and

$$P_\beta P_\alpha = \left(P_\beta P_{\alpha-1} \right) P_\alpha = P_\beta \left(P_{\alpha-1} P_\alpha \right) = P_\beta P_{\alpha-1} = P_\beta .$$

If α is a limit ordinal, we may take as P_α any cluster point in the weak operatorial topology of the net $\{P_\beta\}_{\beta < \alpha}$ and use the fact that α is a limit ordinal to obtain $P_\beta P_\alpha = P_\alpha P_\beta = P_\beta$ for $\beta < \alpha$. It is now an easy matter to verify that P_α has also the remained properties.
Thus the construction of the set $\{P_\alpha\}_{\alpha < \mu}$ is acheived.
By the inductive hypothesis, for each $\omega \leq \alpha < \mu$ there exists a set Γ_α and a continuous linear injective operator $T_\alpha : W_\alpha \to c_0(\Gamma_\alpha)$ with $\| T_\alpha \| \leq 1$. We can assume that the sets Γ_α are pairwise disjoint.

Let $\Gamma = N \cup \left[\bigcup_{\omega \leq \alpha + 1 < \mu} \Gamma_{\alpha+1} \right]$ and define $T : X \to c_0(\Gamma)$ by

$$(Tx)(n) = \left(T_\omega P_\omega x \right)(n) \quad \text{for} \quad n \in N = \Gamma_\omega$$

$$(Tx)(\gamma) = \frac{T_{\alpha+1}\left(P_{\alpha+1} x - P_\alpha x \right)(\gamma)}{2} \quad \text{for } \gamma \in \Gamma_{\alpha+1},\ \alpha < \mu .$$

It is obvious that T is linear and $\| T \| \leq 1$.
Let be $\varepsilon > 0$ and suppose that $|(Tx)(\gamma)| \geq \varepsilon$ for $x \in X$; then it is clear that $\| P_{\alpha+1} x - P_\alpha x \| \geq 2\varepsilon$, $\alpha < \mu$.
By Lemma 2.8. there are only a finite number of such γ hence the range of T is contained in $c_0(\Gamma)$.
To end the proof we have only to show that T is injective.
Let be $x \in X$ with $Tx = 0$; then as all T_α, $\omega \leq \alpha \leq \mu$, are injective, it follows

$$P_\omega x = 0 \text{ and } P_{\alpha+1} x = P_\alpha x \text{ for all } \omega \leq \alpha < \mu. \tag{2.5}$$

We shall prove by transfinite induction that $P_\alpha x = 0, \forall \omega \le \alpha < \mu$.
Suppose that $P_\beta x = 0$ for any $\beta < \alpha$; if α is not a limit ordinal, then by by (2.5), $P_\alpha x = P_{\alpha-1}x = 0$.
If α is a limit ordinal, then we use (v) to obtain that also in this case $P_\alpha x = 0$. Finally, since by (iv), $x \in \overline{\bigcup_{\omega \le \alpha \le \mu} W_\alpha}$, then it is of the form $x = \text{weak}\lim_{\alpha' < \mu, \alpha' \uparrow \mu} P_{\alpha'} x$, hence $x = 0$. ■

Using now the Corollary 1.6. and Theorem 1.7., we obtain the

2.9. THEOREM. On each reflexive Banach space there is an equivalent norm such that both X and X^* are strictly convex.

2.10. THEOREM. (Trojanski) Every reflexive Banach space admits an equivalent locally uniformly convex norm.

PROOF. Firt part. We shall prove by transfinite induction on the density character λ of X that there is a well ordered set $\Delta (=\Delta_X)$ and a set of bounded linear operators $T_\delta: X \to X, \delta \in \Delta$, with the following properties:

a) the set $\Delta(x,\varepsilon) = \{\delta \in \Delta; \| T_{\delta+1}x - T_\delta x \| \ge \varepsilon(\| T_{\delta+1} \| + \| T_\delta \|)\}$ is finite for each $x \in X$ and $\varepsilon > 0$;

b) let be $\Delta(x) = \bigcup_{\varepsilon > 0} \Delta(x,\varepsilon)$, 1 the smallest element of Δ and

$$Y_x = \text{closed linear span of } \left\{ \| T_1 x \| T_1 X \cup \left[\bigcup_{\delta \in \Delta(x)} (T_{\delta+1} - T_\delta) \right] \right\};$$

then $x \in Y_x$, for each $x \in X$.

c) the density character of the closed linear span of $(T_{\delta+1} - T_\delta)X$ is $\le \chi_0$.

Indeed, if $\lambda = \chi_0$, then $\Delta = \{1\}$ and $T_1 = I$ (the identity operator) fullfill the conditions.
Let the density character λ of X be $> \chi_0$ and suppose that for any reflexive Banach space Y of density character $< \lambda$, the set Δ_Y and the operators $\{T_\delta\}_{\delta \in \Delta_Y}$ can be constructed.
Let μ denote again the first ordinal number whose cardinality is λ.

By the method presented in the proof of the Theorem 2.1., there are projections $\{P_\alpha\}_{\omega\le\alpha<\mu}$ satisfying (i) to (v). From this construction we have that the density character of $P_\omega X$ is χ_0. Observe that the subspace $Y_\alpha=(P_{\alpha+1}-P_\alpha)X$ is closed, hence reflexive, and that the density character of Y_α is $<\lambda$; then by the inductive hypothesis, for each α with $\omega\le\alpha<\mu$ there exists a set $\{S_\beta^\alpha\}_{\beta\in\Delta_\alpha}$ of continuous linear operators from Y_α into itsself satisfying a), b) and c) for $\Delta_\alpha=\Delta_{Y_\alpha}$.
Let $\Delta=\{(\alpha,\beta);\ \beta\in\Delta_\alpha\cup\{0\}\}$. For $\delta\in\Delta$ denote by δ' and δ'' the first and the second component of δ, respectively. Agree that $\delta_1<\delta_2$ whenever $\delta_1'<\delta_2'$ or $\delta_1'=\delta_2'$ and $\delta_1''<\delta_2''$ and observe that with this ordering Δ is well-ordered. For $\delta\in\Delta$, define $T_\delta: X\to X$ by $T_\delta=S_{\delta''}^{\delta'}(P_{\delta'+1}-P_{\delta'})+P_{\delta'}$, where $S_0^{\delta'}=0$.
We shall prove that $\{T_\delta\}_{\delta\in\Delta}$ satisfies the requirements a), b), c).
It is clear that the density character of the closed linear span of $(T_{\delta+1}-T_\delta)X$ is $\le\chi_0$, thus c) is true.
Since $T_\delta x=S_{\delta''}^{\delta'}x, \forall x\in(P_{\delta'+1}-P_{\delta'})(X)$ it follows that $\|T_\delta\|\ge\|S_{\delta''}^{\delta'}\|$.
Remark next that by Lemma 2.8. and by the inductive hypothesis, respectively, for each $\varepsilon>0$, $x\in X$ and $\omega\le\alpha<\mu$, the sets

$$\{\omega\le\beta<\mu;\|P_{\beta+1}x-P_\beta x\|>\varepsilon\}$$

and

$$\left\{\beta\in\Delta_\alpha;\left\|\left(S_{\beta+1}^\alpha-S_\beta^\alpha\right)(P_{\alpha+1}-P_\alpha)x\right\|>\varepsilon\left(\|S_{\beta+1}^\alpha\|+\|S_\beta^\alpha\|\right)\right\}$$

are finite. Then

$$\frac{\|T_{\delta+1}x-T_\delta\|}{\|T_{\delta+1}\|+\|T_\delta\|}\le\max\left(\|P_{\delta'+1}x-P_{\delta'}x\|,\ \frac{\left\|\left(S_{\delta''+1}^{\delta 1}-S_{\delta''}^{\delta'}\right)(P_{\delta'+1}-P_{\delta'})x\right\|}{\|S_{\delta''+1}^{\delta'}\|+\|S_{\delta''}^{\delta'}\|}\right).$$

It follows that the set $\Delta(x,\varepsilon)$ is finite for each $x\in X$, hence a) is satisfied.
We finally prove by transfinite induction that $P_\alpha x\in Y_x$ for each $x\in X$ and $\omega\le\alpha<\mu$; this yields b) since by property (iv), $X=\overline{\bigcup_{\omega\le\alpha<\mu}P_\alpha X}$.
It is clear that $P_\omega x\in Y_x$.
Suppose that $P_\beta x\in Y_x$ for all $\beta>\alpha$.
Remark that for $\delta=(\alpha,0)$, $\delta'=(a+1,0)$, $\omega\le\alpha<\mu$, we have:

$$T_\delta=S_0^\alpha(P_{\alpha+1}-P_\alpha)+P_\alpha=P_\alpha,\text{ hence }P_{\alpha+1}x-P_\alpha x=T_{\delta'}x-T_\delta x\in Y_x.$$

If α is not a limit ordinal, then $P_\alpha x=(P_\alpha x-P_{\alpha-1}x)+P_{\alpha-1}x\in Y_x$.

If α is a limit ordinal, then the property (v) of the set $\{P_\alpha\}_{\omega\le\alpha<\mu}$ implies that $P_\alpha x = \text{weak}\lim_{\beta'<\alpha,\beta'\uparrow\alpha} P_{\beta'}x$ and hence, by the induction hypothesis, $P_\alpha x \in Y_x$.

Second part. For each $\delta\in\Delta$ and $x\in X$ denote

$$t_\delta(x) = \left(\| T_{\delta+1} \| + \| T_\delta \|\right)^{-1} \cdot \| T_{\delta+1}x - T_\delta x \|.$$

Then by a), the map $\delta \to t_\delta(x)$ belongs to $c_0(\Delta)$. Moreover for each $\delta\in\Delta$, t_δ is a semi-norm on X with $t_\delta(x) \le \| x \|$.

For each finite subset A of Δ, denote $F_A(x) = \sum_{\delta\in A} t_\alpha(x)$. Then each F_A is a semi-norm on X so that $F_A(x) \le$ (cardinality of A)$\cdot \| x \|$.

Consider $\{\ell_n^\delta\}_{n\in N}$ a dense set in $(T_{\delta+1}-T_\delta)X$, $\delta\in\Delta$; for each $n\in N$ and $A\subset\Delta$ finite, denote $Y_{n,A}$ = linear span of $\bigcup_{\delta\in A}\{\ell_1^\delta,\dots,\ell_n^\delta\}$.

It is obvious that $Y_{n,A}$ is finite dimensional.

Let $E_A^n(x) = \inf_{y\in Y_{n,A}} \| x-y \|$. It is clear that E_A^n is a semi-norm on X with $E_A^n(x) \le \| x \|$. Finally, for each $n\in N$ let

$$G_n(x) = \sup\{E_A^n(x) + nF_A(x), A\subseteq\Delta,\ \text{card}\ A\le n\}.$$

It is not difficult to see that G_n is a semi-norm on X with $G_n(x) \le (1+n^2)\| x \|$.

By Theorem 2.1. there exists an injective continuous linear operator $T: X\to c_0(\Gamma)$ for some set Γ.

Consider then the disjoint union $\Lambda = \{0\}\cup N\cup\Delta\cup\Gamma$ and define $Q: X\to c_0(\Gamma)$ by

$$(Qx)(\lambda) = \begin{cases} \| x \| & \text{if } \lambda = 0 \\ 2^{-n}G_n(x) & \text{if } \lambda = n\in N \\ t_\delta(x) & \text{if } \lambda = \delta\in\Delta \\ (Tx)(\gamma) & \text{if } \lambda = \gamma\in\Gamma \end{cases}$$

Note that by the definitions of $t_\delta(x)$, $G_n(x)$ and $c_0(\Gamma)$, we have that $Qx\in c_0(\Gamma)$, $\forall x\in X$; moreover $\| x \| \le \| Qx \|_\infty \le K\| x \|$ for some constant $K>0$.

Define: $|\| x \||=p(Qx)$ where p is the Day's norm on $c_o(\Gamma)$; then $|\|\cdot\||$ is a norm on X, equivalent to $\|\cdot\|$.

Consider now $|\| x_n \||=1=|\| x \||$ and $|\| x_n+x \|| \xrightarrow[n]{} 2$.

Remark that $p(Q(x_n+x)) \xrightarrow[n]{} 2$ implies $\| Q(x_n)-Q(x) \|_\infty \xrightarrow[n]{} 0$ (use to this purpose the proof of Theorem 1.4.). This yields

$$|G_k(x_n)-G_k(x)| \xrightarrow[n]{} 0 \qquad \text{for each } k \in N \tag{2.6}$$

$$|t_\delta(x_n)-t_\delta(x)| \xrightarrow[n]{} 0 \qquad \text{for each } \delta \in \Delta \tag{2.7}$$

$$\| Tx_n - Tx \|_\infty \xrightarrow[n]{} 0. \tag{2.8}$$

We shall prove that $|\| x-x_n \|| \xrightarrow[n]{} 0$.

To this purpose it is sufficient to show that $\{x_n\}_n$ is totally bounded. Indeed, then each of its subsequence will have a convergent subsequence. Since T is injective and continuous, then (2.8) will imply that these subsequencial limits are all equal to x, so that $|\| x_n-x \|| \xrightarrow[n]{} 0$.

Let be $\varepsilon > 0$; we will show that $\{x_n\}_n$ lies within ε of a bounded subset of $Y_{k,A}$, for some k and A. To do this we will find k and A with $E_A^k(x_n)<\varepsilon$, $\forall n \in N$ and this will finish the proof.

By condition b) on the operators $\{T_\delta\}_{\delta\in\Delta}$, there exists an integer m and a finite subset $B \subseteq \Delta(x)$ so that $E_B^m(x)<\varepsilon/3$. This implies

$$E_A^k(x)<\varepsilon/3, \text{ for } k \geq m,\ A \geq B. \tag{2.9}$$

Now we note that the set $\Delta(x)=\{\delta\in\Delta;\ t_\delta(x)>0\}$ is countable and thus we can write $\Delta(x)\subseteq\{\delta_1,\delta_2,\ldots\}\subseteq\Delta$ with $t_\delta(x)\geq t_{\delta_2}(x)\geq\ldots$. Moreover we may assume (by enlarging B if necessary) that $B=\{\delta_1,\ldots,\delta_m\}$ and that $t_{\delta_{m+1}}(x)<t_{\delta_m}(x)$.

Let $b=t_{\delta_m}(x)-t_{\delta_{m+1}}(x)$ and select $k\geq\max\{m,\ (\varepsilon+3\| x \|)/3b\}$.

By the definition of $G_k(x)$, there exists $A\subseteq\Delta$ wutg card $A\leq k$ so that

$$G_k(x)-\left[E_A^k(x)+kF_A(x)\right]<\frac{\varepsilon}{3}. \tag{2.10}$$

We assert that $B \subseteq A$. In effect, suppose $B \not\subseteq A$; then there exists $k_1 \in A$ with $k_1 > m$. Let $D=\{\delta_1,\delta_2,\ldots,\delta_k\}$; we see that

$$G_k(x)-\left[E_A^k(x)+kF_A(x)\right] \geq E_D^k(x)+kF_D(x)-E_A^k(x)-kF_A(x)$$

$$\geq k\left[\sum_{i=1}^{k} t_{\delta_i}(x)-\sum_{\delta\in A} t_\delta(x)\right]-E_A^k(x)\geq k\left(t_{\delta_m}-t_{\delta_{k_1}}\right)-E_A^k(x)$$

$$\geq kb-\|x\|>\varepsilon/3.$$

This contradicts (2.10)
Using (2.9) and (2.10) we obtain

$$G_k(x)-kF_A(x)\leq 2\varepsilon/3. \tag{2.11}$$

Now (2.6) and (2.7) imply that there exists $n_0 \geq 0$ so that for $n > n_0$ $G_k(x_n)-kF_A(x_n)<\varepsilon$. Consequently

$$E_A^k(x_n)+kF_A(x_n)\leq G_k(x_n)<\varepsilon+k\,F_A(x_n),\ n>n_0.$$

Thus $\quad E_A^k(x_n)<\varepsilon, n>n_0.$

By (b) (enlarging k and A if necessary) we see t hat $E_A^k(x_n)<\varepsilon$ for all $n\in N$ and thus the proof is complete. ■

Using Theorem 1.8 we get the

2.11 THEOREM. On each reflexive Banach space there is an equivalent norm such that both X and X* are locally uniformly convex.

EXERCISES

1. Let $\|\cdot\|_1$ and $\|\cdot\|_2$ be two equivalent norms on the Banach space X; then

   $$\|x\|_0=\left(\|x\|_1^2+\|x\|_2^2\right)^{1/2} \text{ and } \|\!|x|\!\|=\inf_{y\in X}\left(\|x+y\|_1^2+\|x-y\|_2^2\right)^{1/2}$$

   are equivalent norms on X.

2. Every separable Banach space is isometric to a quotient of ℓ^1. (see Theorem 1. § 2, Ch. IV, Part 2, Diestel [1]).

3. Every separable Banach space is isometric to a subspace of ℓ^∞. (see Theorem 2, § 2, Ch. IV, Part 2, Diestel [1]).

4. ℓ^∞ admits an equivalent strictly convex norm (Klee).

   Hint. Define T: $\ell^\infty \to c_0 = c_0(N)$ by $(Tx)_n=\left(\frac{x_n}{n}\right)_n$.

5. ℓ^∞ cannot be equivalently renormed in a weakly locally uniformly convex manner (Lindenstrauss).
(see Theorem 1, § 5, Ch. IV, Diestel [1]).

6. If Γ is an uncountable set, then $\ell^\infty(\Gamma)$ cannot be renormed in an equivalent strictly convex manner (Day [4]).
(see Theorem 2, § 5, Ch. IV, Diestel [1]).

7. Every separable Banach space is isomorph to a locally uniformly convex Banach space (Kadec [1]).

8. Every super-reflexive space admits an equivalent uniformly convex norm (Enflo [1]).
(see Theorem, Ch. IV, Part IV, Beauzamy [1]).

9. Give the complete proof of the Theorem 1.8.

BIBLIOGRAPHICAL COMMENTS

§ 1. The norm p introduced in the proof of the Theorem 1.4 was defined by Day [4] who proved that it is strictly convex. Theorem 1.4. is due to Klee [1] but the proof follows Rainwater [1]. Theorem 1.5. is also due to Klee [1]. The method of averaging norms exposed in the proofs of Theorem 1.7. and 1.8. is due to Asplund [1]. Connected results are in Enflo [1] and John-Zisler [1] [2].
§ 2. The basic results of Theorems 2.1. and 2.9. are obtained by Lindenstrauss [2]; they are true in a larger class of Banach spaces, namely in weakly compact generated Banach spaces, in short WCG-Spaces; see Amir-Lindenstrauss [1].
In [1] Kadec proved that on a separable Banach space there is an equivalent locally uniformly convex norm. This result was generalized by Trojanski first to Banach spaces with unconditional basis [1] and then to WCG-spaces [2]; see also Trojanski [3]; accurate proofs are in Distel [1]. In Theorem 2.10. we present the reflexive case and the proof is taken from Cioranescu [1]. For the elementary properties of cardinal numbers used in the proof of Proposition 2.7. we send to Kaplansky [1].

CHAPTER IV

ON THE TOPOLOGICAL DEGREE IN FINITE AND INFINITE DIMENSIONS

We give a short (analytical) presentation of Brouwer's theory of the topological degree of continuous mappings in finite dimensional Banach spaces and a generalization of it to A proper mappings in Banach spaces. The connection with the Leray-Schauder degree is commented and some applications to the topological degree of normalized duality mappings are made.

§1. BROUWER'S DEGREE

Consider $R^n = \{x = (x_i)_{1 \le i \le n}; x_i \in R, 1 \le i \le n\}$ endowed with the norm $|x| = (\sum_{i=1}^{n} x_i^2)^{1/2}$; let $\Omega \subset R^n$ be open and bounded; we denote by $\bar{\Omega}$ and $\partial\Omega$ the closure and the boundary of Ω, respectively.

Let $f : \Omega \to R^n$; if $f'(x_0)$ exists, then $I_f(x_0) = \det(\frac{\partial f_i}{\partial x_j}(x_0))$ $(1 \le i, j \le n)$ is the Jacobian of f at x_0.

We denote by $A_{ij}(x_0)$ the cofactor of $\frac{\partial f_j}{\partial x_i}(x_0)$ in $I_f(x_0)$, namely $A_{ij}(x_0)$ is $(-1)^{i+j}$ times the determinant which we obtain from $I_f(x_0)$ cancelling the j-th row and the i-th column.

1.1. LEMMA. Let $\Omega \subset R^n$ be open and $f \in C^2(\Omega)$; then

$$\sum_{i=1}^{n} \frac{\partial A_{ij}(x)}{\partial x_i} = 0 \ ; \quad 1 \le j \le n \ ; \ x \in \Omega. \tag{1.1}$$

PROOF. Fix j and let f_{x_k} denote the column

$$\left(\frac{\partial f_1}{\partial x_k}, \dots, \frac{\partial f_j}{\partial x_k}, \frac{\partial f_{j-1}}{\partial x_k}, \dots, \frac{\partial f_n}{\partial x_k}\right).$$

Then $A_{ij} = (-1)^{i+j} \det(f_{x_1}, \dots, f_{x_{i-1}}, f_{x_{i+1}}, \dots, f_{x_n})$.

Since a determinant is linear in each column, we have

$$\frac{\partial A_{ij}}{\partial x_i} = (-1)^{i+j} \sum_{k=1}^{n} \det\left(f_{x_1}, \dots, f_{x_{i-1}}, f_{x_{i+1}}, \dots, \frac{\partial f_{x_k}}{\partial x_i}, \dots, f_{x_n}\right).$$

Let be $d_{ki} = \det\left(\frac{\partial f_{x_k}}{\partial x_i}, f_{x_1}, \dots, f_{x_{i-1}}, f_{x_{i+1}}, \dots, f_{x_{k-1}}, f_{x_{k+1}}, \dots, f_{x_n}\right)$.

Since $f \in C^2(\Omega)$, then $\frac{\partial^2 f_m}{\partial x_i \partial x_k} = \frac{\partial^2 f_m}{\partial x_k \partial x_i}$, $1 \le i,k,m \le n$; therefore $d_{ki} = d_{ik}$ and this yields

$$(-1)^{i+j} \frac{\partial A_{ij}}{\partial x_i} = \sum_{k<i} (-1)^{k-1} d_{ki} + \sum_{k>i} (-1)^{k-2} d_{ki} = \sum_{k=1}^{n} (-1)^{k-1} \sigma_{ki} d_{ki}$$

with

$$\sigma_{ki} = \begin{cases} 1 & \text{for} \quad k < i \\ 0 & \text{for} \quad k = i \\ -1 & \text{for} \quad k > i \ . \end{cases}$$

But then $\sigma_{ki} = -\sigma_{ik}$ for all $1 \le i, k \le n$ and we get

$$(-1)^j \sum_{i=1}^{n} \frac{\partial A_{ij}}{\partial x_i} = \sum_{i,k=1}^{n} (-1)^{k-1+i} \sigma_{ki} d_{ki} =$$

$$= \sum_{k,i=1}^{n} (-1)^{i-1+k} \sigma_{ik} d_{ik} = - \sum_{i,k=1}^{n} (-1)^{k-1+i} \sigma_{ki} d_{ki} \ .$$

Hence (1.1) is proved. ■

1.2. LEMMA . Let $f \in C^1(\Omega) \cap C(\bar{\Omega})$ and $\varepsilon > 0$ such that $|f(x)| > \varepsilon$ for every $x \in \partial\Omega$; let φ be a real continuous function with $\operatorname{supp}\varphi \subset (0,\varepsilon)$ and $\int_0^\infty t^{n-1}\varphi(t)dt = 0$; then

$$\int_\Omega \varphi(|f(x)|) . I_f(x)dx = 0. \tag{1.2}$$

PROOF. By Weierstrass Approximation Theorem we can suppose $f \in C^2(\Omega)$. Denote

$$\psi(t) = \begin{cases} t^{-n}\int_0^t s^{n-1}\varphi(s)ds & \text{for} \quad t > 0 \\ 0 & \text{for} \quad t < 0 . \end{cases}$$

Then $\operatorname{supp}\psi \subset (0,\varepsilon), \psi \in C^1$ and satisfies the following differential equation

$$t\psi'(t) + n\psi(t) = \varphi(t), \quad t \geq 0. \tag{1.3}$$

Consider now the functions $g_i : R^n \to R$ defined as follows

$$g_i(y) = \psi(|y|) . y_i, \qquad 1 \leq i \leq n, \qquad y = (y_1, \ldots, y_n) \in R^n .$$

Then $g_i \in C^1 R^n$ and $g_i(y) = 0$ for $|y| \geq \varepsilon$; hence the functions $R^n \ni x \to g_i(f(x))$ are of class C^1 on Ω and zero on a neighborhood of $\partial\Omega$. Using now (1.1) and (1.3) we get

$$\sum_{i=1}^n \frac{\partial}{\partial x_i} \sum_{j=1}^n A_{ij}(x) g_j(f(x))$$

$$= \sum_{j=1}^n \Big(\sum_{i=1}^n \frac{\partial A_{ij}(x)}{\partial x_i} \Big) g_j(f(x)) + \sum_{j=1}^n \sum_{i=1}^n A_{ij}(x) \Big[\sum_{k=1}^n \frac{\partial f_k(x)}{\partial x_i} . \frac{\partial g_j(f(x))}{\partial y_k} \Big]$$

$$= \sum_{j=1}^n \sum_{k=1}^n \Big[\Big(\sum_{i=1}^n A_{ij}(x) . \frac{\partial f_k(x)}{\partial x_i} \Big) . \frac{\partial g_j(f(x))}{\partial y_k} \Big]$$

$$= \sum_{j=1}^n I_f(x) . \frac{\partial g_j(f(x))}{\partial y_j} = I_f(x) . \sum_{j=1}^n \frac{\partial g_j(f(x))}{\partial y_j}$$

$$= I_f(x) . \Big[n\psi(|f(x)|) + \sum_{j=1}^n \psi'(|f(x)|) . I_f(x) . \frac{f_j^2(x)}{|f(x)|} \Big] = I_f(x) . \varphi(|f(x)|).$$

We apply now the theorem of Gauss-Ostrogradski to get the result.■

1.3. DEFINITION. Let $f \in C^1(\Omega) \cap C(\bar{\Omega})$ and $y \in R^n \setminus f(\partial\Omega)$; let $0 < \varepsilon < \min_{x \in \partial\Omega} |f(x) - y|$ and consider a real continuous function φ with $\operatorname{supp}\varphi \subset (0,\varepsilon]$ and $\int_{R^n} \varphi(|x|)dx = 1$. We define Brouwer's degree to be

$$d(f,\Omega,y) = \int_\Omega \varphi(|f(x) - y|) I_f(x)dx.$$

1.4. PROPOSITION. The definition of $d(f,\Omega,y)$ is independent of the function φ.

PROOF. Let be

$$\mathcal{G} = \{\varphi;\ \text{real continuous},\ \operatorname{supp}\varphi \subset (0,\varepsilon],\ 0 < \varepsilon < \min_{x \in \partial\Omega} |f(x) - y|\};$$

we can define on $\mathcal{G}$ the following three linear functionals

$$L(\varphi) = \int_0^\infty t^{n-1}\varphi(t)dt;\ M(\varphi) = \int_{R^n} \varphi(|x|)dx \text{ and } N(\varphi) = \int_\Omega \varphi(|f(x) - y|) I_f(x)dx.$$

Consider a $\varphi \in \mathcal{G}$ with $L(\varphi) = 0$ and apply Lemma 1.2. to the function $id: S_r \to R^n$ where $S_r = \{x; |x| < r\} \supseteq \Omega$, $id(x) = x$ and $r > \varepsilon$; we obtain

$$M(\mathcal{G}) = \int_{R^n} \varphi(|x|)dx = 0.$$

Apply again the same Lemma to the function $\Omega \ni x \to f(x) - y$; we get

$$N(\varphi) = \int_\Omega \varphi(|f(x) - y|) I_f(x)dx = 0.$$

Choose $\varphi_1, \varphi_2 \in \mathcal{G}$ with $M(\varphi_1) = M(\varphi_2) = 1$; then

$$L[L(\varphi_2)\varphi_1 - L(\varphi_1)\varphi_2] = 0$$

and by the above considerations we obtain

$$0 = M[L(\varphi_2)\varphi_1 - L(\varphi_1)\varphi_2] = L(\varphi_2)M(\varphi_1) - L(\varphi_1)M(\varphi_2)$$

$$= L(\varphi_1) - L(\varphi_2) = L(\varphi_1 - \varphi_2)$$

Then $N(\varphi_1 - \varphi_2) = 0$, i.e. $N(\varphi)$ is independent of $\varphi \in \mathcal{G}$ with $M(\varphi) = 1$. ■

1.5. REMARK. If $y \notin f(\bar{\Omega})$, then $d(f,\Omega,y) = 0$.

In effect, we can find an $\varepsilon > 0$ with $|f(x) - y| > \varepsilon$ for every $x \in \bar{\Omega}$ and then $\varphi(|f(x) - y|) = 0$, $\forall\, x \in \bar{\Omega}$ and $\varphi \in \mathcal{G}$.

1.6. PROPOSITION. $d(id, \Omega, y) = 1$ for $y \in \Omega$.

PROOF. For $y \in \Omega$, let $0 < 2\varepsilon = \operatorname{dist}(y, \partial\Omega)$; then $S_\varepsilon(y) = \{x; |x-y| < \varepsilon\} \subseteq \Omega$ hence

$$d(id, \Omega, y) = \int_\Omega \varphi(|x-y|)dx = \int_{S_\varepsilon} \varphi(|x-y|)dx = \int_{R^n} \varphi(|x|)dx = 1 . \qquad \blacksquare$$

In order to extend the definition of the degree to continuous functions, we give the

1.7. LEMMA . Let $f, g \in C^1(\Omega) \cap C(\bar{\Omega}), y_1, y_2 \in R^n$ and $\varepsilon > 0$ so that $\min\limits_{x \in \partial\Omega}\{|f(x) - y_1|, |g(x) - y_2|\} > 5\varepsilon$ and $\max\limits_{x \in \bar{\Omega}}|f(x) - g(x) - y_1 + y_2| < \varepsilon$. Then $d(f, \Omega, y_1) = d(g, \Omega, y_2)$.

PROOF. It is clear from the definition of the degree that $d(f, \Omega, y) = d(f(\cdot) - y, \Omega, 0)$ so that we can suppose $y_1 = y_2 = 0$.

Consider a function $\psi: R^n \to R$ of class C^1 so that

$$\psi(x) = \begin{cases} 1 & \text{for } |x| \le 2\varepsilon \\ 0 & \text{for } |x \ge 3\varepsilon| \end{cases} \qquad \text{and} \qquad |\psi(x)| \le 1, x \in R^n .$$

Define

$$h(x) = [1 - \psi(f(x))]f(x) + \psi(f(x))g(x), \qquad x \in \bar{\Omega} .$$

It is clear that $h \in C^1(\Omega) \cap C(\bar{\Omega})$ and that

$$|f(x) - h(x)| < \varepsilon, |g(x) - h(x)| < 2\varepsilon \quad \text{for every } x \in \bar{\Omega} .$$

Observe that

$$h(x) = \begin{cases} f(x) & \text{if } |f(x)| \ge 3\varepsilon \\ g(x) & \text{if } |f(x)| \le 2\varepsilon \end{cases}$$

Then it is clear that

$$\max_{x \in \partial\Omega}\{|f(x)|, |g(x)|, |h(x)|\} > 5\varepsilon . \tag{1.4}$$

Consider now $\varphi_1, \varphi_2 : R_+ \to R$, continuous with

$\operatorname{supp}\varphi_1 \subseteq [4\varepsilon,5\varepsilon]$, $\operatorname{supp}\varphi_2 \subset (0,\varepsilon]$ and $\int_{R^n} \varphi_1(|x|)dx = \int_{R^n} \varphi_2(|x|)dx = 1$

For every $x \in \Omega$ we have:

$$\varphi_1(|h(x)|)I_h(x) = \varphi_1(|f(x)|)I_f(x) \text{ and } \varphi_2(|h(x)|)I_h(x) = \varphi_2(|g(x)|)I_g(x). \quad (1.5)$$

In effect, if $|f(x)| \geq 3\varepsilon$, then $h(x) = f(x)$ and if $|f(x)| < 3\varepsilon$ then $|h(x)| \leq |f(x)| + |h(x) - f(x)| < 4\varepsilon$. Hence $\varphi_1(|h(x)|) = \varphi_1(|f(x)|) = 0$.

Analogously,if $|f(x)| \leq 2\varepsilon$, then $h(x) = g(x)$; if $|f(x)| > 2\varepsilon$, then $|h(x)| \geq |f(x)| - |h(x) - f(x)| \geq \varepsilon$ and $|g(x)| \geq |f(x)| - |g(x) - f(x)| > \varepsilon$ so that $\varphi_2(|h(x)|) = \varphi_2(|g(x)|) = 0$.

Taking in account (1.4) as well as the location of the supports of φ_1 and φ_2, we get by integration of (1.5) over Ω:

$$d(h,\Omega,0) = d(f,\Omega,0) = d(g,\Omega,0). \qquad \blacksquare$$

1.8. DEFINITION. Let $f \in C(\bar{\Omega})$ and $y \in R^n \setminus f(\partial\Omega)$; consider a sequence $\{f_k\}_{k\in N} \subset C^1(\Omega) \cap C(\bar{\Omega})$, so that $y \notin f_k(\partial\Omega)$, $k \in N$ and $\lim_{k\to\infty} f_k = f$ uniformly on Ω; then we define

$$d(f,\Omega,y) = \lim_{k\to\infty} d(f_k,\Omega,y).$$

1.9. REMARK. The existence of the sequence $\{f_k\}_{k\in N}$ is assured by the Weierstrass approximation Theorem.

If $y \notin f(\partial\Omega)$, then $\min_{x\in\partial\Omega} |f(x) - y| > \varepsilon$ for some $\varepsilon > 0$ and by the uniform convergence there is k_ε so that $y \notin f_k(\partial\Omega)$ for $k \geq k_\varepsilon$. The Lemma 1.7 implies that for k sufficiently large all $d(f_k,\Omega,y)$ coincide, thus the above limit exists and is independent of the sequence $\{f_k\}_{k\in N}$ with the properties of the definition.

1.10. PROPOSITION. Let Ω_1,Ω_2 be open disjoint subsets of Ω, $f \in C(\bar{\Omega})$ and $y \notin f(\bar{\Omega}\setminus(\Omega_1 \cup \Omega_2))$; then

$$d(f,\Omega,y) = d(f,\Omega_1,y) + d(f,\Omega_2,y).$$

PROOF. It is sufficient to prove the assertion for $f \in C^1(\Omega)$.

Let be $0 < \varepsilon < \min_{x\in\Omega\setminus\Omega_1\cup\Omega_2} |f(x) - y|$ and $\varphi: R_+ \to R$ continuous with $\operatorname{supp}\varphi \subset (0,\varepsilon]$ and $\int_{R^n} \varphi(|x|)dx = 1$; then

$$d(f,\Omega,y) = \int_{\Omega_1 \cup \Omega_2} \varphi(|f(x)-y|) I_f(x)dx$$

$$= \int_{\Omega_1} \varphi(|f(x)-y|) I_f(x)dx + \int_{\Omega_2} \varphi(|f(x)-y|) I_f(x)dx = d(f,\Omega_1,)(y) + d(f,\Omega_2,y).$$

■

1.11. COROLLARY. Let be $\Omega_1 \subseteq \Omega$ and $y \notin f(\bar{\Omega} \backslash \Omega_1)$; then $d(f,\Omega,y) = d(f,\Omega_1,y)$. Moreover for $\Omega_1, \Omega_2 \subseteq \Omega$ open, disjoint sets and $y \notin f(\partial\Omega_1 \cup \partial\Omega_2)$ we have

$$d(f,\Omega_1 \cup \Omega_2,y) = d(f,\Omega_1,y) + d(f,\Omega_2,y).$$

PROOF. Apply the above proposition to Ω_1 and $\Omega_2 = \phi$ to obtain the first part of the assertion. For the second part denote $\Omega = \Omega_1 \cup \Omega_2$ and observe that $\bar{\Omega} \backslash \Omega_1 \cup \Omega_2 = \partial\Omega_1 \cup \partial\Omega_2$; then one can apply again the Proposition 1.10. ■

1.12. THEOREM. Let $\Omega \subset R^n$ be open and bounded, $F:[0,1] \times \bar{\Omega} \to R^n$ and $y:[0,1] \to R^n$ be continuous s.t. $y(t) \notin F(t,\partial\Omega)$ for every $t \in [0,1]$; then $d(F(t,.),\Omega,y(t))$ is independent of $t \in [0,1]$..

PROOF. It is clear that we can find $\varepsilon > 0$ so that $|F(t,x) - y(t)| > 6\varepsilon$ for every $(x,t) \in [0,1] \times \partial\Omega$. Moreover there is $\delta > 0$ so that for $|t_1 - t_2| < \delta$ and $x \in \bar{\Omega}, |F(t_1,x) - F(t_2,x)| < \varepsilon/4$ and $y(t_1) - y(t_2)| < \varepsilon/4$. For $t_1, t_2 \in [0,1]$ fixed with $|t_1 - t_2| < \delta$ we approximate uniformly the functions $F(t_1,.)$ and $F(t_2,.)$ by sequences $\{f_k^1\}_{k \in N}$, $\{f_k^2\}_{k \in N}$ as in the Definition 1.8; then we can find $k_\varepsilon \in N$ such that

$$\left|f_k^1(x) - y(t_1)\right| > 5\varepsilon \text{ and } \left|f_k^2(x) - y(t_2)\right| > 5\varepsilon \text{ for } k \geq k_\varepsilon \quad x \in \partial\Omega$$

and

$$\left|f_k^1(x) - f_k^2(x) - y(t_1) + y(t_2)\right| < \varepsilon \text{ for } x \in \bar{\Omega}.$$

By Lemma 1.7. we have that

$$d\left(f_k^1,\Omega,y(t_1)\right) = d\left(f_k^2,\Omega,y(t_2)\right), \; k \geq k_\varepsilon$$

so that

$$d(F(t_1,.),\Omega,y(t_1)) = d(F(t_2,.),\Omega,y(t_2)).$$

Since the points t_1, t_2 were arbitrary in $[0,1]$ with $|t_1 - t_2| < \delta$, it follows that $d(F(t,.),\Omega,y(t))$ is constant for $t \in [0,1]$. ■

1.13. COROLLARY. If $f,g \in C(\bar{\Omega})$ and f and g coincide on $\partial\Omega$, then for every $y \notin f(\partial\Omega)$ we have $d(f,\Omega,y) = d(g,\Omega,y)$.

PROOF. Consider $F(t,x) = tf(x) + (1-t)g(x), x \in \bar{\Omega}, t \in [0,1]$; then $F(t,x) = f(x) = g(x)$ for $(t,x) \in \partial\Omega \times [0,1]$. Hence $y \notin F([0,1] \times \partial\Omega)$. By the above Theorem we have

$$d(F(0,.),\Omega,y) = d(F(1,.),\Omega,y)\text{ , i.e. } d(f,\Omega,y) = d(g,\Omega,y). \quad ■$$

1.14. COROLLARY. For $f \in C(\bar{\Omega}), d(f,\Omega,.)$ is constant on every connected component of $R^n \setminus f(\partial\Omega)$.

PROOF Since $R^n \setminus f(\partial\Omega)$ is open, its connected components are open; for open sets in R^n connectedness is the same as arcwise connectedness. Therefore if C is a component of $R^n \setminus f(\partial\Omega)$ and $y_1, y_2 \in C$, there is a continuous curve $y:[0,1] \to R^n$ with $y(0) = y_1$ and $y(1) = y_2$.

The result is now a consequence of the Theorem 1.12. for $F(t,x) = f(x)$. ■

1.15. THEOREM. Let be $f \in C(\bar{\Omega})$ and $y \notin f(\partial\Omega)$; if $d(f,\Omega,y) \neq 0$, then there exists at least a point $z \in \Omega$ so that $f(z) = y$.

PROOF. Suppose that $f(x) \neq y$ for every $x \in \Omega$; then $f(x) \neq y$ on $\bar{\Omega}$ and this implies the existence of an $\varepsilon > 0$, such that $|f(x) - y| > \varepsilon$ for every $x \in \bar{\Omega}$. Consider a sequence $\{f_k\}_{k \in N} \subset C^1(\Omega) \cap C(\bar{\Omega})$ approximating uniformly f; then $|f_k(x) - y| > \varepsilon/2$ for $x \in \bar{\Omega}$ and k sufficiently large, i.e. $y \notin f_k(\bar{\Omega})$. By the Remark 1.5 it follows that for k sufficiently large $d(f_k,\Omega,y) = 0$.

Hence $d(f,\Omega,y) = \lim_k d(f_k,\Omega,y) = 0$, which is a contradiction. ■

1.16. THEOREM. (Brouwer). Let f be a continuous mapping from the closed unit ball $\bar{S}_1(0)=\{x;|x|\le 1\}$ into itself; then f has fixed point.

PROOF. Suppose that $f(x)\ne x$ for every $x\in\bar{S}_1(0)$ and denote by $F(t,x)=x-tf(x)$, $t\in[0,1]$. Then $0\notin F([0,1]\times\partial S_1(0))$; indeed if $(t,x)\in[0,1]\times\partial S_1(0)$ then $|F(t,x)|\ge|x|-t|f(x)|\ge 1-t>0$ and if $t=1$, then $F(1,x)=x-f(x)\ne 0$ on $\partial S_1(0)$.

Hence by Theorem 1.12 and Proposition 1.6 we have;

$$d\big((F(1,.),S_1(0),0\big)=d(f(0,.),S_1(0),0)=d(id,S_1(0),0)=1.$$

The existence of a fixed point is now a consequence of the Theorem 1.15. ■

1.17. DEFINITION. Let $f\in C(\bar{S}_r(x_0))$ and $f(x)\ne f(x_0)$ in $\bar{S}_r(x_0)\setminus\{x_0\}$; we call the index of f at x_0 at the number $d(f,S_r(x_0),f(x_0))$ and denote it by $i(f,x_0)$.

1.18. REMARK. By Corollary 1.11. it follows that the index is independent of $r>0$ with the above properties.

1.19. PROPOSITION. Let be $f\in C(\bar{\Omega})$ and $x_0\in\Omega$. If f is of class C^1 on a neighbourhood of x_0 and $I_f(x_0)\ne 0$, then $i(f,x_0)=\text{sign } I_f(x_0)$.

PROOF. Consider a ball $S_{r_0}(x_0)$ on which f is of class C^1 and denote $A=f'(x_0)$; then

$$f(x)-f(x_0)=A(x-x_0)+0(|x-x_0|) \quad \text{as} \quad |x-x_0|\to 0$$

where the remainder if of class C^1 on $S_{r_0}(x_0)$.

Since $\det A=I_f(x_0)\ne 0$, then A^{-1} exists; we can write A under its polar form $A=TP$, where T is an orthogonal matrix and P is positive definite and symmetric; both T and P are invertible and $|Tx|=|x|$ for every $x\in R^n$.

For $t\in[0,1]$ and $x\in S_{r_0}(x_0)$ let be

$$F(t,x)=f(x_0)+T[(1-t)P(x-x_0)+t(x-x_0)]+(1-t)0(|x-x_0|)$$

Since P has no negative eigenvalues, then $\det[(1-t)P + t\,id] \neq 0$ in $[0,1]$; therefore $|(1-t)Px + tx| \geq M_t|x|$ for some $M_t > 0$ and $x \in R^n$, $t \in [0,1]$. Thus

$$|F(t,x) - f(x_0)| \geq |T[(1-t)P(x-x_0) + t(x-x_0)]| - (1-t)O(|x-x_0|)|$$
$$= |(1-t)P(x-x_0) + t(x-x_0)| - (1-t)O(|x-x_0|)| \geq$$
$$\geq M_t \cdot |x-x_0| - |O(|x-x_0|)| > 0$$

for all $t \in [0,1]$ and $x \in S_r(x_0)\backslash\{x_0\}$ provided that $r > 0$ is sufficiently small.
Hence by Theorem 1.12

$$d(f, S_r(x_0), f(x_0)) = d(F(0,.), S_r(x_0), f(x_0))$$

$$= d(F(1,.), S_r(x_0), f(x_0)) = d(F(1,.) - f(x_0), S_r(x_0), 0)$$

$$= d(T(x-x_o), S_r(x_o), 0) = \int_{S_r(x_o)} \varphi(|T(x-x_0)|) I_T(x-x_0)dx$$

$$= \int_{S_r(x_0)} \varphi(|x-x_0|) \det T dx = \det T . \int_{S_r(0)} \varphi(|y|)dy$$

$$= \det T . \int_{R^n} \varphi(|y|)dy = \text{sign } \det A = \text{sign } I_f(x_0) \cdot \qquad \blacksquare$$

1.20. COROLLARY. Let A be a linear map on R^n with $\det A \neq 0$; then $i(A,x) = \text{sign } \det A$, for every $x \in R^n$.

1.21. COROLLARY. Let be $f \in C(\bar{\Omega})$, $y \notin f(\partial\Omega)$ and suppose that in Ω the equation $f(x) = y$ has only the solutions $x_1, \ldots, x_k$; then if f is of class C^1 in a neighbourhood of every $x_j, 1 \leq j \leq k$ and $I_f(x_j) \neq 0$, we have: $d(f, \Omega, y) = \sum_{j=1}^{k} \text{sign } I_f(x_j)$.

PROOF. Consider the disjoint balls $S_{r_j}(x_j)$, $1 \leq j \leq k$; then the open set $\Omega_0 = \bigcup_{j=1}^{k} S_{r_j}(x_j)$ contains all the zeros of the function $f(x) = y$ and by Corollary 1.11

$$d(f, \Omega, y) = d(f, \Omega_0, y) = \sum_{j=1}^{k} d(f, S_{r_j}(x_j), f(x_j))$$

$$= \sum_{j=1}^{k} i(f, x_j)) = \sum_{j=1}^{k} \text{sign } I_f(x_j).$$

■

1.22. LEMMA. Let be $f \in C^1(\Omega) \cap C(\bar{\Omega})$, $y_0 \in R^n \backslash f(\partial\Omega)$ and $\varepsilon < \min_{x \in \partial\Omega} |f(x) - y_0|$; then there exists $y \in S_\varepsilon(y_0) \backslash f(\partial\Omega)$ so that the equation $f(x) = y$ has at most a finite number of solutions $x_1, \ldots, x_k$ in Ω and $I_f(x_j) \neq 0$, $1 \leq j \leq k$; moreover if $y_0 \notin \{f(x); I_f(x) = 0\}$ we can take $y = y_0$.

PROOF. Let $K = \{x \in \bar{\Omega}; |f(x) - y_0| \leq \varepsilon\}$ and $\Gamma = \{x \in K; I_f(x) = 0\}$. Then $K \subset \Omega$ and Γ is closed, hence measurable; by Sard's lemma we have

$$m(f(\Gamma) \leq \int_\Gamma |I_f(x)| dx = 0.$$

Hence $f(\Gamma)$ has a void interior, so that there exists a $y \in S_\varepsilon(y_0) \backslash f(\Gamma)$. Consider a solution $x \in \bar{\Omega}$ of the equation $f(x) = y$; then $|f(x) - y_0| = |y - y_0| \leq \varepsilon$, that is $x \in K \subset \Omega$; hence $y \notin f(\partial\Omega)$. Since $y \notin f(\Gamma)$, it follows that $x \notin \Gamma$, i.e. $I_f(x) \neq 0$.

Suppose that there are infinitely many distinct $\{x_k\}_{k \in N}$ with $f(x_k) = y, k \in N$. Then there is $x_0 = \lim_{k' \to \infty} x_{k'}$, where $\{x_{k'}\}_{k'}$ is a subsequence of $\{x_k\}_{k \in N}$; consequently , hence $x_0 \in K$ and $I_f(x_0) \neq 0$. We have

$$0 = f(x_k) - f(x_0) = I_f(x_k^*)(x - x_0), \text{ for some } x_k^* \in [x_0, x_k], k \in N.$$

Therefore $I_f(x_k^*) = 0$ and this yields that $I_f(x_0) = 0$, which is a contradiction. ■

1.23. THEOREM. For every $f \in C(\bar{\Omega})$ and $y \notin f(\partial\Omega)$, $d(f, \Omega, y)$ is an integer.

PROOF. It is sufficient to prove the statement for $f \in C^1(\Omega) \cap C(\bar{\Omega})$. By the above Lemma we can construct a sequence $\{y_k\}_{k \in N}$ converging to y so that for each y_k the conditions of the Corollary 1.21 are satisfied; then $d(f, \Omega, y)$ has to be an integer. ■

1.24. THEOREM. (Borsuk) Let $\Omega \subseteq R^n$ be open, bounded, symmetric and $0 \in \Omega$; let $f \in C(\bar{\Omega})$ be odd and $0 \notin f(\partial\Omega)$; then $d(f,\Omega,0)$ is odd.

PROOF. We may assume that $f \in C^1(\Omega) \cap C(\bar{\Omega})$ and $I_f(0) \neq 0$. Indeed, we only have to approximate f uniformly by $f_k \in C^1(\Omega) \cap C(\bar{\Omega})$, to consider the odd functions $f_k^*(x) = \frac{1}{2}[f_k(x) - f_k(-x)]$ and to choose λ_k which are not eigenvalues of $(f_k^*)'(0)$. Then $\tilde{f}_k = f_k^* - \lambda_k . \mathrm{id} \in C^1(\Omega) \cap C(\bar{\Omega})$, $I_{\tilde{f}_k}(0) \neq 0$ and moreover if $\lambda_k \underset{k}{\to} 0$ then $\lim_{k\to\infty} \tilde{f}_k = f$ uniformly on Ω.

We shall construct by induction an odd map $g \in C^1(\Omega) \cap C(\bar{\Omega})$ sufficiently close to f such that $0 \notin \{g(x); x \in \Omega, I_g(x) = 0\}$. Then by Lemma 1.7, Lemma 1.22 and Corollary 1.20 we shall have

$$d(f,\Omega,0) = d(g,\Omega,0) = \sum_{g(x)=0} \text{sign } I_g(x)$$

$$= \text{sign } I_g(0) + \sum_{\substack{g(x)=0 \\ x\neq 0}} \text{sign } I_g(x) .$$

But $g(x) = 0$ iff $g(-x) = 0$ and the function $I_g(\cdot)$ is even, hence the sum $\sum_{\substack{g(x)=0 \\ x\neq 0}} \text{sign } I_g(x)$ is also even and this will end our proof.

For $k \in N$, denote $\Omega_k = \{x \in \Omega; x_i \neq 0 \text{ for some } i \leq k\}$ and for $h:\Omega_k \to R^n$ denote $\Gamma_{k,h} = \{x \in \Omega_k; I_h(x) = 0\}$. Pick an odd function $\psi \in C^1(R)$ such that $\psi'(0) = 0$ and $\psi(t) = 0$ iff $t = 0$, and define on Ω_1 the function $h_1(x) = \dfrac{f(x)}{\psi(x_1)}$.

By Sard's lemma we find $y^1 \in S_r \setminus h(\Gamma_{1,h_1})$, where r is arbitrary small. Define $g_1(x) = f(x) - \psi(x_1).y^1$ on Ω; then $0 \notin \{g_1(x); x \in \Gamma_{1,g_1}\}$. Indeed if $g_1(x) = 0$ for $x \in \Gamma_{1,g_1}$, then $y^1 = \dfrac{f(x)}{\psi(x_1)} = h_1(x) \notin h_1(\Gamma_{1,h_1})$, i.e.

$x \notin \Gamma_{1,h_1}$. Hence $I_{h_1}(x) \neq 0$, and since $I_{g_1}(x) = \psi(x_1)I_{h_1}$, we get that also $I_{g_1}(x) \neq 0$, which is a contradiction.

Now suppose that we have already constructed an odd $g_k \in C^1(\Omega) \cap C(\bar{\Omega})$ close to f on $\bar{\Omega}$ so that $0 \notin \{g_k(x); x \in \Gamma_{k,g_k}\}$ for $k < n$. We define

$$g_{k+1}(x) = g_k(x) - \psi(x_{k+1})y^{k+1} \text{ on } \Omega$$

where $y^{k+1} \in S_r \setminus f\left(\Gamma_{k+1,h_{k+1}}\right)$ and $h_{k+1} = \dfrac{f}{\psi(x_{k+1})}$ for $x_{k+1} \neq 0$.

It is clear that $0 \notin \{g_{k+1}(x); x \in \Omega, x_{k+1} \neq 0, I_{g_{k+1}}(x) = 0\}$. Moreover if $x \in \Omega_{k+1}$ and $x_{k+1} = 0$, then $x \in \Omega_k, g_{k+1}(x) = g_k(x)$ and $g'_{k+1}(x) = g'_k(x)$, hence $I_{g_{k+1}}(x) \neq 0$ and therefore $0 \notin \{g_{k+1}(x); x \in \Gamma_{k+1,g_{k+1}}\}$. Evidently $g_{k+1} \in C^1(\Omega) \cap C(\bar{\Omega})$ is odd and close to f on $\bar{\Omega}$. Define $g = g_n$; then g is odd, close to f on $\bar{\Omega}$ and such that $0 \notin \{g(x); x \in \Omega_n = \Omega \setminus \{0\}, I_g(x) = 0\}$. We see that $g(0) = g_{n-1}(0) = \ldots = g_1(0) = 0$ and that $g'(0) = g'_{n-1}(0) = \ldots = g'_1(0) = f'(0)$; therefore $I_g(0) = I_f(0) \neq 0$ and thus $0 \notin \{g(x); x \in \Omega, I_g(x) = 0\}$. ■

1.25. COROLLARY. Let $\Omega \subset R^n$ be bounded, open, symmetric and $0 \in \Omega$; let $f \in C(\bar{\Omega})$ be such that $0 \notin f(\partial\Omega)$ and $f(-x) \neq \lambda f(x)$ on $\partial\Omega$ for all $\lambda \geq 1$; then $d(f, \Omega, 0)$ is odd.

PROOF. Let $F(t,x) = f(x) - tf(-x)$, $t \in [0,1]$; then $F(t,x) \neq 0$ on $\partial\Omega$. Since $F(1,.)$ is odd, we obtain

$$d(f, \Omega, 0) = d(F(0,.), \Omega, 0) = d(F(1,.), \Omega, 0) = \text{odd}.$$ ■

1.26. PROPOSITION. Let $\Omega \subset R^n$ be open and $f: \Omega \to R^n$ continuous and locally injective; then $f(\Omega)$ is open.

PROOF. It is sufficient to show that to any $x_0 \in \Omega$ there exists a ball with centre $f(x_0)$ included in $f(\Omega)$. Passing to $\Omega - x_0$ and $\tilde{f}(x_0) = f(x + x_0) - f(x_0)$ (defined on $\Omega - x_0$), we may assume $x_o = 0$

and $f(0)=0$. Let us choose $r>0$ such that f is injective on $\bar{S}_r(0)\subset\Omega$. Define

$$F(t,x)=f\left(\tfrac{1}{1+t}x\right)-f\left(-\tfrac{t}{1+t}x\right) \text{ for } t\in[0,1],\ x\in\bar{S}_r(0).$$

Obviously F is continuous, $F(0,.)=f$ and $F(1,x)=f\left(\tfrac{1}{2}x\right)-f\left(-\tfrac{1}{2}x\right)$ is odd.

If $F(t,x)=0$ for some $(t,x)\in[0,1]\times\partial S_r(0)$ then as f is injective, $\frac{x}{1+t}=\frac{tx}{1+t}$, i.e. $x=0$, a contradiction. Therefore by Borsuk's Theorem

$$d(f,S_r(0),0)=d(F(1,0),S_r(0),0)\neq 0.$$

Hence $d(f,S_r(0),y)\neq 0$ for every y in some ball $S_{r'}(0)$ and now Theorem 1.15 yields $S_{r'}(0)\subset f(S_r(0))$. ■

1.27. COROLLARY. Let $f:R^n\to R^n$ be continuous, injective and $|f(x)|\to\infty$ as $|x|\to\infty$; then $f(R^n)=R^n$.

PROOF. By the above proposition, $f(R^n)$ is open. It is also closed; indeed: $f(x_n)\underset{n}{\to}y$ implies that $\{x_n\}_n$ is bounded, hence $x_{n'}\underset{n'}{\to}x_0$ for a subsequence and therefore $y=f(x_0)$. Now the surjectivity of f is a consequence of the connectedness of R^n. ■

1.28. REMARK. Let X be an n-dimensional normed real space; then X is homeomorphic to R^n. For $F:\bar{\Omega}\to X$, $F\in C(\bar{\Omega})$ and $y\notin F(\partial\Omega)$ we can define the degree by the formula: $d(F,\Omega,y)=d(f,h(\Omega),h(y))$, where h is the homeomorphism of X onto R^n corresponding to a given base in X and $f=hFh^{-1}$. This definition of the degree is independent of the base in X. Indeed, let A be the transform matrix from one base into the other, with $\det A\neq 0$; then if $\tilde{x}=Ax, \tilde{\Omega}=A\Omega$, $G(\tilde{x})=AF\left(A^{-1}\tilde{x}\right)$ are the representations of x,Ω and F with respect to the new base, then

$$I_G(\tilde{x})=\det A . I_F\left(A^{-1}\tilde{x}\right)\det A^{-1}=I_F\left(A^{-1}\tilde{x}\right)=I_F(x).$$

Hence we obtain the same value for the degree in the differentiable case, hence in general.

Notice that if we have two real n dimensional vector spaces X and Y and two bases in X and Y respectively, then for $F:\bar{\Omega} \subset \to Y$ and $y \notin F(\partial\Omega)$, we can define $d(F,\Omega,y) = d\left(f,h(\Omega),\tilde{h}(y)\right)$, where h and $\tilde{h}$ are the corresponding homeomorphisms of X, respectively Y onto R^n and $f = \tilde{h} F h^{-1}$. However, if we change the bases, then $h = A\hat{h}, \tilde{h} = B h^*$ and $B^{-1} f A$ is the new f; therefore

$$d\left(B^{-1} fA, \hat{h}(\Omega), h^*(y)\right) = \operatorname{sign}(\det A . \det B) d\left(f,h(\Omega),\tilde{h}(y)\right).$$

Our definition will be independent of the bases only if, for example, $\det A > 0$ and $\det B > 0$. We call X oriented if we consider as admissible only those bases in X for which the transform matrix has a positive determinant. The degree of continuous maps between oriented spaces of the same dimension is then well defined.

1.29 REMARK. Finally suppose that X_n is a real normed space with $\dim X_n = n, X_m$ a subpace with $\dim X_m = m, n > m$, $f:\bar{\Omega} \subset X_n \to X_m$ is continuous and $y \notin g(\partial\Omega)$, where $g = id - f$.
Then $d(g,\Omega,y) = d(g_m,\Omega_m,y)$, where $\Omega_m = \Omega \cap X_m$ and $g_m = g|\bar{\Omega}_m$.

Indeed, by Remark 1.28, we may assume $X_n = R^n$ and $X_m = R^m = \{x \in R^n; x_{m+1} = \ldots = x_n = 0\}$. Then, in the differentiable case we have: $I_{g_m} = \det(I_m - (\frac{\partial f_i}{\partial x_j}(x)))$, $1 \le i,j \le m$, and

$$I_g(x) = \det\begin{pmatrix} I_m - (\frac{\partial f_i}{\partial x_j}(x)) & -(\frac{\partial f_i}{\partial x_j}(x)) \\ (0) & I_{n-m} \end{pmatrix}$$

where I_{n-m} is the $(n-m)\times(n-m)$ identity matrix and (0) the $(n-m)xm$ zero matrix. It is now clear that $I_g(x) = I_{g_m}(x)$.

§2. BROWDER-PETRYSHYN'S DEGREE FOR A-PROPER MAPPINGS

Our aim in the next two paragraphs is to study by means of the topological degree two classes of mappings, namely the A-proper and the P-compact mappings. These classes of mappings include the duality mappings, the mappings with the condition (S) and with the modified condition (S).

We shall make use of most of the notations and results of §5, Chapter II.

We recall that in Chapter II, Definition 5.13. the spaces with projection scheme of type $(\Pi)_1$ were introduced. Now we shall consider a class of mappings between such Banach spaces which play an important role in the "approximation solvability" and for which a "degree can be defined."

2.1 DEFINITION. a) Let X, Y be two real Banach spaces with the projections schemes of type $(\Pi)_1$, $\{X_n, P_n\}$ and $\{Y_n, Q_n\}$ respectively. Assume that for every $n \in N$, $\dim X_n = \dim Y_n$; then $\Pi = \{X_n, P_n; Y_n, Q_n\}$ will be called an operator projection scheme, in short an OP-scheme for the mappings from X into Y.

We shall write $\Pi = \{X_n, P_n\}$ if $X = Y, X_n = Y_n$ and $P_n = Q_n$, $\forall n \in N$.

b) Let $\Pi = \{X_n, P_n; Y_n, Q_n\}$ be an OP-scheme, $D \subset X$ and $T: D \to Y$ be a continuous map; denote $D_n = D \cap X_n$ and $T_n = Q_n T|D_n$; T is called A-proper with respect to Π if given a bounded sequence $\{x_{n'}\}_{n'} \subset D, x_{n'} \in D_{n'}, N \ni n' \to \infty$ and an $y \in Y$ such that $T_{n'} x_{n'} \xrightarrow[n']{} y$, there exists a subquence $\{x_{n''}\}_{n''}$ and an $x \in \Omega$ such that $x_{n''} \xrightarrow[n'']{} x$ and $Tx = y$.

Denote $A_\Pi(D,Y)$ the class of all A-proper mappings and write $A_\Pi(D)$ if $X = Y$.

c) Let $D \subseteq X$ be open; we say that $T: D \to Y$ is locally A-proper if for every $x \in D$ there is a ball $S_r(x) \subset D$ so that $T \in A_\Pi(S_r(x), Y)$.

2.2. REMARK. Notice that $\pm I \in A_\Pi(X)$ but $0 \notin A_\Pi(X)$; thus $A_\Pi(X)$ is not a linear space. However it is clear that $\lambda T \in A_\Pi(D,Y)$ if $T \in A_\Pi(D,Y)$ an $\lambda \neq 0$. Moreover we have

2.3. PROPOSITION. If $T \in A_\Pi(\Omega,Y)$ and $S: D \to Y$ is continuous and relatively compact, then $T + S \in A_\Pi(D,Y)$.

PROOF. Let be $\{x_{n'}\}_{n'} \subset D$ a bounded sequence s.t., $x_{n'} \in D_{n'}, n' \to \infty$ and let be $y \in X$ with $Q_{n'}(T + S)x_{n'} \xrightarrow[n']{} y$; there is a subsequence $\{x_{n''}\}_{n''}$ so that $Sx_{n''} \xrightarrow[n'']{} y'$ for some $y' \in Y$.

Then

$$Q_{n''}Sx_{n''}-y'=(Q_{n''}Sx_{n''}-Q_{n''}y')+(Q_{n''}y'-y'=)\xrightarrow[n'']{}0$$

Hence

$$Q_{n''}Tx_{n''}\xrightarrow[n'']{}y-y'.$$

Since T is A-proper, there is an infinite subsequence $\{x_{n'''}\}_{n'''}$ and $y\in D$ with $x_{n'''}\xrightarrow[n''']{}x$ and $Tx=y-y'$.

But then $Sx_{n'''}\xrightarrow[n''']{}Sx=y'$, i.e. $(T+S)x=y$. ■

2.4. PROPOSITION. Let $D\subset X$ be closed and bounded and $T\in A_{\Pi}(D,Y)$; then T is a closed proper mapping, i.e. $T(M)$ is closed for every closed subset $M\subseteq D$ and $T^{-1}(L)$ is relatively compact for every relatively compact subset L in Y.

PROOF. Let $Tx_n\xrightarrow[n]{}y, x_n\in M$; then from $P_kx_n\xrightarrow[k]{}x_n$ and $TP_kx_n\xrightarrow[k]{}Tx_n$, $\forall n\in N$, we get that for every $n\in N$, there is $k_n\geq n$ such that

$$\|x_n-P_{k_n}x_n\|\leq\frac{1}{n}\text{ and }\|Tx_n-TP_{k_n}x_n\|\leq\frac{1}{n}.$$

Denote $z_n=P_{k_n}x_n$; then

$$x_n-z_n\xrightarrow[n]{}0\text{ and }Tx_n-Tz_n\xrightarrow[n]{}0 \tag{2.1}$$

Hence

$$Q_{k_n}Tz_n-y=(Q_{k_n}Tz_n-Q_{k_n}Tx_n)+(Q_{k_n}Tx_n-Q_{k_n}y)+(Q_{k_n}y-y)\xrightarrow[n]{}0.$$

Since T is A-proper, there is a subsequence $\{z_{n'}\}_{n'}=\{P_{k_{n'}}x_{n'}\}_{n'}\subseteq\{z_n\}_n$ and $x\in D$ so that $P_{k_{n'}}x_{n'}\xrightarrow[n']{}x$ and $Tx=y$. But then (2.1) implies $x_{n'}\xrightarrow[n']{}x$ and since M is closed, $x\in M$, i.e. $T(M)$ is closed.

In order to prove the second assertion, let $\{x_n\}_n\subset\overline{T^{-1}(L)}$; then it is clear that $Tx_n\in\overline{L}, n\in N$ and since L is relatively compact, passing to a subsequence we can assume that $Tx_n\xrightarrow[n]{}y$ for some $y\in L$. By a similar argument as above we find that for a subsequence $\{x_{n'}\}_{n'}$, and an element $x\in D, x_{n'}\xrightarrow[n']{}x$ and $Tx=y$, i.e. $x\in\overline{T^{-1}(L)}$. ■

2.5. COROLLARY. Let be D, M and T as in the above Proposition and $y \in Y$ so that $Tx - y \neq 0$ on ∂M; then there is a constant $c > 0$ with $\|Tx - y\| \geq c$ on ∂M.

PROOF. By the above Proposition $T(\partial M)$ is closed. Suppose that for every $n \in N$ there is $x_n \in \partial M$ s.t. $\|Tx_n - y\| < \frac{1}{n}$. Then $Tx_n \xrightarrow{n} y$, hence $y \in T(\partial M)$, which is impossible. ■

2.6. PROPOSITION. (Petryshyn) Let X,Y be Banach spaces, $\Pi = \{X_n, P_n, Y_n, Q_n\}$ an OP-scheme and $T: X \to Y$ a continuous map; assume that there exists a continuous function $\varphi: R_+ \to R_+$ with $\lim_{t \to \infty} \varphi(t) = \infty$, $\varphi(0) = 0, \varphi(t) > 0$ for $t > 0$ and s.t.

$$\|T_n x - T_n y\| \geq \varphi(\|x - y\|) \text{ on } X_n \times X_n \text{ for } n \in N. \tag{2.2}$$

Then $T \in A_\Pi(X,Y)$ if and only if $T(X) = Y$.

PROOF. Suppose $T \in A_\Pi(X,Y)$. Condition (2.2) implies that $T_n : X_n \to Y_n$ is injective. We apply the Corollary 1.27 to get that T_n is also surjective.

Then for any $y \in Y$ there exists a unique $x_n \in X_n$ such that $T_n x_n = Q_n y, \forall n \in N$; hence $T_n x_n - y \xrightarrow{n} 0$. It is an easy matter to see that by (2.2) $\{x_n\}_n$ is bounded. But as T is A-proper, there is $\{x_{n'}\}_{n'}$ and $x \in X$ with $x_{n'} \to x$ and $Tx = y$. We note that x is unique because (2.2) implies $\|Tx - Ty\| \geq \varphi(\|x - y\|)$ on $X \times X$.

Let T be surjective; consider a bounded sequence $\{x_{n'}\}_{n'} \subset X, x_{n'} \in X_{n'}$ and $y \in X$ so that $T_{n'} x_{n'} - y \xrightarrow{n'} 0$.
There is $x \in X$ with $Tx = y$. As $P_{n'} x \xrightarrow{n'} x$, then $T_{n'} P_{n'} x \xrightarrow{n'} Tx = y$.
Apply now (2.2) to get

$$\varphi(\|x_{n'} - P_{n'} x\|) \leq \|T_{n'} x_{n'} - T_{n'} P_{n'} x\|$$

$$\leq \|T_{n'} x_{n'} - y\| + \|T_{n'} P_{n'} x - A_n y\| + \|Q_n y - y\| \xrightarrow{n'} 0$$

It follows $x_{n'} - P_n x \xrightarrow{n'} 0$, hence $x_{n'} \xrightarrow{n'} x$, i.e. T is A-proper. ■

2.7. COROLLARY. Let X,Y and Π be as in Proposition 2.6 and $T \in L(X,Y)$; the following assertions are equivalent:
i) T is injective and A-proper;
ii) $T(X) = Y$ and $\|T_n x\| \geq c\|x\|$ on X_n for some $c > 0$ and all $n \in N$.

PROOF. Since T satisfies the condition (2.2) for $\varphi(t) = c.t$, it is clear that ii) = i)

Suppose that i) holds and that for every $c = \frac{1}{n}$, there is $x_{n'} \in X_{n'}$, $\|x_{n'}\| = 1$ so that $\|T_{n'} x_{n'}\| < \frac{1}{n}$; then there is an $x \in X$ and a subsequence $\{x_{n''}\}_{n''}$ with $x_{n''} \xrightarrow[n'']{} x$ and $Tx = 0$. The injectivity of T implies $x = 0$, which is nonsense, because $\|x\| = 1$.

Now we can use the above Proposition to obtain $T(X) = Y$. ■

2.8. REMARK. A-proper mappings arrised in conection with the problem to solve an infinite-dimensional equation $Tx = y$ replacing it by finite-dimensional equations $T_n x_n = y_n$ and requiring that at least a subsequence $\{x_{n'}\}_{n'}$ converges to a solution. Using an OP-scheme the finite dimensional equations are $Q_n Tx = Q_n y, x \in X_n$, where $Q_n y \xrightarrow[n]{} y$. In the case of A-proper mappings with condition (2.2), these equations have unique solutions $x_n, n \in N$ and for a subsequence we have $x_{n'} \xrightarrow[n]{} x$, where x is the unique solution of $Tx = y$.
This method to solve equations in Banach spaces is called Galekin's method.

Next we consider a reflexive Banach spaces X with a projection scheme $\{X_n, P_n\}$ so that $P_m P_n = P_{\min(m,n)}$; then by Proposition 5.16., Ch. I, $\Pi = \{X_n, P_n; P_n^* X^*, P_n^*\}$ is an OP-scheme for the mappings from X into X^*.

2.9. PROPOSITION. Let be X reflexive, $D \subset X$ open and $T: D \to X^*$ continuous and bounded with the Property (S); then T is locally A-proper with respect to the OP-scheme $\Pi = \{X_n, P_n; P_n^* X^*, P_n^*\}$.

PROOF. Let $x_0 \in D$ and $\bar{S}_r(x_0) \subset D$; consider $x_{n'} \in \bar{S}_r(x_0) \cap X_{n'}$ (for $n' \to +\infty$) and $x^* \in X^*$ with

$$T_{n'} x_{n'} \xrightarrow[n']{} x^* \tag{2.3}$$

As $\|x_n\| \le r$ and X is reflexive, there is $x \in X$ so that for a subsequence $x_{n''} \xrightarrow[n'']{w} x$.

We shall prove that $x_{n''} \xrightarrow[n'']{} x$ and $Tx = x^*$.

For a fixed $y \in X$ we have

$$\left|< Tx_{n''} - x^*, x_{n''} - P_{n''}y >\right| = \left|< Tx_{n''} - x^*, P_{n''}x_{n''} - P_{n''}y >\right|$$

$$= \left|< P^*_{n''}Tx_{n''} - P^*_{n''}x^*, x_{n''} - y >\right| \le M\left\|P^*_{n''}Tx_{n''} - P^*_{n''}x^*\right\| \xrightarrow[n'']{} 0$$

with $M = \sup_{n''}(\|x_{n''}\| + \|y\|)$. Since $< x^*, x_{n''} - P_{n''}y > \xrightarrow[n'']{} < x^*, x - y >$ then $< Tx_{n''}, x_{n''} - P_{n''}y > \xrightarrow[n'']{} < x^*, x - y >$; whence, using the fact that T is bounded we get

$$< Tx_{n''}, x_{n''} - y > \xrightarrow[n'']{} < x^*, x - y >. \tag{2.4}$$

Take in (2.4) $y = x$ to find that $< Tx_{n''}, x_{n''} - x > \xrightarrow[n'']{} 0$; hence $< Tx_{n''} - Tx, x_{n''} - x > \xrightarrow[n'']{} 0$.

The Property (S) for T yields now $x_{n''} \xrightarrow[n'']{} x$. But then $Tx_{n''} \xrightarrow[n'']{} Tx$, so that

$$P^*_{n''}Tx_{n''} - Tx = (P^*_{n''}Tx_{n''} - P^*_{n''}Tx) + (P^*_{n''}Tx - Tx) \xrightarrow[n'']{} 0$$

Then by (2.3) it follows that $Tx = x^*$. ■

2.10. COROLLARY. Let X be reflexive, $D \subset X$ open and bounded and $T: D \to X^*$ continuous and bounded satisfying the Property (S) on D; then T is A-proper.

2.11. COROLLARY. Let X, D and T be as in the above Corollary and $\varphi: R_+ \to R_+$ continuous, $\varphi(0) = 0$, $\varphi(t) > 0$ for $t > 0$ and $\lim_{t \to \infty} \varphi(t) = \infty$ such that

$$< Tx - Ty, x - y > \ge \varphi(\|x - y\|), x, y \in D. \tag{2.5}$$

Then T is A-proper.

PROOF. The condition (2.5) implies that T has the Property (S). ■

2.12. PROPOSITION. Let X be reflexive and smooth so that both X and X^* have the Property (h); then the normalized duality mapping on X is A-proper.

PROOF. Let us consider $\{x_{n'}\}_{n'} \subset X, x_{n'} \in X_{n'}$ and $x^* \in X^*$ with $P^*_n J x_n x_{n'} \xrightarrow[n']{} x^*$. By Proposition 5.17, Ch. II, $P^*_n J x_{n'} = J x_{n'}$, so that $\|x_{n'}\| = \|J x_{n'}\| = \|P^*_{n'} J x_{n'}\|, n' \to \infty$; whence $\{x_{n'}\}_{n'}$ is bounded. Thus there exists a subsequence $\{x_{n''}\}_{n''} \subseteq \{x_{n'}\}_{n'}$ and $x \in X$ so that $x_{n''} \xrightarrow[n'']{w} x$. The extimate

$$< Jx - Jy , x - y > = (\|x\| - \|y\|)^2 + (\|Jx\|.\|y\| - < Jx, y >) + (\|Jy\|.\|x\| - < Jy , x >)$$
$$\geq (\|x\| - \|y\|)^2, \quad x, y \in X$$

yields

$$< Jx_{n''} - JP_{n''}x, x_{n''} - P_{n''}x > \geq \left(\|x_{n''}\| - \|P_{n''}x\|\right)^2 . \tag{2.6}$$

By proposition 5.2 Ch. II, J is norm-norm continuous so that $Jx_{n''} - JP_{n''}x \xrightarrow[n'']{} x^* - Jx$; then the left side of (2.6) converges to 0, whence $\|x_{n''}\| - \|P_{n''}x\| \xrightarrow[n'']{} 0$.
Consequently

$$\|x_{n''}\| - \|x\| = (\|x_{n''}\| - \|P_{n''}x\|) + (\|P_{n''}x\| - \|x\|) \xrightarrow[n'']{} 0 \text{ i.e. } \|x_{n''}\| \xrightarrow[n'']{} \|x\|.$$

As X has the Property (h), it follows that $x_{n''} \xrightarrow[n'']{} x$. Then $Jx_{n''} \xrightarrow[n'']{} Jx$ and from the fact that

$$P^*_{n''} J x_{n''} - Jx = \left(P^*_{n''} J x_{n''} - Jx\right) + \left(P^*_{n''} Jx - Jx\right) \xrightarrow[n']{} 0$$

it is clear that $Jx = x^*$ and this ends the proof. ■

In order to define the degree for A-proper mappings we need the

2.13. LEMMA. Let X, Y be Banach spaces with an OP-scheme $\Pi, \Omega \subset X$ open and bounded and $T \in A_{\Pi}(\bar{\Omega}, Y)$. If $y \in Y \setminus T(\partial\Omega)$, then there exist $n_0 \in N$ and $c > 0$ so that $\|T_n x - y\| \geq c$ for every $x \in \partial\Omega_n, n \geq n_0$.

PROOF. Suppose that the property in the Lemma does not hold; then there is $x_{n'} \in \partial\Omega_{n'}$, so that $T_{n'} x_{n'} - y \xrightarrow[n']{} 0$. As T is A-proper, there is a subsequence $\{x_{n''}\} \subset \{x_{n'}\}$ and $x \in \bar{\Omega}$ with $x_{n''} \xrightarrow[n'']{} x$ and $Tx = y$. It is clear that $x \in \partial\Omega$ and thus $y \in T(\partial\Omega)$ which is impossible. ■

We can now give the:

2.14. DEFINITION. Let X, Y be Banach spaces, $\Pi = \{X_n, P_n, Y_n, Q_n\}$ an OP-scheme where X_n and Y_n are oriented. Let $\Omega \subset X$ be open and bounded, $T \in A_\Pi(\bar{\Omega}, Y)$, and $y \notin T(\partial\Omega)$; we define the multivalued degree (of Browder-Petryshyn) by
$d(T,\Omega,y) = \{k \in Z \cup \{-\infty,+\infty\}; d(T_{n'},\Omega_{n'}y) \xrightarrow[n']{} k$ for some $n' \to \infty\}$.

2.15. REMARK. First notice that by Lemma 2.13. $Q_n y \notin T_n(\partial\Omega_n)$, for n sufficiently large so that $d(T_n,\Omega_n,Q_n y)$ is defined. Moreover if $k \in d(T,\Omega,y)$ is finite, then $d(T_{n''},\Omega_{n''},Q_{n'}y) = k$ for n' large.

Some usual properties of a degree are valid.

2.16. THEOREM.

i) $d(T,\Omega,y) \neq \emptyset$

ii) if $d(T,\Omega,y) \neq \{0\}$, there is $x \in \Omega$ with $Tx=y$

iii) if $F:[0,1] \times \bar{\Omega} \to Y$ is such that $F(.,x)$ is continuous on $[0,1]$ uniformly with respect to $x \in \bar{\Omega}$ and $F(t,.) \in A_\Pi(\bar{\Omega},y)$ for every $t \in [0,1]$, then for $y \notin F[0,1] \times \partial\Omega$, $d(F(t,.),\Omega,y)$ is independent of $t \in [0,1]$.

iv) If Ω is symmetric, $0 \in \Omega$, T is odd on $\bar{\Omega}$ and $0 \notin T(\partial\Omega)$, then the numbers of the form $2m \notin d(T,\Omega,0)$, $m \in Z$.

PROOF. i) Consider the sequence $\{d(T_n,\Omega_n,Q_n y)\}_{n \geq n_0}$ (with n_0 given by the Lemma 2.13); if it is bounded, there is $k \in Z$ with $k \in d(T,\Omega,y)$; if it is undounded, then $+\infty$ or $-\infty$ are in $d(T,\Omega,y)$,
ii) If $d(T,\Omega,y) \neq \{0\}$, then $d(T_{n''},\Omega_{n''},Q_{n'}y) \neq 0$ for a sequence $n' \to +\infty$. By Theorem 1.15 there is $x_{n'} \in \Omega_{n'}$ with $T_{n'}x_{n'} = Q_{n'}y$. But $Q_{n'}y \xrightarrow[n']{} y$ and since T is A-proper, there is $x \in \bar{\Omega}$ so that $x_{n''} \xrightarrow[n'']{} x$ for a subsequence $\{x_{n''}\}_{n''} \subseteq \{x_{n'}\}_{n'}$, and $Tx = y$. Now it is clear that $x \in \Omega$.
iii) It is sufficient to prove that

$$Q_n y \notin F_n([0,1] \times \partial\Omega) = Q_n F([0,1] \times \partial\Omega_n), \text{ for } n \text{ large,} \tag{2.7}$$

because in this case $d(F_n(t,.),\Omega_n,Q_n y)$ is independent of $t \in [0,1]$. Suppose that (2.7) is not true; then there are sequences $\{x_{n'}\} \subseteq \partial\Omega_{n'}$ and $\{t_{n'}\} \subset [0,1]$ with

$$t_{n'} \xrightarrow[n']{} t_0 \quad \text{and} \quad F_{n'}(t_{n'},x_{n'}) = Q_{n'}y \tag{2.8}$$

Since $F(t_{n'},.) \xrightarrow[n']{} F(t_0,.)$ uniformly on $\bar{\Omega}$, we obtain that $F(t_{n'},x_{n'}) - F(t_0,x_{n'}) \xrightarrow[n']{} 0$.

Using (2.8) we get

$$Q_{n'}y - Q_{n'} - F(t_0,x_{n'}) \xrightarrow[n']{} \text{ i.e. } Q_{n'}F(t_0,x_{n''}) \xrightarrow[n']{} y.$$

As $F(t_0,.)$ is A-proper, there is $x \in \partial\Omega$ and a subsequence $\{x_{n''}\} \subseteq \{x_{n'}\}$ so that $x_{n''} \xrightarrow[n'']{} x$ and $F(t_0,x) = y$; consequently $y \in F(t_0,\partial\Omega)$, which is nonsense.

iv) Observe that T_n is odd, $\forall\, n \in N$, and we can apply Borsuk's Theorem to get that $d(T_n,\Omega_n,0)$ is odd, $\forall\, n \in N$. ■

2.17. PROPOSITION. Let X be reflexive, smooth so that both X and X^* have the Property (h) and J be the normalized duality mapping on X; then for every r>0 and $x^* \in X^*$, we have

$$d(J,S_r(0),x^*) \neq \{0\} \text{ if } \|x^*\| < r \text{ and } d(J,S_r(0),x^*) = \{0\} \text{ if } \|x^*\| > r.$$

PROOF. Let $\|x^*\| < r$ and define $F(t,x^*) = Jx - tx^*$ on $[0,1] \times \bar{S}_r(0)$.

Then by Proposition 2.3. $F(t,.)$ is A-proper for every $t \in [0,1]$. Moreover $F(.,x)$ is continuous on $[0,1]$, uniformly with respect to $x \in \bar{S}_r(0)$.

Let us note that $F(t,x^*) \neq 0$ for $(t,x^*) \in [0,1] \times \partial S_r(0)$. In effect, if for $t_0 \in [0,1]$ and $x_0 \in X$ with $\|x_0\| = r$, we would have $Jx_0 - t_0 x^* = 0$, then $\|x_0\| = \|Jx_0\| = t_0\|x^*\| < r$ which is impossible.
Then by Theorem 2.14. iii)

$$d(J,S_r(0),0] = d(J(.) - x^*,S_r(0),0) = d(J,S_r(0),x^*)$$

But J is odd and then by Theorem 2.14 (iv), $d(J,S(0,r),0) \neq \{0\}$. In the case $\|x^*\| > r$, we have $Jx \neq x^*, \forall\, x \in \partial S_r(0)$; indeed if $Jx = x^*$, then $\|x^*\| = \|x\| = r$, a contradiction. Suppose $d(J,S_r(0),x^*) \neq \{0\}$; then by Theorem 2.14 ii), there is $x_0 \in \partial S_r(0)$ with $Jx_0 = x^*$, which by the above argument is impossible. ■

2.18. DEFINITION. Let be $T:\Omega \to Y$, Ω open; we say that T has a locally A-proper solvable homotopy if to every $x_0 \in \Omega$ there is $\bar{S}_r(x_0) \subset \Omega$ and a map $F(t,x):[0,1]\times \bar{S}_r(x_0) \to Y$ so that

i) $F(0,x) = Tx$ on $\bar{S}_r(x_0)$, $F(t,.) \in A_\Pi(\bar{S}_r(x_0),Y)$ for all $t \in [0,1]$ and $F(.,x)$ is continuous on $[0,1]$, uniformly for $x \in \bar{S}_r(x_0)$.

ii) $d(F(1,.),S_r(x_0),Tx_0) \neq \{0\}$ if $Tx_0 \notin F(1,\partial \bar{S}_r(x_0))$.

iii) If there is $r_0 \le r$ such that $Tx - Tx_0 \neq 0$ on $\partial S_{r_0}(x_0)$, then $F(t,x) - Tx_o \neq 0$ for every $(t,x) \in [0,1]\times \partial S_{r_0}(x_0)$.

2.19. LEMMA. Let $T:\bar{S}_r(x_0) \to Y$ be A-proper and $Tx - Tx_0 \neq 0$ on $\partial S_r(x_0)$; if $d(T,S_r(x_0),Tx_0) \neq \{0\}$, then there is $r' > 0$ so that $S_{r'}(Tx_0) \subseteq TS_r(x_0)$.

PROOF. By Corollary 2.5. there is $c > 0$ with $\|Tx - Tx_0\| \ge c$ for every $x \in \partial S_r(x_0)$.
Let $r' = \inf_{x\in \partial S_r(x_0)} \|Tx - Tx_0\|$ and fix a $y \in S_{r'}(Tx_0)$.
For every $x \in S_r(x_0)$ we have

$$\|Tx - y\| \ge \|Tx - Tx_0\| - \|Tx_0 - y\| \ge r' \|Tx_0 - y\| > 0 .$$

Consider now the map

$$F(t,x) = Tx - t\,Tx_0 - (1-t)y \ ,(t,x) \in [0,1]\times \bar{S}_r(x_0).$$

It is clear that for every $t \in [0,1]$, $F(t,.)$ is A-proper and that $F(.,x)$ is continuous on $[0,1]$, uniformly with respect to x in $\bar{S}_r(x_0)$. In effect we have the estimate

$$\|F(t_1,x) - F(t_2,x)\| \le |t_1 - t_2|(\|Tx_0\| + \|y\|) .$$

Moreover, for $x \in \partial \bar{S}_r(x_0)$ and $t \in [0,1]$, we have

$$\|F(t,,x)\| = \|Tx - tx_0 + (1-t)Tx_0 - (1-t)\|$$

$$\ge \|Tx - tx_0\| - (1-t)\|Tx_0 - y\| \ge r' - (1-t)r' = t.r'$$

We can therefore apply Theorem 2.16. iii) to get that

$$d(T, S_r(x_0), y) = d(T(\cdot) - Tx_0, S_r(x_0), 0) = d(T, S_r(x_0), Tx_0) \neq \{0\}\ .$$

Then again by Theorem 2.16. ii), there is $x \in S_r(x_0)$ so that $Tx = y$. Hence $S_{r'}(Tx_0) \subseteq T\, S_r(x_0)$. ■

We can now give the following domain-invariance result (compare with Proposition 1.26)

2.20. THEOREM. Let be Ω open and $T:\Omega \to Y$ a mapping which has a locally A-proper solvable homotopy ; if T is locally injective, then $T(\Omega)$ is open.

PROOF. Consider $y_0 \in T(\Omega)$ and $x_0 \in \Omega$ so that $Tx_0 = y_0$; there exists $\bar{S}_r(x_0) \subset \Omega$ and $F(t,x): [0,1] \times \bar{S}_r(x_0) \longrightarrow Y$ with the properties i), ii) and iii) from the above definition.

Since T is locally injective, there is a ball $S_{r_0}(x_0), r_0 \leq r$, so that $Tx \neq Tx_0$ on $\partial S_{r_0}(x_0)$.

But then, by iii) it follows that $F(t,x) - Tx_0 \neq 0$ for every $(t,x) \in [0,1] \times \partial S_{r_0}(x_0)$.

Using ii) and the Theorem 2.16 iii) we obtain

$$d\big(T, S_{r_0}(x_0), Tx_0\big) = d\big(F(1,\cdot), S_{r_0}(x_0), Tx_0\big) \neq \{0\}$$

Hence by Lemma 2.18 there is a ball so that $S_{r'}(Tx_0) \subseteq S_{r_0}(x_0)$.■

2.21. THEOREM. Let X be reflexive and smooth so that X and X^* have the Property (h); let $T:\Omega \to X^*$ be a continuous bounded, locally monotone and locally injective mapping with the Property (S); then $T(\Omega)$ is open.

PROOF. By the above Theorem it is sufficient to prove that T has a locally A-proper solvable homotopy. Let $x_0 \in \Omega$ and $\bar{S}_r(x_0) \subset \Omega$ so that T is monotone and injective on $\bar{S}_r(x_0)$; as the monotonicity and the property (S) are invariant to translations, we can suppose $x_0 = 0$ and $T\,0 = 0$. Consider now the map

$$F(t,x) = (1-t)Tx + tJx, \quad (t,x) \in [0,1] \times \bar{S}_r(0)$$

where J is the normalized duality mapping on X.

As T and J are bounded and continuous on $\bar{S}_r(0)$, it is easy to check that $F(t,.)$ is continuous on $\bar{S}_r(0)$ and $F(.,x)$ is continuous on $[0,1]$, uniformly for $x \in S_r(x_0)$. Moreover $F(0,.) = T$ is A-proper on $\bar{S}_r(0)$ by Proposition 2.9 and $F(t,.) = J$ is A-proper on $S_r(0)$ by Proposition 2.12. In order to prove that $F(t,.) \in A_\Pi(\bar{S}_r(0),Y)$ for $t \in (0,1)$, it is sufficient to show that $F(t,.)$ has the property (S) on $\bar{S}_r(0)$.

Fix $t \in (0,1)$ and consider $\{x_n\}_n \subset \bar{S}_r(0)$, $x_n \xrightarrow[n]{w} x \in \bar{S}_r(0)$ with $< F(t,x_n) - F(t,x), x_n - x > \xrightarrow[n]{} 0$.
Then the relation

$$< F(t,x_n) - F(t,x), x_n - x > \\ = (1-t) < Tx_n - Tx, x_n - x > + t < Jx_n - Jx, x_n - x >$$

shows that $< Tx_n - Tx, x_n - x > \xrightarrow[n]{} 0$.
Since T has the property (S), then $x_n \xrightarrow[n]{} x$.
We further note that by Proposition 2.17 $F(1,x) = Jx \neq 0$ for $x \in \partial S_r(0)$ and $d(J, S_r(0), 0) \neq \{0\}$.
We only need to prove that for F the property iii), from the Definition 2.18. is valid.
Since T is injective on $\bar{S}_r(0)$ and $T0 = 0$, then $Tx \neq 0$ on $\partial S_r(0)$. We assert that $F(t,x) \neq 0$ for every $(t,x) \in (0,1] \times \partial S_r(0)$. Indeed, assume that $F(t_0,x_0) = 0$ for some $(t_0,x_0) \in (0,1] \times \partial S_r(0)$; then

$$0 = < F(t_o,x_o) - F(t_o,0), x_o - 0 >$$

$$= (1-t_o) < Tx_o - T0, x_o - 0 > + t_o < Jx_o - J0, x_o - 0 >$$

As T and J are monotone, we obtain

$$0 = < Jx_o - J0, x_o - 0 > = < Jx_o, x_o > = \|x_o\|^2$$

i.e. $x_0 = 0$, which is a contradiction. ■

2.22. REMARK. We end our considerations by comparing the Browder-Petryshyn degree with the Leray-Schauder one.
Recall first the following

2.23. DEFINITION. Let X be a real Banach space, $\Omega \subset X$ open, bounded, $T : \bar{\Omega} \to Y$ continuous and compact and $y \notin (I-T)(\partial\Omega)$; then the Leray-Schauder degree is defined as

$$d_{L.S}(I-T,\Omega,y)=d((I-T_1)|_{\bar{\Omega}_1}\Omega_1,y)$$

where T_1 is any continuous compact finite-dimensional operator on $\bar{\Omega}$ such that $\sup_{x\in\bar{\Omega}}\|Tx-T_1x\|<\mathrm{dist}(y,(I-T)(\partial\Omega))$, X_1 is any finite dimensional subspace of X with $T_1(\bar{\Omega})\cup\{y\}\subset X_1$, $\Omega_1=\Omega\cap X_1$ and d is the Brouwer degree on X_1.
The existence of a T_1 and X_1 as above is a consequence of the fact that every compact operator can be aproximated uniformly by finite-dimensional operators.
We note that if T_2 and X_2 satisfy the same conditions as T_1 and X_1, we let X_o be the span of X_1 and X_2 and $\Omega_0=\Omega\cap X_o$. Then, by Remark 1.30

$$d(I-T_i)|_{\bar{\Omega}_o},\Omega_0,y)=d(I-T_i)|_{\bar{\Omega}_i},\Omega_i,y) \qquad \text{for } i=1,2.$$

Let $F(t,x)=x-t\,T_1x-(1-t)T_2x$, $(t,x)\in[0,1]\times\Omega_0$; then for $x\in\partial\Omega_0$

$$\|y-F(t,x)\|=\|y-(I-T)x+t(T_1x-Tx)+(1-t)(T_2x-Tx)\|$$

$$\geq \mathrm{dist}(y,(I-T)(\partial\Omega))-t\sup_{\bar{\Omega}}\|T_1x-Tx\|-(I-T)\sup_{\bar{\Omega}}\|T_2x-Tx\|>0$$

hence by Theorem 1.12.

$$d(I-T_1)|_{\bar{\Omega}_o},\Omega_0,y)=d((I-T_2)|_{\bar{\Omega}_o},\Omega_0,y)\ .$$

Therefore the above definition of the degree is consistent.

All the properties of the degree given in §1 are valid; it is only to be remarked that Theorem 1.12. has the form

the degree $d(I-F(t,.),\Omega,y(t))$ is independent of $t\in[0,1]$ whenever $F:[0,1]\times\bar{\Omega}\to X$ is compact, continuous and $y:[0,1]\to X$ is continuous and such that $y(t)\notin(I-F(t,.))(\partial\Omega)$ for $t\in[0,1]$.
The analogue of Theorem 1.16. can also be proved; this is

Schauder's fixed point Theorem. Let $T:\bar{S}_r(0)\to\bar{S}_r(0)$ be continuous and compact; then T has a fixed point.

We also have a "domain-invariance Theorem" similar to Proposition 1.25, namely

Let $\Omega \subset X$ be open, $T:\Omega \to X$ continuous compact, and locally injective; then $(I-T)(\Omega)$ is open.
Suppose now that X has an OP-scheme, $\Pi = \{X_n, P_n\}$; we know by Proposition 2.3. that if Ω is open bounded and $T:\bar{\Omega} \to X$ is continuous and compact, then $I - T \in A_\Pi(\bar{\Omega}, Y)$. We can prove that the Leray-Schauder degree for $I-T$ coincides with the degree of Browder-Petryshyn.
Indeed, in this case $P_n Tx \xrightarrow[n]{} Tx$ uniformly on $\bar{\Omega}$: given $\varepsilon > 0$, we have $T(\bar{\Omega}) \subset \bigcup_{i=1}^{m} S_\varepsilon(Tx_i)$ and $\|P_n Tx_i - Tx_i\| \le \varepsilon$ for $n \ge n_o(\varepsilon)$, $i = 1, \dots, n$; hence

$$\sup_{x\in\bar{\Omega}} \|P_n Tx - x\| \le \sup_{x\in\bar{\Omega}} \left(\|P_n Tx - P_n Tx_{i_o}\|\right) + \|P_n Tx_{i_o} - Tx_{i_o}\| + \|Tx_{i_o} - Tx\| \le 3\varepsilon$$

for i_o such that $Tx \in S_\varepsilon(Tx_{i_o})$.
Then for n sufficiently large

$$d(I_n - T_n, \Omega_n, P_n y) = \text{constant} = d_{L.S}(I - T, \Omega, y) .$$

We conclude our considerations with the remark that the duality mapping plays a similar role in the Browder-Petryschyn's degree theory to the identity in the Leray-Schauder theory for compact operators; Proposition 2.17. underlines this analogy.

§3. P-COMPACT MAPPINGS

In this section X will be a Banach space with an OP-scheme $\Pi\{X_n, P_n\}$. We apply the results of the previous paragraph to study the following class of mappings

3.1. DEFINITION. A mapping $T:\bar{\Omega} \subseteq X \to X$ is called P-compact if for any $\lambda \ge 0$, $T_\lambda = T + \lambda I \in A_\Pi(\Omega)$.
We say that $T:\Omega \to X$ is locally P-compact if for every $x \in \Omega$, there is a ball $S_r(x_o) \subset \Omega$ so that T is P-compact on $\bar{S}_r(x_o)$.

3.2. THEOREM. Let $\Omega \subset X$ be open and bounded with $0 \in \Omega$ and $T:\bar{\Omega} \to X$ a bounded P-compact mapping; if for all $\lambda \geq 0$, $0 \notin T_\lambda(\partial\Omega)$, then $d(T,\Omega,0) = \{1\}$; in particular, there is $x \in \Omega$ with $Tx = 0$.

PROOF. For $t \in [0,1]$ and $x \in \bar{\Omega}$, define $F(t,x) = (1-t)Tx + tx$. We have

$$\|F(t,x) - F(s,x)\| = |t-s|\,\|Tx - x\| .$$

Since T is bounded it follows that $F(.,x)$ is continuous on $[0,1]$, uniformly with respect to $x \in \bar{\Omega}$.

It is clear that $F(t,.)$ is A-proper, $\forall\, t \in [0,1]$. Moreover $F([0,1] \times \partial\Omega) \neq 0$; indeed, suppose that $F(t_0,x_0) = 0$ for $t_0 \in [0,1]$ and $x_0 \in \partial\Omega$. If $t_o \in [0,1)$ then $t_0 x_0 + \frac{t_0}{1-t_0} x_0 = 0$, which is impossible; if $t_0 = 1$, then $x_0 = 0 \in \partial\Omega$ which again is a contradiction. We can hence apply Theorem 2.16 iii) to obtain that

$$d(T,\Omega,0) = d(I,\Omega,0) = \{1\}$$

The last part is a consequence of ii) in the above mentioned Theorem 2.16. ∎

3.3. THEOREM. Let $\Omega \subseteq X$ be open and $T:\Omega \to X$ bounded and locally P-compact such that for every $x \in \Omega$, there is a ball $\bar{S}_r(x) \subset \Omega$ on which T_λ is injective, for all $\lambda \geq 0$; then $T_\lambda(\Omega)$ is open in X, for all $\lambda \geq 0$.

PROOF. It suffices to prove that T has a locally A-proper solvable homotopy and make use of the Theorem 2.20.

For $x_0 \in \Omega$ we can choose a ball $\bar{S}_r(x_0) \subset \Omega$ on which T is P-compact all T_λ are injective and so that $T\left(\bar{S}_r(x_0)\right)$ is bounded. For $\lambda \geq 0$ and $(t,x) \in [0,1] \times \bar{S}_r(x_0)$ define

$$F_\lambda(t,x) = (1-t)T_\lambda x + t\,T_\lambda x_0 + t(x - x_0) .$$

Then $F_\lambda(0,.) = T_\lambda$ and $F_\lambda(1,.) = I + T_\lambda x_0 - x_0$ are A-proper on $\bar{S}_r(0)$. Moreover, for $t \in (0,1)$, we have

$$\frac{1}{1-t}F_\lambda(t,.) = T + (\lambda + \frac{t}{1-t})I + \frac{t}{1-t}(T_\lambda x_0 - x_0).$$

Then it is an easy matter to see that $F_\lambda(t,.)$ is A-proper for all $\lambda \geq 0$. We also have

$$F_\lambda(t,x) - F_\lambda(s,x) = (s-t)[T_\lambda x - T_\lambda x_0 - (x - x_0)];$$

Since $T(\bar{S}_r(x_0))$ is bounded, then $F_\lambda(t,x)$ is continuous in t, uniformly for $x \in \bar{S}_r(x_0)$. Hence the condition i) from Definition 2.18 is fullfilled. Define now on $[0,1] \times \bar{S}_r(x_0)$

$$H_\lambda(t,x) = F_\lambda(t,x) - T_\lambda x_0.$$

Then if $T_\lambda x_0 \notin F_\lambda(1, \partial S_r(x_0))$ we have

$$d(F_\lambda(1,.), S_r(x_0), T_\lambda x_0) = d(H_\lambda(1,.), S_r(x_0), 0)$$

$$= d(I - x_0, S_r(x_0), 0) = \{1\}$$

and thus also the condition ii) from the Definition 2.18 is satisfied. In order to end the proof we have only to prove that $H_\lambda(t,x) \neq 0$ for $x \in \partial S_r(x_0)$ and $t \in [0,1]$. Suppose that $H_\lambda(t_1, x_1) = 0$ for some $x_1 \in \partial S_r(x_0)$ and $t_1 \in [0,1]$. If $t_1 = 1$, then we have that $0 = H_\lambda(t_1, x_1) = x_1 - x_0$, i.e. $x_1 = x_0 \in \partial S_r(x_0)$, which is impossible. If $t \in [0,1)$, then

$$Tx_1 + (\lambda + \frac{1}{1-t_1})x_1 = Tx_0 + (\lambda + \frac{1}{1-t_1})x_0$$

relation which contradicts the fact that for any $\lambda \geq 0, T_\lambda$ is injective on $\bar{S}_r(x_0)$. ■

3.4. DEFINITION. A mapping $T: \Omega \subseteq X \to X$ is called locally accretive if for every $x_0 \in X$ there is a ball $\bar{S}_r(x_0) \subset \Omega$ so that for every $x, y \in \bar{S}_r(x_0)$ there is $x^* \in J(x-y)$ with

$$< x^*, Tx - Ty > \geq 0 \qquad (3.1)$$

(J is the normalized duality mapping.)

3.5. COROLLARY. Let $\Omega \subseteq X$ be open and $T:\Omega \to X$ bounded, locally accretive, locally inyective and locally P-compact; then $T_\lambda(\Omega)$ is open, for all $\lambda \geq 0$.

PROOF. For $x_0 \in \Omega$ consider a ball $\bar{S}_r(x_0)$ on which T is accretive, P-compact injective and such that $T(\bar{S}_r(x_0))$ is bounded. Then T_λ is injective on $S_r(x_0)$, $\forall \lambda \geq 0$.
Indeed, suppose that there are $\lambda > 0$ and $x \neq y \in S_r(x_0)$ with $Tx + \lambda x = Ty + \lambda y$; then for $x^* \in J(x-y)$ from the Definition 3.4 we have

$$<x^*, Tx - Ty> + \|x-y\|^2 = 0$$

Since $<x^*, Tx - Ty> \geq 0$ it follows that $\|x-y\| = 0$, i.e. $x = y$
We can apply now the above Theorem to obtain the result. ∎

3.6. DEFINITION. We say that the continuous mapping $T:\Omega \subseteq X \to X$ satisfies the modified condition (S) if
for $x_n \in X_n \cap \Omega$ and $x \in \Omega$ so that $P_n x \in \Omega$ for all $n \in N$ the conditions $x \xrightarrow[n]{w} x$ and $<J(x_n - P_n x), TP_n x> \xrightarrow[n]{} 0$ imply $x_n \xrightarrow[n]{} x$.

3.7. THEOREM. Let X be a reflexive Banach space with the property (I), and $T:\Omega \subseteq X \to X$ be a continuous, locally accretive mapping satisfying the modified condition (S); then T is locally P-compact.

PROOF. Let $x_0 \in \Omega$ and $\bar{S}_r(x_0) \subseteq \Omega$ on which (3.1) is satisfied.

First we shall prove that T is A-proper on $\bar{S}_r(x_0)$. Consider $x_n \in X_n \cap \bar{S}_r(x_0)$, $n \in N$, so that $P_n T x_n \xrightarrow[n]{} y$. Since X is reflexive, there is $x \in \bar{S}_r(x_0)$ with the property that for a subsequence $\{x_{n'}\}_{n'}$ we have

$$P_{n'} x \in \bar{S}_r(x_0) \text{ and } x_{n'} \xrightarrow[n']{w} x.$$

Then the property (1) of the space X yields

$$<J(x_{n'} - P_{n'} x), Tx_{n'} - TP_{n'} x>$$

$$= < P^*_{n'} J(x_{n'} - P_{n'}x), Tx_{n'} - TP_{n'}x >$$

$$= < J(x_{n'} - P_{n'}x), P_{n'}Tx_{n'} - P_{n'}TP_{n'}x > \xrightarrow[n']{} 0$$

(We recall that by Proposition 5.17 Ch. II, $P^*_n Tx = Tx$ for every $x \in X_n$).
By the modified condition (S) it follows that $x_{n'} \xrightarrow[n']{} 0$. Then we have

$$\|P_{n'}Tx_{n'} - Tx_{n'}\| \le \|P_{n'}Tx_{n'} - P_{n'}Tx\| + \|P_{n'}Tx - Tx\|$$

$$\le \|P_{n'}\| \, \|Tx_{n'} - Tx\| + \|P_{n'}Tx - Tx\| \xrightarrow[n']{} 0 .$$

If follows that $\|Tx - y\| = \lim_{n'} \|P_{n'}Tx_{n'} - Tx_{n'}\| = 0$, i.e. $Tx = y$.

Consider now $\lambda > 0$ and the mapping T_λ; for $x, y \in \bar{S}_r(x_0)$ we have

$$< J(x - y), T_\lambda x - T_\lambda y > = < J(x - y), Tx - Ty > +$$
$$+ \lambda < J(x - y), x - y > \ge \lambda \|x - y\|^2 \tag{3.2}$$

hence T_λ is locally accretive. Moreover, the relation (3.2) shows also that T_λ satisfies the modified condition (S) on $\bar{S}_r(x_0)$. Then we can use the same argument as above to get that T_λ is A-proper on $\bar{S}_r(x_0)$. ■

3.8. COROLLARY. Let X be a Hilbert space and $T:\Omega \subseteq X \to X$ continuous, locally monotone, locally injective and satisfying the modified condition (S); then for every $\lambda \ge 0, T_\lambda(\Omega)$ is open.

3.9. COROLLARY. (Browder-Minty). Let X be a Hilbert space and $T:\Omega \subseteq X \to X$ be continuous and such that

$$< Tx - Ty, x - y > \ge \text{const.} \|x - y\|^2, \quad \text{for all } \forall x,y \in \Omega;$$

then $T(\Omega)$ is open.

As a consequence of the Theorem 3.3, 3.7 and the Remark 5.10, Ch. II we also have the

3.10. COROLLARY. Let $T:\Omega \subseteq l^p \to l^p$ be continuous, locally injective, locally accretive and satisfying the modified condition (S); then $T_\lambda(\Omega)$ is open in l^p for all $\lambda \geq 0$.

EXERCISES

1. a) Let $\Omega \subset R$ be an open interval with $0 \in \Omega$ and let $f(x) = \alpha x^k$, $\alpha \neq 0$, then

$$d(f,\Omega,0) = \begin{cases} 0 & \text{if } k \text{ is even} \\ \operatorname{sign}\alpha & \text{if } k \text{ is odd} \end{cases}$$

b) Let $g(x) = f(x) + \sum_{i=0}^{k-1} \alpha_i x^i$, $x \in R$ and f as in a); then for r sufficiently large we have:

$$d(g,(-r,r),0) = d(f,(-r,r),0).$$

2. Prove that for $m \in Z$ there exists $\Omega \subset R$ open and bounded and $f \in C(\bar{\Omega})$ with $0 \notin f(\partial\Omega)$ so that $d(f,\Omega,0) = m$.

3. Let $f: R^2 \to R^2$ be defined by

$$f_1(x,y) = x^3 - 3xy^2, \qquad f_2(x,y) = -y^3 + 3x^2 y$$

and let a = (1,0); then $d(f, S_2(0), a) = 3$.

4. Let $f: R^2 \to R^2$ be continuous and so that $\langle f(x), \frac{x}{|x|} \rangle \xrightarrow[|x|\to\infty]{} \infty$; then $f(R^n) = R^n$.

<u>Hint</u>. Consider $F(t,x) = tx + (1-t)f(x)$ and prove that for $y \in R^n$ and r sufficently large one has $F(t,x) - y \neq 0$ for $|x| = r$.

5. There is no continuous $f: \bar{S}_1(0) \to \partial S_1(0)$ such that $f(x) = x$ for all $x \in \partial S_1(0)$.

<u>Hint</u>. See Example 3.3. Ch. I. §3, Deimling [2].

6. Let $\Omega \subset R^n$ be open and bounded with $0 \in \Omega$ and let $f \in C(\bar{\Omega})$ with $0 \notin f(\partial\Omega)$; if the space dimension n is odd then there exists $x \in \partial\Omega$ and $\lambda \neq 0$ such that $f(x) = \lambda x$.

Hint. Consider the functions $F(t,x)=(1-t)f(x) \pm tx, t \in [0,1], x \in \bar{\Omega}$ and use the fact that if n is odd then $d(-id, \Omega, 0) = -1$ (Theorem 3.4, Ch. I, §3 in Deimling [2]).

7. Let $\Omega \subset R^n$ be open and bounded, $f \in C(\bar{\Omega}), g \in C(\bar{\Omega})$ and $|g(x)| < |f(x)|$ on $\partial\Omega$; then $d(f+g, \Omega, 0) = d(f, \Omega, 0)$.

Hint. Consider $F(t,x) = t(f+g)(x) + (1-t)f(x), (t,x) \in [0,1] \times \bar{\Omega}$.

8. The system $2x+y+\sin(x+y)=0, x-2y+\cos(x+y)=0$ has a solution in $S_r(0) \subset R^2$, where $r > 1/\sqrt{5}$.
Hint. Apply exercise 7 to the functions $f_1(x,y)=2x+y$, $f_2(x,y)=x-2y$ and $g_1(x,y)=\sin(x+y)$, $g_2(x+y)$.

9. Let A be a real $n \times n$ matrix with $\det A \neq 0$ and $f \in C(R^n)$ such that $|x - Af(x)| \leq \alpha|x| + \beta$ on R^n for some $\alpha \in [0,1)$ and $\beta \geq 0$; then $f(R^n) = R^n$.

Hint. Use the exercise 4.

10. Let $X = c_0$ and $T(x) = \frac{1}{3}\sum_{i=1}^{\infty} \alpha_i^3 e_i$, for $x = \sum_{i=1}^{\infty} \alpha_i e_i \in c_0$ and $\{e_i\}_{i \in N}$ the canonical basis in c_0; then T is A-proper but $T'(\sum_{i=1}^{\infty} i^{-1/2} e_i)$ is not.

11. Let $X = c_0$ and $Tx = -\sum_{i=1}^{\infty}(1+\frac{1}{i})\alpha_i e_i$ for $x = \sum_{i=1}^{\infty} \alpha_i e_i \in c_0$ and $\{e_i\}_{i \in N}$ the basis in c_0; prove that T is A-proper and $d(T, S_1(0), 0) = \{-1, 1\}$.

Hint. See Ch. II §4, in Sburlan [1].

12. Let $\Omega \subset R^n$ be bounded, open, symmetric and $0 \in \Omega$. Consider a continuous function $f: \partial\Omega \to R^m$, where $m < n$; then $f(x) = f(-x)$ for some $x \in \partial\Omega$. (the Borsuk-Ulam's Theorem).

Hint. See Corollary 4.2. Ch. I §4 in Deimiling [2].

BIBLIOGRAPHICAL COMMENTS

§1. We present in this section Heinz' analytical approach to the construction of Brouwer's degree [1]. Other analytical versions of the theory have been given by Nagumo [1] and Schwartz [1]. A proof of Sard's Lemma which is crucial in the treatment can be found in Deimling [2] (Proposition 1.4. §1, Chapter 1) or in Schwartz [1] . For the uniqueness of the degree see Amann-Weiss [1] . The proof of Borsuk's Theorem is due to Gromes [1].

§2. The A-proper mappings were introduced by Browder and Petryshyn in 1968, [1]. The generalization to this class of operators of the notion of degree has the advantage to unify a series of results on compact perturbations of the identity or of monotone operators. The theory works only in spaces with an projection scheme but as this class of spaces includes separable Banach spaces which are very frequent in applications, this restriction is not so severe.
For the effective construction of projectional schemes on functional spaces, such as Sobolev spaces, we send to Aubin [1]; more about projectional schemes can be found in Browder [13], [15] and Petryshyn [5].
Proposition 2.6 is taken from Petryshyn [3]; more properties of the A-proper maps are given in Petryshyn [3], [4], [9].
A (multivalued) degree for A-proper mappings was defined by Browder- Petryshyn [1], [2]. It has the inconvenient that it is not additive with respect to the domain Ω; this difficulty was removed in the definition of Wong [1], [2]. Since for fixed points theorems and for the invariance of domains the degree of Browder-Petryshyn is "good" enough (see Petryshyn [8], [9]) we adopted here this approach and send to the monograph of Deimling [2] for Wong's degree as for a throughgoing study of the Leray-Schauder degree ([1]); See also Sburlan [1]. Nice applications of the generalized degree for A-proper mappings to semilinear equations are made by Petryshyn in [11].

§3. The P-compact operators were introduced and studied by Petryshyn [1] to [9].Most of the results in this sections are to be found in [9]; see also Petryshyn-Tucker [1] and Tucker [1]. In selecting the topics in the vaste field of the degree theory we were guided by the idea to obtain the most possible informations on the classes of operators which are the object of our presentation: the acretive operators, the monotone operators, and in particular the duality mappings. We hope that the Proposition 2.12., Proposition 2.17, Theorem 2.21. and the Corollaries to Theorem 3.7 illustrate and motivate these efforts.

CHAPTER V

NONLINEAR MONOTONE MAPPINGS

In this chapter we shall present various results on nonlinear monotone mappings in Banach spaces, pointing out further properties of duality mappings. Applications are made to some nonlinear functional equations.

§1. DEMICONTINUITY AND HEMICONTINUITY FOR MONOTONE OPERATORS

In what follows X and Y are two real Banach spaces and X^*, Y^* their duals and $A:D(A) \subseteq X \longrightarrow Y^*$ a (possibly) nonlinear operator.

1.1. DEFINITION. a) A is called demicontinous at $x \in D(A)$ if $x_n \in D(A)$, $x_n \xrightarrow[n]{} x$ implies $Ax_n \xrightarrow[n]{w^*} Ax$.

b) A is called hemicontinuous at $x \in D(A)$ if $x + t_n y \in D(A)$, for $y \in X$ and $t_n \xrightarrow[n]{} 0_+$ implies $A(x + t_n y) \xrightarrow[n]{w^*} Ax$.

c) A is called locally bounded at $x \in D(A)$ if the sequence $\{Ax_n\}_{n \in N}$ is bounded in Y^* whenever $x_n \in D(A)$ and $x_n \xrightarrow[n]{} x$.

d) A is called locally hemibounded at $x \in D(A)$ if the sequence $\{A(x+t_n y)\}_{n\in N}$ is bounded in Y^* for every $y \in X$ and $t_n \xrightarrow[n]{} 0_+$ with $x+t_n y \in D(A)$.

We say that A has any of the above properties on a set $D \subseteq D(A)$ if it has this property at every $x \in D$.

We used the notation $y_n \xrightarrow[n]{w^*} y$ for the weak* convergence on Y^*.

1.2. REMARK. If Y is reflexive, then in a) and b) the weak* convergence can be replaced by the week convergence. Some authors are using the weak convergence in the definiton of the demicontinuity and hemicontinuity.

1.3. REMARK. It is clear that A is locally bounded at $x \in D(A)$ iff there is an open neighbourhood U of x such that $A(U \cap D(A))$ is bounded in Y.

If A is linear and demicontinuous, then it is continuous (Exercise 1).

In general we have the following relations

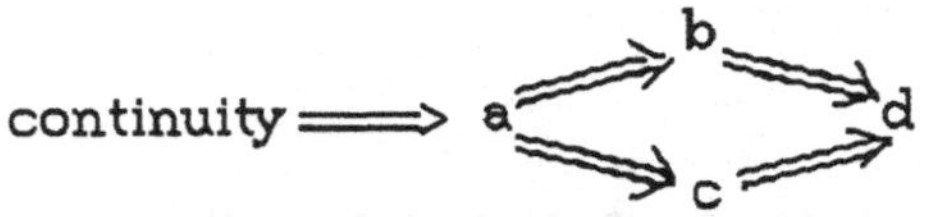

More exact relations between the introduced notions can be obtained in the case of the following class of operators

1.4. DEFINITION. Consider the operators $F:X \longrightarrow Y$, $D(F)=X$ and $A:D(A) \subseteq X \longrightarrow Y^*$; then A is called F-monotone if

$$< Ax - Ay\,,\, F(x-y) > \geq 0, \quad \forall\,(x,y) \in D(A).$$

If X = Y and F = I, the an I-monotone operator is called montone.

In the case of a linear operator $A:X \longrightarrow X^*$ the monotonicity is equivalent to the non-negativity, i.e. $< Ax, x > \geq 0, \ \forall\, x \in D(A)$.

1.5. PROPOSITION. Let $A:D(A) \subseteq X \longrightarrow Y^*$ be F-monotone, where $F:X \longrightarrow Y$ is positively homogeneous, surjective and uniformly continuous on the unit ball of X; then A is locally bounded at $x \in \mathrm{Int}\, D(A)$ if and only if A is locally hemibounded at x.

PROOF. We only have to prove that d) $\Rightarrow$ b). Suppose that A is locally hemibounded at $x \in \operatorname{Int} D(A)$ and that there exists a sequence $\{x_n\}_n \subseteq D(A)$ with $x_n \xrightarrow[n]{} x$ and $0 < r_n = \|Ax_n\| \xrightarrow[n]{} +\infty$. Define

$$\Phi(t) = \sup\{\|F(y) - F(z)\|; \|y\|, \|z\| \le 1, \|y - z\| \le t\}, t > 0;$$

then Φ is nondecreasing and by the uniform continuity of F on $\bar{S}_1(0)$, $\lim_{t \to 0} \Phi(t) = 0$. Moreover $\Phi(t) < +\infty$, $\forall\ t > 0$. Indeed if $\Phi(t_0) = +\infty$ for some $t_0 > 0$, then $\Phi(\lambda t_0) = +\infty$, $\forall \lambda > 0$. Thus $\lim_{\lambda \to 0} \Phi(\lambda t_0) = +\infty$ which is impossible. It is an easy matter to see that

$$\|F(y) - F(z)\| \le \Phi(\|y - z\|), \|y\|, \|z\| \le 1 \tag{1.1}$$

For $n \in N$ define $t_n = \max\left\{r_n^{-1}, \|x_n - x\|^{1/2}, \Phi(\|x_n - x\|)^{1/2}\right\}$; then it is clear that $t_n \xrightarrow[n]{} 0_+$.
Let be $y \in X$ and $y_n = x + t_n y, n \in N$; since $x \in \operatorname{Int} D(A)$, then $y_n \in D(A)$ for n sufficiently large; so that $< Ay_n - Ax_n, F(y_n - x_n) > \ge 0$.
This yields

$$\begin{aligned} t_n < Ax_n, Fy > &= <Ax_n, F(y_n - x_n) > + <Ax_n, F(t_n y) - F(y_n - x_n)) \\ &\le < Ay_n, F(y_n - x_n) > + <Ax_n, F(t_n y) - F(y_n - x_n) >. \end{aligned}$$

Consequently

$$<Ax_n, Fy> \le t_n^{-1} <Ay_n, F(y_n - x_n)> + t_n^{-1} < Ax_n, F(t_n y) - F(y_n - x_n)>. \tag{1.2}$$

We shall estimate the right part of (1.2). By the definition of t_n,

$$t_n^{-1}(y_n - x_n) = y - t_n^{-1}(x_n - x) \xrightarrow[n]{} y$$

so that by the continuity of F

$$t_n^{-1} F(y_n - x_n) = F\left(t_n^{-1}(y_n - x_n)\right) \xrightarrow[n]{} Fy . \tag{1.3}$$

In particular, the sequence $\left\{t_n^{-1} F(y_n - x_n)\right\}_{n \in N}$ is bounded. Since also $\{Ay_n\}_{n \in N}$ is bounded, we have that

$$t_n^{-1} <Ay_n, F(y_n - x_n) > \le C, \text{ for some } C > 0 . \tag{1.4}$$

We note now that (for n sufficiently large)

$$\|F(t_n y) - F(y_n - x_n)\| \le \Phi(\|t_n y - y_n + x_n\|) = \Phi(\|x_n - x\|) \le t_n^2 .$$

Hence

$$t_n^{-1} <Ax_n, F(t_n y) - F(y_n - x_n) > \le t_n . r_n . \tag{1.5}$$

It follows then from (1.2), (1.4) and (1.5) that

$$<Ax_n, Fy> \le C + y_n . r_n .$$

By definition, $t_n . r_n \ge 1$, hence $\limsup_n < (t_n r_n)^{-1} Ax_n, Fy > < +\infty$. Since F is surjective we see that the sequence $\{< (t_n r_n)^{-1} Ax_n, z >\}_{n\in N}$ is bounded, for any $z \in Y$; then, by the uniform boundedness principle, $\{(t_n r_n)^{-1} Ax_n\}_n$ is bounded. But this is in contradiction with the fact that $\|(t_n r_n)^{-1} Ax_n\| = t_n^{-1} \xrightarrow[n]{} +\infty$. ■

1.6. THEOREM. (Kato) Let be A and F as in the above Proposition; then A is demicontinuous on Int D(A) if and only if it is hemicontinuous on Int D(A).

PROOF. Suppose that A is hemicontinuous at $x \in \text{Int}\, D(A)$; then by Proposition 1.4. it is locally bounded at x. Let $x_n \xrightarrow[n]{} x$; then the sequence $\{r_n\}_n = \{\|Ax_n\|\}_n$ is bounded. We shall prove that $Ax_n \xrightarrow[n]{w^*} Ax$.

Define again Φ by (1.1) and put $\{\|x_n - x\|^{1/2}, \Phi(\|x_n - x\|)^{1/2}\}$. Then $t_n \xrightarrow[n]{} 0_+$. For $y \in Y$, let $y_n = x + t_n y$. By a similar argument, as in the previous proposition, we can see that (1.2) and (1.3) are true.
Since $Ay_n \xrightarrow[n]{w^*} Ax$ we have $t_n^{-1} < Ay_n, F(y_n - x_n) > \xrightarrow[n]{} < Ax, Fy >$.
For the second term in the right side of (1.2) we obtain again the inequality (1.5), but now $t_n . r_n \xrightarrow[n]{} 0$.
It follows that $\limsup_n <Ax_n, Fy> \le <Ax, Fy>$. The sujertivity of F yields now $\lim_n <Ax_n, z> = <Ax, z>$, $\forall z \in Y$.
Hence $Ax_n \xrightarrow[n]{w^*} Ax$. ■

1.7. COROLLARY. Any monotone hemicontinuous operator is demicontinuous in the interior of its domain of definition. ■

§.2. MONOTONE AND MAXIMAL MONOTONE MAPPINGS

Let X and Y be two real Banach spaces and $A: X - 2^Y$ (possibly) nonlinear mapping; we recall some usefull notations, namely: the (effective) domain of A is $D(A) = \{x \in X; Ax \ne \emptyset\}$, the range of A is

$R(A) = \{u \in Ax; x \in D(A)\}$ and its graph is the subset of $X \times Y$ defined as $G(A) = \{(x, u) \in X \times Y; x \in D(A), u \in Ax\}$.
We note that one can identify each subset $G \subseteq X \times Y$ with a mapping $A:X \longrightarrow 2^{Y}$ by defining $Ax = \{u \in Y; (x, u) \in G\}$. The inverse $A^{-1}:R(A) \longrightarrow 2^{X}$ is then defined by $A^{-1}u = \{x \in D(A);\ u \in Ax\}$ and it is clear that $G(A^{-1}) = \{(u, x); (x, u) \in G(A)\}$.
We say that a mapping $B:X \longrightarrow 2^{Y}$ extends A if $Ax \subseteq Bx$, $\forall x \in X$ and will denote this by $A \subseteq B$.
We define the sum of the mappings A and B on X by

$$(A+B)(x) = \begin{cases} Ax + Bx & \text{for} \quad x \in D(A) \cap D(B) \\ \emptyset & \text{elsewhere} \end{cases}$$

Single valued mappings are called (in this context) operators. For $U \subseteq X$ we shall denote $A(U) = \bigcup_{x \in U} Ax$.

2.1. DEFINITION. Let $A:X \longrightarrow 2^{X^*}$ be a (possibly) nonlinear mapping; then
i) A is called locally bounded at $x \in D(A)$ if there exists a neighbourhood U of x such that A(U) is bounded in Y. A is locally bounded on $D \subseteq D(A)$ if it is locally bounded at every $x \in D$.
ii) A is called bounded if it maps bounded subsets of X into bounded subsets of Y.

2.2. DEFINITION. A mapping $A:X \longrightarrow 2^{X^*}$ is said to be
a) monotone if for any $x, y \in D(A)$ and $u \in Ax, v \in Ay$, we have

$$< u - v, x - y > \geq 0. \tag{2.1}$$

A is called strictly monotone if equality in (2.1) implies x = y.
b) maximal montone if it is monotone and for $(x, u) \in X \times X^*$ the inequalities $< u - v, x - y > \geq 0$, for all $(y, v) \in G(A)$, imply $(x, u) \in G(A)$.

2.3. REMARK. A set $G \subseteq X \times X^*$ is called monotone provided that $< u - v, x - y > \geq 0$ for any $(x, u), (y, v) \in G$, i.e. if the associated mapping is monotone. Hence a mapping $A:X \longrightarrow 2^{X^*}$ is maximal monotone if and only if its graph G(A) is a maximal monotone set (for the inclusion relation) in $X \times X^*$

By Zorn's Lemma, every monotone map has a maximal monotone extension.

2.4. THEOREM. Any monotone mapping $A:X \longrightarrow 2^{X^*}$ is locally bounded on $\operatorname{Int} D(A)$.

PROOF. Let $x_0 \in \operatorname{Int} D(A)$; we shall prove that A is locally bounded at x_0. Defining $\tilde{A}x = A(x+x_0)$ then $D(\tilde{A}) = D(A) - x_0$ and we may assume $x_0 = 0$. Let $U = S_\rho(0) \subseteq D(A)$; if $y \in U$ and $v \in Ay$, are fixed, then $<u-v, x-y> \geq 0$ for every $x \in U$ and $u \in Ax$. Since there exists $n \in N$ such that $|<v, x-y>| \leq n$, for all $x \in U$, we also have

$$<u, x-y> \geq -n, \quad \forall x \in U, \ u \in Ax.$$

Define $U_n = \{y \in U; <u, x-y> \geq -n, \ \forall x \in U, \ u \in Ax\}$; then $U = \bigcup_{n \geq 1} U_n$ and $U_n = \bar{U}_n$, $\forall n \in N$.

By the Baire category Theorem, $\operatorname{Int} \bar{U}_m \neq \emptyset$ for some $m \in N$ and therefore we can find $r > 0$ and $z_0 \in U_m$ with $z_0 + \bar{S}_r(0) \subset U_m$. Hence

$$<u, x - z_o - z> \geq -m \quad \text{for all} \quad x \in U, u \in Ax \text{ and } z \in \bar{S}_r(0). \tag{2.2}$$

Since $-z_0 \in U$, there is $p \in N$ such that $-z_0 \in U_p$ and this implies

$$<u, x + z_o> \geq -p \quad \text{for all } x \in U \text{ and } u \in Ax. \tag{2.3}$$

Summing (2.2) and (2.3) we obtain

$$<u, 2x - z_o> \geq -(m+p) \quad \text{for all } x \in U, u \in Ax, z \in \bar{S}_r(0). \tag{2.4}$$

We note now that $z = 2x - y \in \bar{S}_r(0)$ if $x \in \bar{S}_{r/4}(0)$ and $y \in \bar{S}_{r/2}(0)$ and consequently (2.4) yields

$$<u, y> \geq -(m+p) \quad \text{for all } x \in \bar{S}_{r/4}(0), u \in Ax, y \in \bar{S}_{r/2}(0) \tag{2.5}$$

Therefore, if $u \in Ax$, where $x \in \bar{S}_{r/4}(0)$ then by (2.5)

$$\|u\| = 2r^{-1}. \sup_{\|y\| \leq r/2} <u, y> \leq 2r^{-1}.(m+p).$$

Hence A is bounded on $\bar{S}_{r/4}(0)$. ■

2.5. THEOREM. Let be $A:X \longrightarrow 2^{X^*}$ monotone, then
i) If A is maximal monotone, then Ax is convex and w^*-closed for all $x \in D(A)$ and A is norm-to- weak* upper semicontinuos.
ii) If for each $x \in X$, Ax is non void convex, w^*-closed subset of X^* and A is norm-to- weak*-upper semicontinuous, then A is maximal monotone.

PROOF. i) As A is maximal monotone, we have

$$Ax = \bigcap_{(y,v)\in G(A)} \{u \in X^*; < u-v, x-y > \geq 0\}, \quad \forall x \in D(A).$$

Then Ax is convex and w^*-closed because in the above intersection each set has these two properties.
Suppose that there is $x_0 \in \operatorname{Int} D(A)$ where A is not norm-to- weak* upper semicontinuous; then there exists a w^*-neighbourhood V of $0 \in X^*$, $x_n \in D(A)$ with $x_n \xrightarrow[n]{} x_0$ and $u_n \in Ax_n$ so that $u_n \notin Ax_0 + V$.
By Theorem 2.4., A is locally bounded at x, so that $\{u_n\}_n$ is bounded, hence w^*-relatively compact. By the finite intersection property we find $u_0 \in \cap \overline{\{u_n; n \geq k\}}^{w^*}$ and obviously $u_0 \notin Ax_0 + V$.
Let be $\varepsilon > 0$ and $(x, u) \in G(A)$; then $< u_n - u, x_n - x > \geq 0$, $n \in N$, and $< u_n - u, x_o - x_n > \geq -\varepsilon$, for $n \geq n_0$; this yields

$$< u_n - u, x_o - x > \geq -\varepsilon \quad \text{for} \quad n \geq n_0. \tag{2.6}$$

Consider the w^*-neighbourhood V_ε of u_0 defined by

$$V_\varepsilon = u_0 + \{v \in X^*; < v, x_o - x > < \varepsilon\}.$$

As $u_0 \in \overline{\{u_n, n \geq n_0\}}^{w^*}$, there exists $n_\varepsilon \geq n_0$ so that $u_{n_\varepsilon} \in V_\varepsilon$. Hence

$$< u_{n_\varepsilon} - u_o, x_o - x > < \varepsilon. \tag{2.7}$$

From (2.6) and (2.7) we obtain

$$< u_o - u, x_o - x > = u_o - u_{n_\varepsilon}, x - x > + < u_{n_\varepsilon} - u, x_o - x > \geq -2\varepsilon.$$

Since $\varepsilon > 0$ and $(x, u) \in G(A)$ were arbitrary, it follows that $< u_o - u, x_o - x > \geq 0$, $\forall (x, u) \in G(A)$. As A is maximal monotone, $u_0 \in Ax_0$, which is a contradiction.
ii) Suppose that there exists $(x_0, u_0) \in S(X) \times X^*$ such that

$$< u_o - v, x_o - y > \geq 0, \quad \forall (y, v) \in G(A) \text{ 'but } u_0 \notin Ax_0 \tag{2.8}$$

Since Ax_0 is convex and w^*-closed in X^*, using a well-known separation theorem we conclude that there exists $y_0 \in X$ with

$$<u_o, y_o> > \sup_{v \in Ax_o} <v, y_o>.$$

Consider the w^*-neighbourhood of Ax given by

$$V = \{v \in X^*; <v, y_o> < <u_o, y_o>\}$$

and define $y_t = x_0 + ty_0$, $t > 0$; take $v_t \in Ay_t$. Then by the upper semicontinuity of A, $v_t \in V$, for $t > 0$ sufficiently small, i.e.

$$<v_t, y_o> < <u_o, y_o>.$$

Finally we observe that if we rewrite (2.8) for y_t and v_t, then $<u_o - v_t, y_o> \le 0$, $t > 0$ and this is a contradiction. ■

2.6. COROLLARY. Every duality mapping on X is maximal monotone.

PROOF. This is a consequence of Proposition 4.7. and Theorem 4.2., §4, Ch. I.

2.7. COROLLARY. Let $A: X \longrightarrow X^*$ be monotone and semicontinuous with $D(A) = X$; then A is maximal monotone.

2.8. EXAMPLES. 1. Let $f: X \longrightarrow \tilde{R}$ be a proper subdifferentiable function; then the map $\partial f: X \longrightarrow 2^{X^*}$ is monotone. Indeed, for any $x^* \in \partial f(x)$, $y^* \in \partial f(y)$, we have

$$f(y) - f(x) \ge <x^*, y - x>, \quad f(x) - f(y) \ge <y^*, x - y>.$$

Summing these inequalities we obtain

$$<x^* - y^*, x - y> \ge 0, \quad \forall x, y \in D(\partial f).$$

2. If X = H is a Hilbert space and $T: H \longrightarrow H$ is a contraction, i.e.

$$\|Tx - Ty\| \le \|x - y\|, \quad \forall x, y \in D(T)$$

then the operator A = I - T is monotone. Indeed

$$<Ax - Ay, x - y> = \|x - y\|^2 - <Tx - Ty, x - y> \ge 0.$$

3. Let $\Omega \subset R^n$ be a bounded domain, $p \ge 2$ and consider the Sobolev space $H^{1,p}(\Omega) = \{f; D_i f \in L^p(\Omega), 0 \le i \le n\}$ with the norm

$$\|f\|_{1,p} = \left(\sum_{i=0}^{n} \|D_i f\|_p^p \right)^{1/p}$$

(we denoted $D_0 f \equiv f$ and $D_i f$ the derivative of f with respect to the i-th variable).

Let $H_o^{1,p}(\Omega)$ denote the closure of $C_o^\infty(\Omega)$ in the norm $\|\cdot\|_{1,p}$ and endowe it with the equivalent norm $\|f\|_{1,p}^o = \left(\sum_{i=1}^n \|D_i f\|_p^p\right)^{1/p}$. Consider the pseudo-laplacian operator $A: H_o^{1,p}(\Omega) \longrightarrow H_o^{1,p}(\Omega)^*$ defined by

$$Af = -\sum_{i=1}^n D_i\left(|D_i f|^{p-2}.D_i f\right), f \in H_o^{1,p}(\Omega)$$

Then for $f, g \in H_o^{1,p}$, we have

$$|<Af, g>| = \left|\sum_{i=1}^n \int_\Omega |D_i f|^{p-2} D_i f.D_i g\, dx\right|$$

$$\le \sum_{i=1}^n \left\| |D_i f|^{p-1} \right\|_q . \|D_i g\|_p \le \left(\sum_{i=1}^n \left\| |D_i f|^{p-1} \right\|_q^q\right)^{1/q} . \left(\sum_{i=1}^n \left\| |D_i g|^p \right\|\right)^{1/p}$$

$$= \left(\sum_{i=1}^n \|D_i f\|_p^p\right)^{1/q} . \left(\sum_{i=1}^n \|D_i f\|_p^p\right)^{1/p} = \left(\|f\|_{1,p}^o\right)^{p-1} . \|g\|_{1,p}^o .$$

consequently if is well defined and A is bounded. Moreover

$$<Af - Ag, f - g>$$

$$\ge \left(\|f\|_{1,p}^o\right)^p + \left(\|g\|_{1,p}^o\right)^p - \left(\|f\|_{1,p}^o\right)^{p-1} . \|g\|_{1,p}^o - \left(\|g\|_{1,p}^o\right)^{p-1} . \|f\|_{1,p}^o$$

$$= \left[\left(\|f\|_{1,p}^o\right)^{p-1} - \left(\|g\|_{1,p}^o\right)^{p-1}\right]\left(\|f\|_{1,p}^o - \|g\|_{1,p}^o\right) \ge 0.$$

Hence the operator A is strictly monotone.

4. More generally consider the second order differential operator $A: H_o^{1,p}(\Omega) \longrightarrow H_o^{1,p}(\Omega)^*$, $p > 1$ ($\Omega \in R^n$ a domain) given by

$$Af = a_0(x, f, \text{grad } f) - \sum_{i=1}^n D_i\, a_i(x, f, \text{grad } f), f \in H_o^{1,p}(\Omega)$$

where $a_i(x, \xi)$ are defined on $\Omega \times R^{n+1}$ and satisfy (for all $0 \le i \le n$) the following conditions

i) a_i are measurable in x, continuous in ξ and there is $\varphi \in L^q(\Omega)$ such that

$$|a_i(x, \xi)| \le C\left(|\xi|^{p-1} + \varphi(x)\right), \text{ a.e. on } \Omega, \forall\, \xi \in R^{n+1}.$$

ii) $\sum_{i=0}^n [a_i(x, \xi) - a_i(x, \eta)](\xi_i - \eta_i) \ge 0$, a.e. on Ω, $\forall\, (\xi, \eta) \in R^{n+1} \times R^{n+1}$,

From i) we obtain

$$\left|a_i(x, f(x), \text{grad } f(x))\right| \le C\left(\sum_{j=0}^{n}\left|D_j f(x)\right|^{p-1} + \varphi(x)\right), \ x \in \Omega.$$

Hence $a_i(x, f(x), \text{grad } f(x)) \in L^q(\Omega)$ and

$$\left\|a_i(x, f, \text{grad } f)\right\|_q \le C'\left(\sum_{j=0}^{n}\left\|\left|D_j f\right|^{p-1}\right\|_q + \|\varphi\|_q\right)$$

$$\le C''\left[\left(\sum_{j=0}^{n}\left\|\left|D_j f\right|^{p-1}\right\|_q\right)^{q\,1/q} + 1\right] = C''\left(\|f\|_{1,p}^{p-1} + 1\right).$$

Then for $f, g \in H_o^{1,p}(\Omega)$ we have

$$\left|<Af, g>\right| = \left|\sum_{i=0}^{n}\int_\Omega a_i(x, f(x), \text{grad } f(x)) D_i g(x) dx\right|$$

$$\le \sum_{i=0}^{n}\left\|a_i(x, f(x), \text{grad } f)\right\|_q \cdot \left\|D_i g\right\|_p$$

$$\le C''\left(\|f\|_{1,p}^{p-1} + 1\right)\left(\sum_{i=0}^{n}\left\|D_i g\right\|_p\right) \le C_0\left(\|f\|_{1,p}^{p-1} + 1\right) . \|g\|_{1,p} .$$

Thus Af is well defined and A is bounded. Moreover A is monotone

$$<Af - Ag, f - g> = <Af, f - g> - <Ag, f - g>$$

$$= \int_\Omega \sum_{i=0}^{n}\left[a_i(x, f, D_1 f, \ldots, D_n f) - a_i(x, g, D_1 g, \ldots, D_n g)\right]\left[D_i f - D_i g\right] dx \ge 0.$$

We can prove that A is demicontinuous; indeed let $f_n \xrightarrow[n]{} f_o$ in $H_o^{1,p}(\Omega)$. Since A is bounded and $H_o^{1,p}(\Omega)$ is reflexive, then passing eventually to a subsequence, we can suppose that $Af_n \xrightarrow[n]{w} g_o$ for some $g \in H^{1,p}(\Omega)^*$. We only need to prove that $g_o = Af_o$. Since $f_n \xrightarrow[n]{} f_o$ in $H_o^{1,p}(\Omega)$, there exists a subsequence $\{f_{n'}\}_{n'}$ such that $D_i f_{n'} \xrightarrow[n']{} D_i f_o$, a.e. on Ω, $0 \le i \le n$. Then by the properties of the functions a_i, we have

$$a_i(x, f, \text{grad } f_{n'}(x)) \xrightarrow[n']{} a_i(x, f_o(x), \text{grad } f_o(x)) \text{ a.e. on } \Omega, 0 \le i \le n.$$

By the Lebesgue convergence Theorem it follows that

$$<Af_{n'}, g> = \sum_{i=0}^{n}\int_\Omega a_i(x, f_{n'} \text{ grad } f_{n'}) D_i g dx \xrightarrow[n']{}$$

$$\sum_{i=0}^{n}\int_\Omega a_i(x, f_o, \text{grad } f_o) D_i g \, dx = <Af_o, g>, \ \forall g \in H^{1,p}(\Omega).$$

Hence A is demicontinuous. Then by Corollary 2.7. it is maximal monotone. Finally we observe that for $a_0(x,\xi)=0$ and $a_i(x,\xi)=|\xi_i|^{p-2}.\xi_i, 1\le i\le n$, we obtain the operator in the Example 3. If the functions a_i are of the from $a_i(x,\xi)=D_i\Phi(x,\xi)$, a.e. on $\Omega, 0\le i\le n$, and if we denote $F(f)=\int_\Omega \Phi(x,f,\operatorname{grad} f)dx$, then we have:

$$\frac{d}{dt}F(f+tg)\Big|_{t=0}=\frac{d}{dt}\int_\Omega \Phi(x,f+tg,\operatorname{grad}(f+tg)dx\Big|_{t=0}$$

$$=\sum_{i=0}^{n}\int_\Omega D_{\xi_i}\Phi(x,f,\operatorname{grad} f).D_i g\,dx=<Af,g>.$$

In particular, for the function

$$F(f)=\frac{1}{p}\left(\|f\|^{o}_{1,p}\right)^p=\frac{1}{p}\sum_{i=1}^{n}\|D_i f\|_p^p$$

the corresponding operator A obtained as the differential of F is the duality mapping on $\overset{o}{H}{}^{1,p}$ (see Proportion 4.12., Ch. II).

§.3. THE ROLE OF THE DUALITY MAPPING IN SURJECTIVITY AND MAXIMALITY PROBLEMS

3.1. DEFINITION. A mapping $A:X\longrightarrow 2^{X^*}$ is called coercive if there is a function $c:R_+\longrightarrow R$ with $\lim_{t\to+\infty} c(t)=+\infty$ and such that

$$<u,x>\ge c(\|x\|).\|x\|,\quad \forall\ (x,u)\in G(A).$$

3.2. REMARK. An example of a coercive mapping is the duality mapping, as one can easily check. Further we note that all mapings $A:X\longrightarrow 2^{X^*}$ with D(A) bounded are coercive; indeed we can define

$$c(t)=\begin{cases}\inf\left\{\dfrac{|<u,x>|}{\|x\|};\ \|x\|=t,\ u\in Ax\right\} & \text{if } D(A)\cap S_t(0)\ne\varnothing\\ +\infty & \text{else where}\end{cases}$$

It is also clear that an operator $A:X\longrightarrow X^*$, with D(A) unbounded, it coercive if and only if $\lim_{\|x\|\mapsto\infty}\dfrac{<Ax,x>}{\|x\|}=\infty,\ \forall x\in D(A)$.

The following result is basic in the study of variational inequalities (i.e. infinite systems of nonlinear inequalities).

3.3. PROPOSITION. Let X be a real reflexive Banach space, $C \subseteq X$ a closed convex set, $A: X \longrightarrow 2^{X^*}$ a monotone mapping with $D(A) \subseteq C$ and $B: C \longrightarrow X^*$ a monotone, bounded, coercive and demicontinuous operator; then there exists $x_0 \in C$ with

$$< u + Bx_0, x - x_0 > \geq 0, \quad \forall\, (x, u) \in G(A). \tag{3.1}$$

PROOF: We can assume without loss of generality that $(0,0) \in G(A)$; indeed we only need to change A and B with $\tilde{A}x = A(x + x_0) - u_0$, $\tilde{B}x = B(x + x_0) + u_0$, for some fixed $(x_0, u_0) \in G(A)$, and to observe that all the properties are preserved.
We also may suppose that A is maximal monotone on $D(A)$, using the Zorn's Lemma to obtain such an extension.
We part the proof into two steps.
1. We shall first consider the case when X is finite dimentional; then it is sufficient to prove the result for $D(A)$ bounded.
Indeed if the result is true for bounded domains, then for every $n \in N$ we can find $x_n \in C$ such that

$$< u + Bx_n, x - x_n > \geq 0, \quad \forall\, (x, u) \in G(A|S_n(0)). \tag{3.2}$$

In particular, since $(0,0) \in G(A|S_n(0))$, we have $< Bx_n, x_n > \leq 0$, $\forall\, n \in N$, and the coercivity of B implies that the sequence $\{x_n\}_n$ is bounded. Then there is a subsequence $\{x_{n'}\}_n \subseteq \{x_n\}_n$ and $x_0 \in C$ with $x_{n'} \xrightarrow[n']{} x_0$. As by hypothesis B is demicontinuous, (3.2) yields

$$< u + Bx_0, x - x_0 > \geq 0, \quad \forall\, (x, u) \in G(A).$$

So let $D(A)$ be bounded and K a compact and convex set so that $D(A) \subseteq K \subseteq C$ $(K = \operatorname{conv} D(A))$. Suppose that (3.1) does not hold; then to every $z \in K$ there exists $(x, u) \in G(A)$ such that $< u + Bz, x - z > < 0$. Hence $K = \bigcup_{(x,u) \in G(A)} \{z \in K; < u + Bz, x - z > < 0\}$.
Moreover the covering sets are all open in K since the function $K \ni z \longrightarrow < u + Bz, x - z >$ is continuous.
The compacity of K implies that there exist $(x_i, u_i) \in G(A)$, $1 \leq i \leq m$, with $K = \bigcup_{i=1}^{m} K_i$, where $K_i = \{z \in K; < u_i + Bz, x_i - z > < 0\}$.
Let $\{\varphi_1, \dots, \varphi_m\}$ be a partition of the unity subordinated to this convering and define the map $K \ni z \longrightarrow f(z) = \sum_{i=1}^{m} \varphi_i(z) x_i$. Since f is

continuous and maps K into K, by Brower's fixed point (Theorem 1.16., Ch. IV) there is $z_0 \in K$ so that $f(z_0) = z_0$.
On the other hand, for every $z \in K$, we may write

$$g(z) = < \sum_{i=1}^{m} \varphi_i(z) u_i + Bz, f(z) - z >$$

$$= < \sum_{i=1}^{m} \varphi_i(z) u_i + Bz, \sum_{j=1}^{m} \varphi_j(z).(x_j - z) >$$

$$= \sum_{i,j}^{m} \varphi_i(z).\varphi_j(z) < u_i + Bz, x_j - z >.$$

If in the last sum $i = j$ and $\varphi_i^2(z) \neq 0$, it follows that $z \in K_i$ and thus $< u_i + Bz, x_i - z > < 0$. If $i \neq j$ and $\varphi_i(z)\varphi_j(z) \neq 0$, then $z \in K_i \cap K_j$; using then the monotonicity of A we have

$$< u_i + Bz, x_j - z > + < u_j + Bz, x_i - z >$$
$$= < u_j + Bz, x_j - z > + < u_i + Bz, x_i - z > + < u_j - u_i, x_i - x_j > < 0.$$

Hence g(z)<0 on K; this is in contradiction with the fact that

$$g(z_0) = < \sum_{i=1}^{m} \varphi_i(z_0) u_i + Bz_0, f(z_0) - z_0 > = 0.$$

II. In order to prove the result in infinite dimensions, we consider a directed family $\{X_\alpha\}_{\alpha \in I}$ of finite dimensional subspaces of X; we can suppose that also I is directed, namely $\alpha < \beta \Rightarrow X_\alpha \subset X_\beta$, and that $X = \bigcup_{\alpha \in I} X_\alpha$. Denote P_α the injection of X_α in X, P_α^* its dual map, and $A_\alpha = P_\alpha^* A P_\alpha$, $B_\alpha = P_\alpha^* B P_\alpha$, $C_\alpha = C \cap X_\alpha$, $\forall \alpha \in I$. It is clear that $A_\alpha : X_\alpha \longrightarrow 2^{X_\alpha^*}$ is monotone and that $B_\alpha : C_\alpha \longrightarrow X_\alpha^*$ is monotone, coercive, bounded and continuous.
By the first step of the proof, for each $\alpha \in I$ there is $x_\alpha \in C_\alpha$ such that $< \tilde{u} + B_\alpha x_\alpha, x - x_\alpha > \geq 0$, $\forall (x, \tilde{u}) \in G(A_\alpha)$, i.e.

$$< u + Bx_\alpha, x - x_\alpha > \geq 0, \quad (x, u) \in G(AP_\alpha). \tag{3.3}$$

In particular, $< Bx_\alpha, x_\alpha > \leq 0$, $\forall \alpha \in I$ and since B is coercive and bounded, this implies that there exists $r > 0$ so that

$$\|x_\alpha\| \leq r \text{ and } \|Bx_\alpha\| \leq r, \quad \forall \alpha \in I.$$

As X is reflexive, there is a subsequence $\{x_{\alpha_n}\}_{n \in N} \subset \{x_\alpha\}_{\alpha \in I}$ and $(x_0, u_0) \in C \times X^*$ such that

$$x_{\alpha_n} \xrightarrow[n]{w} x_0 \text{ and } Bx_{\alpha_n} \xrightarrow[n]{w} u_0.$$

From (3.3) we obtain

$$\limsup_{n\to+\infty} < Bx_{\alpha_n}, x_{\alpha_n} > \le < u, x - x_0 > + < u_0, x >, \ \forall\ (x, u) \in G(A). \quad (3.4)$$

We also note that there is $(x_1, u_1) \in G(A)$ such that

$$< u_1, x_1 - x_0 > + < u_0, x_1 > \le < u_0, x_0 >. \quad (3.5)$$

Indeed, otherwise

$$< u + u_0, x - x_o > > 0, \quad \forall\ (x, u) \in G(A), \quad (3.6)$$

and the maximality of A yields $(x_0, -u_0) \in G(A)$.
Setting now in (3.6) $x = x_0, u = -u_0$, we immediately obtain $0 = < u_0 - u_0, x_0 - x_0 > > 0$ which is a contradiction.
Hence from (3.4) and (3.5) we have $\limsup_{n\to\infty} < Bx_{\alpha_n}, x_{\alpha_n} > \le u_0, x_0 >$; consequently

$$\limsup_{n\to\infty} < Bx_{\alpha_n}, x_{\alpha_n} - x_0 > \le 0. \quad (3.7)$$

Let $x \in D(A)$; for $t \in [0,1]$ denote $x_t = tx_0 + (1-t)x$. Then $x_t \in C$ and the monotonicity of B yields

$$< Bx_{\alpha_n} - Bx_t, x_{\alpha_n} - x_t > \ge 0, \quad n \in N, \quad t \in [0,1].$$

Hence

$$(1-t) < Bx_{\alpha_n}, x_{\alpha_n} - x > + t < Bx_{\alpha_n}, x_{\alpha_n} - x_0 >$$
$$\ge (1-t) < Bx_t, x_{\alpha_n} - x > + t < Bx_t, x_{\alpha_n} - x_0 >.$$

Letting $n \to \infty$ and using (3.7) we have

$$\liminf_{n\to\infty} < Bx_{\alpha_n}, x_{\alpha_n} - x > \ge < Bx_t, x_0 - x >, \quad t \in [0,1].$$

But B is in particular hemicontinuous, so that

$$\liminf_{n\to\infty} < Bx_{\alpha_n}, x_{\alpha_n} > - < u_0, x > \ge < Bx_0, x_0 - x >. \quad (3.8)$$

Finally by (3.4) and (3.8) we have

$$< Bx_0, x_0 - x > \le u, x - x_0 >, \quad \forall.(x, u) \in G(A).$$

Therefore (3.1) is proved. ■
We shall apply this result to solve some functional equations for monotone operators.

3.4. THEOREM. Let X be a real reflexive Banach space, $C \subseteq X$ a closed convex set, $A:X \longrightarrow 2^{X^*}$ a maximal monotone mapping with $D(A) \subseteq C$ and $B:C \longrightarrow X^*$ a monotone, bounded, coercive and demicontinuous operator; then A + B is surjective.

PROOF. For a $u_0 \in X^*$, define $\tilde{A}x = Ax - u_0$; then all the conditions of the above theorem are fullfilled for $\tilde{A}$ and B. Hence there exists $x_0 \in C$ such that

$$< \tilde{u} + Bx_0, x - x_0 > \geq 0, \quad \forall\, (x, \tilde{u}) \in G(\tilde{A})$$

i.e.

$$< u - (u_0 - Bx_0), x - x_0 > \geq 0, \quad \forall\, (x, u) \in G(A).$$

Since A is maximal monotone, we have that $(x_0, u_0 - Bx_0) \in G(A)$ and this yields $u_0 \in Ax_0 + Bx_0$. ■

In what follows we consider the real reflexive Banach space X endowed with an equivalent norm such that X and X^* are both strictly convex; by Theorem 2.9., Ch. III, this is always possible. Then if J denote the normalized duality mapping relative to this norm, it is clear that it is monotone, bounded, demicontinuous and coercive. We shall present two surjectivity results for maximal monotone mappings involving such a duality mapping.

3.5. THEOREM. Let X be a real reflexive Banach space and $A:X \longrightarrow 2^{X^*}$ is a maximal monotone and coercive mapping, then A is surjective.

PROOF. It is sufficient to prove that $0 \in R(A)$. To this purpose, let $\varepsilon_n \to 0$ be a sequence of positive numbers; we apply the above theorem to A and $\varepsilon_n J, n \in N$, to obtain vectors $x_n \in D(A)$ with $0 \in (A + \varepsilon_n J)x_n$. Consequently there are $u_n \in Ax_n$ such that $u_n + \varepsilon_n J x_n = 0$. Let $c:R_+ \longrightarrow R$ be the function from the definition of the coercivity of A; then

$$0 = < u_n + \varepsilon_n Jx_n, x_n > \geq c(\|x_n\|)\cdot\|x_n\| + \varepsilon_n \cdot \|x_n\|^2.$$

Consequently the sequence $\{x_n\}_n$ is bounded. As $\varepsilon_n \longrightarrow 0$, we have

$$\|u_n\| = \varepsilon_n \cdot \|Jx_n\| = \varepsilon_n \|x_n\| \xrightarrow[n]{} 0, \text{ i.e. } u_n \xrightarrow[n]{} 0.$$

By the reflexivity of X, there exist $x_0 \in X$ and a subquence $\{x_{n'}\}_{n'}$ of $\{x_n\}_n$ such that $x_{n'} \xrightarrow[n']{w} x_0$. Since A is monotone, we have

$$<u-u_{n'}, x-x_{n'}> \geq 0, \quad \forall (x,u) \in G(A).$$

Letting $n' \longrightarrow +\infty$, we obtain

$$<u, x-x_{0'}> \geq 0, \quad \forall (x,u) \in G(A).$$

Since A is maximal monotone, it follows that $0 \in Ax_o$. ■

3.6. COROLLARY. If X is a real reflexive Banach space and $A:X \longrightarrow X^*$ is a monotone coercive hemicontinuos mapping with $D(A) = X$, then A is surjective.

PROOF. By Corollary 2.7., A is maximal monotone and we can apply the above Theorem.

3.7. REMARK. Consider the operator A on $H_0^{1,p}(\Omega)$ from example 2.8, 4 and suppose that A is coercive. Then by the above Corollary A is surjective, i.e. for every $g^* \in H^{1,p}(\Omega)^*$, the equation $Af = g^*$ has at least a solution.

3.8. THEOREM. Let X be a real reflexive Banach space and $A:X \longrightarrow 2^{X^*}$ be a maximal monotone mapping; then A is surjective if and only if A^{-1} is locally bounded.

PROOF. Since A is maximal monotone and surjective, A^{-1} is maximal monotone; moreover $D(A^{-1}) = X^*$. Then by Theorem 2.4. A^{-1} is locally bounded on X^*.

In order to prove the converse statement, we show that R(A) is both closed and open in X^*.

Let $u_n \in Ax_n$, $n \in N$, be a sequence in X^* such that $u_n \xrightarrow[n]{} u_o \in X^*$; then $x_n \in A^{-1}$, $\forall n \in N$ and since A^{-1} is locally bounded, $\{x_n\}_{n \in N}$ is bounded. Passing to a subsequence we have $x_{n'} \xrightarrow[n']{w} x_o \in X$. By the monotonicity of A, $<u_{n'} - u, x_{n'} - x> \geq 0$, $\forall (x,y) \in G(A)$, hence $<u_o - u, x_o - x> \geq 0$, $\forall (x,u) \in G(A)$. Consequently $(x_o, u_o) \in G(A)$, i.e. $R(A)$ is closed.

In order to prove that $R(A)$ is open, consider $u_o \in R(A)$ and $r > 0$ so that $A^{-1}S_r(u_o)$ is bounded; we shall show that $S_{r/2}(u_o) \subseteq R(A)$.

There is $x_o \in D(A)$ with $u_o \in Ax_o$. Consider on X^* the normalized duality mapping corresponding to the equivalent norm such that X and X^* are strictly convex and apply Theorem 3.4. to get that $A+\lambda J(.-x_0)$ is surjective for every $\lambda>0$. Then for a fixed $u\in S_{r/2}(u_0)$, there are $x_\lambda\in D(A)$ and $u_\lambda\in Ax_\lambda, \lambda>0$ such that $u_\lambda+\lambda J(x_\lambda-x_0)=u$. We have

$$<u-\lambda J(x_\lambda-x_o)-u_o, x_\lambda-x_o>=<u_\lambda-u_o, x_\lambda-x_o>\geq 0.$$

Consequently

$$<u-u_o, x_\lambda-x_o>\geq\lambda<J(x_\lambda-x_o), x_\lambda-x>=\lambda\|x_\lambda-x\|^2,$$

and thus $\lambda\|x_\lambda-x\|\leq\|u-u_0\|<r/2$. Then

$$\|u-u_\lambda\|=\|J(x_\lambda-x_0)\|=\lambda\|x_\lambda-x\|<\frac{r}{2}$$

and this yields $\|u_\lambda-u_0\|\leq\|u_\lambda-u\|+\|u-u_o\|<r$. Since A^{-1} is bounded on $S_r(u_o)$, the set $\{x_\lambda\}_{\lambda>0}$ is bounded so that

$$\|u-u_\lambda\|=\lambda\|x_\lambda-x\|\xrightarrow[\lambda\to 0]{} 0.$$

As $R(A)$ is closed, it follows that $u\in R(A)$. ■

3.9. COROLLARY. Let $A:X\longrightarrow X^*$ be a hemicontinuous operator with $D(A)=X$ which is strongly monotone, i.e. there is a continuous increasing function $c:R_+\longrightarrow R_+$ with $c(0)=0$ such that

$$<Ax-Ay, x-y>\geq c(\|x-y\|)\cdot\|x-y\|, \quad \forall\, x,y\in X; \tag{3.3}$$

then A is surjective.

PROOF. By Corollary 2.7. A is maximal; we shall show that A^{-1} is locally bounded.
By (3.9) $c(\|x-y\|)\leq\|Ax-Ay\|$ for all $x,y\in D(A)$ so that

$$\|A^{-1}u-A^{-1}v\|\leq c^{-1}(\|u-v\|) \text{ for } u=Ax, v=Ay,$$

where c^{-1} is the inverse of the function c. Since c^{-1} has the same properties as c, for all $u\in R(T)$ with $\|u\|\leq M$ and $u_0\in R(T)$ fixed, we have $\|A^{-1}u_0\|\leq\|A^{-1}u_0\|+c^{-1}(M+\|u_0\|)$. Hence A^{-1} is bounded and the result is a consequence of the above theorem. ■

3.10. PROPOSITION. Let X be a real reflexive Banach space and $A:X\longrightarrow 2^{X^*}$ a maximal monotone mapping with $D(A)=X$; then for every $\lambda>0$, $A+\lambda J$ is maximal monotone and surjective and $(A+\lambda J)^{-1}:X^*\longrightarrow X$ is a demicontinuous maximal monotone operator. Moreover, if X has the property (h), then $(A+\lambda J)^{-1}$ is continuous.

PROOF. The surjectivity of $A+\lambda J$ is a consequence of Theorem 3.4, hence $(A+\lambda J)^{-1}$ is defined on all X^*. We also note that it is single valued. Indeed, let $x_1, x_2 \in (A+\lambda J)^{-1}u$; then there are $u_1 \in Ax_1$ and $u_2 \in Ax_2$ such that $u = u_1 + \lambda Jx_1 = u_2 + \lambda Jx_2$. Consequently
$0 = \langle u-u, x_1 - x_2\rangle = \langle u_1 - u_2, x_1 - x_2\rangle + \lambda\langle Jx_1 - Jx_2, x_1 - x_2\rangle$.
Since each of the terms at the right side of the above equality is positive, we have $\langle Jx_1 - Jx_2, x_1 - x_2\rangle = 0$. By Theorem 1.8., Ch. II. J is strictly monotone, thus $x_1 = x_2$.

We shall prove that $(A+\lambda J)^{-1}$ is demicontinuous. Let $u_n \longrightarrow u_o$ in X^* and $x_n = (A+\lambda J)^{-1}$, $n \in N_o$; then there are $u'_n \in Ax_n$ such that $u_n = u'_n + Jx_n$, $n \in N_o$, and we may write

$$\langle u_n-u_o, x_n-x_o\rangle = \langle u'_n-u'_o, x_n-x_o\rangle + \lambda\langle Jx_n-Jx_o, x_n-x_o\rangle,\ n\in N. \quad (3.10)$$

We note that $(A+\lambda J)^{-1}$ is monotone, hence locally bounded on X^*. Then the sequence $\{x_n\}_n$ is bounded so that $\langle u_n-x_o, x_n-x_o\rangle \xrightarrow[n]{} 0$.
As both terms at the right side of the equality (3.10) are positive, it follows that $\langle Jx_n-Jx_o, x_n-x_o\rangle \xrightarrow[n]{} 0$. But then in virtue of Proposition 3.6, Ch. II, $x_n \xrightarrow[n]{w} x_o$.
If X has the property (h), then by Proposition 5.6., Ch. II, we have $x_n \xrightarrow[n]{} x_o$. Finally $(A+\lambda J)^{-1}$ is maximal monotone by Corollary 2.7. and this yields also the maximality of $A+\lambda J$. ■

3.11. THEOREM. Let X be a real reflexive Banach space $A: X \longrightarrow 2^{X^*}$ a monotone mapping; then A is maximal monotone if and only if A + J is surjective.

PROOF. The necessity of the condition is a consequence of Theorem 3.4. Suppose that $R(A+J) = X^*$ and that A is not maximal montone; then there exists $(x_o, u_o) \in X \times X^*$ such that $(x_o, u_o) \notin G(A)$ but

$$\langle u-u_o, x-x_n\rangle \geq 0,\quad \forall\, (x,u) \in G(A). \quad (3.11)$$

By hypothesis there exists $(x_1, u_1) \in G(A)$ so that

$$u_0 + Jx_0 = u_1 + Jx_1. \tag{3.12}$$

Taking in (3.11) $x = x_1$, $u = u_1$ we obtain $< u_1 - u_0, x_1 - x_0 \geq 0$. Hence $< Jx_1 - Jx_0, x_1 - x_0 > = 0$. Since J is strictly monotone it follows that $x_0 = x_1 \in D(A)$. Thus by (3.12) $u_0 = u_1 \in Ax_1 \in G(A)$, i.e. $(x_0, u_0) \in G(A)$ and this is a contradiction. ■

3.12. COROLLARY. If $A{:}X \longrightarrow 2^{X^*}$ is maximal monotone and $B{:}X \longrightarrow X^*$ is a monotone bounded and hemicontinuous operator with $D(B) = X$, then A + B is maximal monotone.

PROOF. Since $D(B) = X$, by Theorem 1.6. B is demicontinuous. Then also B + J is demicontinuous; moreover it is monotone, bounded and coercive. Then, by Theorem 3.4, A + B + J is surjective and we only have to apply the above theorem to get that A + B is maximal. ■

The fact that the sum of two maximal monotone operators is not always maximal monotone is illustrated in Exercise 13. A more general situation in which A + B is maximal monotone is described in Exercise 14.
We end this section with the following convexity result

3.13. THEOREM. Let be $A{:}X \longrightarrow 2^{X^*}$ a maximal mapping on the real reflexive Banach space X ; then $\overline{D(A)}$ and $\overline{R(A)}$ are convex.

PROOF. It is sufficient to prove that $\overline{D(A)}$ is convex; indeed, as A^{-1} is maximal monotone too, it follows that $\overline{D(A^{-1})} = \overline{R(A)}$ is convex.
In order to prove the convexity of $\overline{D(A)}$ we define for every $\lambda > 0$ two operators on X in the following way
By Theorem 3.4. the equation $0 \in \lambda A\tilde{x} + J(\tilde{x} - x)$ has a solution $x_\lambda \in D(A)$ for every $x \in X$; consequently $0 = \lambda u_\lambda + J(x_\lambda - x)$, for some $u_\lambda \in Ax_\lambda$. Moreover the solution x_λ is unique.
Indeed, suppose that $0 = \lambda u_\lambda + J(x_\lambda - x) = \lambda u'_\lambda + J(x'_\lambda - x)$, for some $u_\lambda \in Ax_\lambda$, $u'_\lambda \in Ax'_\lambda$. Then

$$0 \leq < J(x_\lambda - x) - J(x'_\lambda - x), x_\lambda - x'_\lambda > = - \lambda < u_\lambda - x'_\lambda, x_\lambda - x'_\lambda > \leq 0$$

This yields $x_\lambda = x'_\lambda$.

Denote $J_\lambda x = x_\lambda$ and $A_\lambda x = \frac{1}{\lambda} x_\lambda J(x_\lambda - x)$, $x \in X$, $\lambda > 0$; then the maps $J_\lambda{:}X \longrightarrow X$ and $A_\lambda{:}X \longrightarrow X^*$ are single-valued.

Since $A_\lambda x \in Ax_\lambda$, the monotinicity of A yields

$$<v, J_\lambda x - y> = <v, x_\lambda - y> \le <A_\lambda x, J_\lambda x - y>$$

$$= -\frac{1}{\lambda} <J(J_\lambda x - x), J_\lambda x - x> - \frac{1}{\lambda} <J(J_\lambda x - x), x - y>, (y, v) \in G(A)$$

Whence, for every $x \in X$, $(y, v) \in G(A)$, $\lambda > 0$, we have

$$\|J_\lambda x - x\|^2 \le <J(J_\lambda x - x), x - y> + \lambda <v, J_\lambda x>. \qquad (3.13)$$

It follows that

$$\|J_\lambda x - x\|^2 \le \|J_\lambda x - x\| \, \|x - y\| + \lambda\|v\| \, \|J_\lambda x - x\| + \lambda\|x\| . \|x - y\|$$

and it is an easy matter to see now that $\{J_\lambda x\}_{\lambda>0}$ is bounded, for all $x \in X$. Then there exists a sequence $\lambda_n \xrightarrow{n} 0_+$ such that $J(J\lambda_n x - x) \xrightarrow{w} u_o \in X^*$. From (3.13) we obtain

$$\limsup_n \|J_{\lambda_n} x - x\|^2 \le <u_o y - x>, \forall y \in D(A). \qquad (3.14)$$

It is clear that (3.14) is also true for any $y \in \mathrm{conv}D(A)$. In particular, if $x \in \overline{\mathrm{conv}D(A)}$, we can take in (3.14) y = x to get that $J_{\lambda_n} x \xrightarrow{\lambda_n \to 0_+} x$. Since $J_{\lambda_n} x \in D(A)$, $\forall n \in N$, we easily see that $x \in \overline{D(A)}$, i.e. $\overline{D(A)} = \overline{\mathrm{conv}D(A)}$. ■

3.12. REMARK. From (3.13) it follows that J_λ and thus also A_λ are bounded on $D(A)$. The operators A_λ are called Yosida approximants and J_λ the resolvent of A; this terminology is motivated by the fact that in the Hilbert case it is easy to check that $J_\lambda = (1 + \lambda A)^{-1}$ and $A_\lambda = \frac{1}{\lambda}(1 - J_\lambda)$. More properties of these operators are given in Exercise 18.

§.4. AGAIN ON SUBDIFFERENTIALS OF CONVEX FUNCTIONS

The subdifferential of a convex function has some remarkable properties and therefore play a central role in the class of monotone mappings.

4.1. PROPOSITION. Let $f: X \longrightarrow R$ be a proper convex l.s.c function on the reflexive real Banach space X; then $\partial f: X \longrightarrow 2^{X^*}$ is maximal monotone.

PROOF. As it was shown in Example 1, 2.8, ∂f is monotone. We shall prove that $\partial f + J$ is surjective and then the maximality is a consequence of Theorem 3.11.

Consider $u \in X^*$ and define the function $\varphi: X \longrightarrow \tilde{R}$ by

$$\varphi(x) = f(x) + \frac{1}{2}\|x\|^2 - \langle u, x \rangle, x \in X.$$

Then φ is proper, convex and l.s.c.; moreover, by Theorem 2.8., Ch. I. we have

$$\partial\varphi(x) = \partial f(x) + \partial\left(\frac{1}{2}\|x\|^2\right) - \partial \langle u, x \rangle, = (\partial f + J)\, x - u.$$

Then $u \in (\partial f + J)(x)$ if and only if $0 \in \partial\varphi(x)$ and this is equivalent to the fact that the function φ has a minimum value at x. We shall show that

$$\lim_{\|x\| \to \infty} \varphi(x) \equiv +\infty \tag{4.1}$$

and then apply Corollary 1.8., Ch. I to obtain that φ attains its minimum. To this purpose suppose that (4.1) is not true; then there is $\{x_n\} \subseteq D(\varphi)$ with $\|x_n\| = t_n \xrightarrow[n]{} \infty$ but $\varphi(x_n) \le M$. It follows that

$$f(x_n) - \langle u, x_n \rangle \le M - \frac{1}{2}t_n^2, \quad n \in N.$$

Let be $y_n = t_n^{-1}x_n$; then $\|y_n\| = 1$ and we may write

$$\begin{aligned} f(y_n) - \langle u, y_n \rangle &= f\left(t_n^{-1}.x_n + \left(1 - t_n^{-1}\right).0 - \langle u, t_n^{-1}x_n\right) \rangle \\ &\le t_n^{-1}.f(x_n) + \left(1 - t_n^{-1}\right)f(0) - t_n^{-1} \langle u, x_n \rangle \\ &\le t_n^{-1}.M - \frac{1}{2}.t_n + \left(1 - t_n^{-1}\right)f(0) \xrightarrow[n]{} -\infty \end{aligned}$$

This is in contradiction with the fact that the proper convex l.s.c. function $\psi(x) = f(x) - \langle u, x \rangle$ is bounded from below on the unit ball. This completes our proof. ■

In order to characterize those maximal monotone mappings on X which are subdifferentials of convex l.s.c. functions, we give the:

4.2. DEFINITION. A mapping $A: X \longrightarrow 2^{X^*}$ is said to be cyclically monotone if for any subset $\{x_0, x_1, \ldots, x_n\} \subseteq D(A), n \in N$ and $u_i \in Ax_i$, $0 \le i \le n$, we have

$$\sum_{i=0}^{n} \langle u_i, x_i - x_{i+1} \rangle \ge 0, \text{ where } x_{n+1} = x_0.$$

A is called cyclically maximal monotone if it has no cyclically monotone extension.

4.3. EXAMPLE. Every monotone mapping $A:R \longrightarrow 2^R$ is cyclically montone.
Indeed, let $\{x_0, \dots, x_n\} \subseteq D(A)$ and $u_i \in Ax_i, 0 \le i \le n$; we can suppose $x_i \le x_{i+1}$ for $0 \le i \le n-1$ and this implies $u_i \le u_{i+1}$, $0 \le i \le n-1$. Hence

$$\sum_{i=0}^{n} < u_i, x_i - x_{i+1} > = \sum_{i=1}^{n} < u_i, x_i - x_{i+1} > + < u_o, x_o - x_1 >$$

$$\sum_{i=1}^{n} < u_i - u_o, x_i - x_{i+1} > \ge 0.$$

The main result of this paragraph is

4.4. THEOREM. A mapping is cyclically maximal monotone if and only if A is the subdifferential of a proper convex l.s.c. function.

PROOF. Let $f:X \longrightarrow \tilde{R}$ be a proper convex l.s.c. function such that $A = \partial f$ and $\{x_0, x_1, \dots, x_n\} \subseteq D(A)$, $u_i \in Ax_i, 0 \le i \le n$; we have

$$f(x_i) - f(x_{i+1}) \le < u_i, x_i - x_{i+1} >, \quad 0 \le i \le n, \quad x_{n+1} = x_0.$$

Consequently

$$\sum_{i=0}^{n} < u_i, x_i - x_{i+1} > \ge \sum_{i=o}^{n} f(x_i) - f(x_{i+1}) = 0.$$

Conversely, let A be a cyclically maximal monotone mapping and $(x_0, u_0) \in G(A)$; define

$$f(x) = \sup\left\{ \sum_{i=o}^{n} < u_i, x_{i+1} - x_i > + < u_n, x - x_n >; (x_i, u_i) \in G(A), 0 \le i \le n, n \in N \right\}.$$

Since f is the upper bound of affine functions, then f is convex and l.s.c.; moreover as $f(x) = \sum_{i=0}^{n} < u_i, x_{i+1} - x_i > \le 0$, f is proper. Let $(x, u) \in G(A)$ and $y \in X$; consider in the definiton of f at y the n + 1 pairs in $G(A):(x_0, u_0), (x_1, u_1), \dots, (x_n, u_n), (x, u)$, to obtain that

$$f(y) \ge \le u_o, x_1 - x_o > + u_1, x_2 - x_1 > + \dots + u_n, x - x_n > + < u, y - x >.$$

Hence,

$$f(y) \ge f(x) + < u, y - x >, \text{ i.e. } u \in \partial f(x).$$

Consequently $G(A) \subseteq G(\partial f)$, so that by the maximality of A we have $A = \partial f$. ■

4.5. COROLLARY. Any maximal monotone map $A:R \longrightarrow 2^R$ is the subdifferential of a proper convex l.s.c. function.

4.6. COROLLARY. Let H be a Hilbert space and $A:H \longrightarrow H$ a linear maximal monotone operator; then A is cyclically maximal monotone if and only if A is sefadjoint.

PROOF. Suppose A selfadjoint and define

$$f(x) = \begin{cases} 1/2\left\|A^{1/2}x\right\|^2, & x \in D\left(A^{1/2}\right) \\ +\infty & , \quad \text{elsewhere} \end{cases}.$$

Then since $A^{1/2}$ is closed, f is proper, convex and l.s.c. Moreover, for $x \in D(A)$ and $y \in D\left(A^{1/2}\right)$, we have

$$1/2\left\|A^{1/2}x\right\|^2 + 1/2\left\|A^{1/2}y\right\|^2 \geq \left\|A^{1/2}x\right\|.\left\|A^{1/2}y\right\|$$
$$\geq < A^{1/2}x, A^{1/2}y > = < Ax, y >.$$

This yields

$$1/2\left\|A^{1/2}x\right\|^2 - 1/2\left\|A^{1/2}y\right\|^2 \geq < Ax, y > - \left\|A^{1/2}x\right\|^2$$
$$= < Ax, y >. - < Ax, x > = < Ax - y - x >.$$

Hence $Ax \in \partial f(x)$, i.e. $A \subseteq \partial f$ and the maximality of A implies $A = \partial f$.
Suppose A cyclically maximal monotone; then $A = \partial f$ for some $f : H \longrightarrow \tilde{R}$, convex, proper and l.s.c.

It is an easy matter to see that $\overline{D(A)} = H$; indeed, if $\overline{D(A)} \neq H$, there is $u \neq 0$ such that $< u, x > = 0$, $\forall x \in D(A)$. Now the maximality of A implies $u = Ao = 0$.

We shall show further that A^* is monotone, i.e. positive. In effect, if $x \in D(A) \cap D(A^*)$, then $< A^*x, x > = < x, Ax > \geq 0$. If $x \in D(A) \cap D(A^*)$, but $x \notin D(A)$, then there is $x_0 \in D(A)$ with $< Ax_o + A^*x, x_o - x > < 0$.
Hence

$$< Ax_o, x_o > < A^*x, x_o > + < A^*x, x > - < Ax_o, x > = < A^*x, x >.$$

Consequently $< A^*x, x > \geq 0$.

For $x \in D(A)$, let be $\varphi(t) = f(tx), t \in (0,1)$; then φ is differentiable $\varphi'(t) = t < Ax, x >$ and we have

$$\varphi(1) - \varphi(0) = \int_0^1 \varphi'(t)dt = \int_0^1 t < Ax, x > dt = 1/2 < Ax, x >.$$

Hence $f(x) = f(0) + 1/2 < Ax, x >$. A simple computation yields

$$< \partial f(x), y > = 1/2(< Ax, y > + < x, Ay >), \qquad \forall y \in D(A).$$

Since $< \partial f(x), y > = < Ax, y >$, we have $< Ax, y > = < x, Ay >$, $\forall y \in D(A)$.
Thus $x \in D(A^*)$ and $Ax = A^*x$, i.e. $A \subseteq A^*$. The fact that A^* is monotone and the maximality of A imply $A = A^*$. ■

4.7. THEOREM. Let X be a (real) reflexive Banach space, and $f: X \longrightarrow \tilde{R}$ a proper l.s.c. convex function with $0 \in D(\partial f)$; then the following statements are equivalent:

i) $\lim\limits_{\|x\| \to \infty} \dfrac{f(x)}{\|x\|} = +\infty$;

ii) $A = \partial f$ is coercive;

iii) $R(A) = X^*$ and A^{-1} is bounded on X^*.

PROOF. i) $\Rightarrow$ ii) Let $(x, u) \in G(\partial f)$; then $f(x) - f(y) \le \langle u, x - y \rangle$, $\forall y \in X$. In particular, for $y = 0$ we have $\langle u, x \rangle \ge f(x) - f(0)$.

If we denote $c(t) = \inf\limits_{\|x\| = t} \dfrac{f(x) - f(0)}{\|x\|}$, then $c(t) \xrightarrow[t \to +\infty]{} +\infty$ and $\langle u, v \rangle \ge c(\|x\|).\|x\|$, i.e. A is coercive.

ii) $\Rightarrow$ iii) Since A is coercive and maximal monotone, then by Theorem 3.5. $R(A) = X^*$. Since $c(\|x\|).\|x\| \le \langle u, x \rangle$, $\forall (x, u) \in G(A)$, then $c(\|x\|) \le \|u\|$, $\forall (x, u) \in G(A)$. It is clear now that A^{-1} is bounded.

iii) $\Rightarrow$ i) We can suppose that $f \ge 0$. Indeed for $x_0 \in D(\partial f)$, we select $u_0 \in \partial f(x_0)$; then $f(x) - f(x_0) \ge \langle u_0, x - x_0, \forall x \in D(f)$.

Let be $\tilde{f}(x) = f(x) - f(x_0) - \langle u_0, x - x_0 \rangle$; it is clear that $\tilde{f}(x) \ge 0$ and $\partial \tilde{f}(x) = \partial f(x) - u_0 = Ax - u_0$. Hence we only have to make a convenient translation of A in order to consider positive f.

Consider $r > 0$; as $A^{-1}[S_r(0)]$ is bounded, there exists $M > 0$ such that for every $u \in X^*, \|u\| \le r$, there is $x \in D(A)$ with $\|x\| \le M$ and $u \in Ax = \partial f(x)$. For $(x, u) \in G(A)$ as above we have

$$f(y) - f(x) \ge \langle u, y - x \rangle, \forall y \in D(f),$$

i.e.

$$\langle u, y \le f(y) + Mr - f(x) \le f(y) + M.r.$$

We can now easily see that

$$r.\|y\| = \sup_{\|u\| \le r} \langle u, y \rangle \le f(y) + M.r, \quad y \in D(f).$$

Hence $\dfrac{f(y)}{\|y\|} \ge r - \dfrac{M.r}{\|y\|}$, $\forall y \in D(f)$, so that $\liminf\limits_{\substack{\|y\| \to \infty \\ y \in D(f)}} \dfrac{f(y)}{\|y\|} = +\infty$. ■

We end our considerations with the following results concerning some variational inequalities:

4.8. THEOREM. Let $A:X \longrightarrow X^*$ be a monotone, hemicontinuous, coercive and bounded operator with $D(A) = X$ and $f:X \longrightarrow \tilde{R}$ a proper, convex, l.s.c. function; then for any $u \in X^*$ there is $x \in D(f)$ so that

$$< Ax - u, x - y >, \leq f(y) - f(x), \quad \forall\, y \in X \tag{4.2}$$

Moreover the solution x is unique if A is strictly monotone.

PROOF. We note that (4.2) is equivalent to the functional equation $u - Ax \in \partial f(x)$ i.e. $u \in (A + \partial f)(x)$. Since ∂f is maximal monotone, they by Theorem 3.4., $u \in R(A + \partial f)(x_1) \cap (A + \partial f)(x_2)$; then there are $u_1 \in \partial f(x_1)$, $u_2 \in \partial f(x_2)$ so that $u = Ax_1 + u_1 = Ax_2 + u_2$.
Hence

$$< Ax_1 - Ax_2, x_1 - x_2 > = < u_2 - u_1, x_2 - x_2 > \leq 0.$$

This obviously yields $x_1 = x_2$. ■

4.9. COROLLARY. Let $A:X \longrightarrow X^*$ be a monotone, hemicontinuous, coercive, bounded operators with $D(A) = X$ and let $K \subseteq X$ be a convex, closed subset of X; then for every $u \in X^*$ there exists $x \in K$ such that

$$< Ax, y - x > \geq < u, y - x > \quad \forall\, y \in K.$$

PROOF. Consider $f = I_K$, where I_K is the indicator function of K (see Remark 1.4, Ch. I); then I_K is proper, convex, l.s.c. and $\partial I_K = \{x^* \in X^*; < x^*, x - y > \geq 0, \ \forall\, y \in K\}$.
The result is now a direct consequence of the above theorem.

EXERCISES

1. Let X be a Banach space and $A:X \longrightarrow X$ a nonlinear operator on X; define $|A| = \sup\limits_{x,\, y \in D(A),\, x \neq y} \|Ax - Ay\| / \|x - y\|$, $\text{Lip } X = \{A; |A| < +\infty\}$ and $\text{Lip}_o X = \{A \in \text{Lip } X; D(A) = X\}$.
   i) Prove that $|A|$ is seminorm on Lip X and that $\text{Lip}_o X$ is a Banach space with the norm $\|A\| = |A| + \|Ax_o\|$, for a fixed $x_o \in D(A)$.

ii) Let $A \in \text{Lip } X$ with $|A| < 1$; then $(I - A)^{-1} \in \text{Lip } X$ and $\left|(I - A)^{-1}\right| \leq (1 - |A|)^{-1}$.

iii) Let $\rho(A) = \{\lambda \in R; (\lambda I - A) - 1 \in \text{Lip } X\}$; prove that $\rho(A)$ is open.

Hint. iii) We note that

$\lambda I - A = \left[I + (\gamma - \lambda)(\lambda I - A)^{-1}\right](\lambda I - T), \quad \lambda \in \rho(A), \quad \gamma \in R$

and we can use ii) to prove that $\rho(A)$ is open.

2. Prove that any locally bounded linear operator $A:X \longrightarrow X^*$ is continuous.

Hint. Suppose that there exists $\{x_n\}_n \subset D(A)$ such that $x_n \xrightarrow[n]{} 0$ and $\|Ax_n\| > \varepsilon$, for some $\varepsilon > 0$; write $t_n = \|x_n\|^{-1/2}$ and $y_n = t_n x_n$; then $y_n \xrightarrow[n]{} 0$ and $\|Ay_n\| = t_n\|Ax_n\| \geq \varepsilon t_n \xrightarrow[n]{} +\infty$ which is a contradiction.

3. Let $A:X \longrightarrow X^*$ be a linear monotone operator; then:

i) A is maximal monotone if and only if $G(A)$ is maximal among all linear monotone graphs.

ii) A is maximal monotone if and only if A is closed and A^* is monotone.

Hint. i) Suppose $G(A)$ is maximal among all linear monotone graphs and that there exists $(x, u) \notin G(A)$ so that $\langle u - v, x - y \rangle \geq 0$, for every $(y, v) \in G(A)$; then the linear space spanned by $G(A) \cup (x, u)$ is monotone too.

ii) See Theorem 2.7. Ch. III, Pascali-Sburlan [1].

4. i) Let $A:R^m \longrightarrow R^m$ be a monotone surjective map; then $\lim\limits_{\|x\| \to \infty} \|Ax\| = +\infty$.

ii) Let $A:R^m \longrightarrow R^m$ be monotone on $S_{r+\varepsilon}(0) \subset R^m$; prove that $A(\bar{S}_r(0))$ is bounded.

Hint. i) Suppose that there are $\{x_n\}_n \subset R^m$, $x_0, y_0 \in R^m$ such that $\|x_n\| \xrightarrow[n]{} \infty, x_n/\|x_n\| \xrightarrow[n]{} x_0$ and $Ax_n \xrightarrow[n]{} y_0$; then

$< Ax_n - Ax, x_n/\|x_n\| - x/\|x_n\| > \geq 0$ implies $< y_0, x_0 > \geq Ax, x_0 >$, $\forall\, x \in R^n$. We use the surjectivity to obtain $x_0 = 0$.

ii) Use a similar argument as in i).

5. Let $\{x_n\} \subset \partial S_1(0)$ be such that $x_n \neq x_{n'}$ for $n \neq n'$ and $x_n \xrightarrow[n]{} x_0$; define the mapping $A:\bar{S}_1(0) \subset R^m \longrightarrow R^m$:

$$Ax = \begin{cases} x & \text{for} \quad x \neq x_n \\ (n+1)x_n & \text{for} \quad x = x_n ; \end{cases}$$

then A is monotone on $\bar{S}_1(0)$ but it is unbounded at x_0.

6. A mapping $A:H \longrightarrow 2^H$ on a Hilbert space is monotone iff

$$\|x - y + t(u - v)\| \geq \|x - y\|, \quad \forall\, u \in Ax, v \in Ay \text{ and } t \geq 0.$$

Hint. Suppose A monotone; then the equality

$$\|(x + y) + t(u - v)\|^2 = \|x - y\|^2 + 2t < u - v, x - y > + t^2\|u - v\|^2 \qquad (*)$$

yields $\|(x - y) + t(u - v)\| \geq \|x - y\|$.

Conversely, from $(*)$ we obtain $2 < u - v, x - y + t\|u - v\|^2 \geq 0$. Letting now $t \longrightarrow 0$, we get that A is monotone.

7. Let $A:X \longrightarrow 2^{X^*}$ be maximal monotone, $\{x_n\}_n \subset D(A)$, $u_n \in Ax_n$ and $(x, u) \in X \times X$ such that $x_n \xrightarrow[n]{} x$ (resp. $x_n \xrightarrow[n]{w} x$) and $u_n \xrightarrow[n]{w^*} u$ (resp. $u_n \longrightarrow u$); then $(x, u) \in G(A)$.

8. Let $A:X \longrightarrow 2^{X^*}$ and $B:X \longrightarrow X^*$ be monotone and such that $R(A + B) = X^*$ and $(A + B)^{-1}$ is continuous; then A is maximal monotone.

Hint. Suppose that for $(x, u) \in X \times X$ we have

$$< u - v, x - y > \geq 0, \quad \forall\, (y, v) \in G(A);$$

then

$$< u + Bx - (v + By), x - y > \geq 0, \quad \forall\, (y, v) \in G(A). \qquad (*)$$

Let be $w \in X^*$, $t>0$, $u_t = u + Bx + tw, x_t = (A + B)^{-1} u_t$ and $v_t \in A x_t$ such that $v_t + B x_t = u_t$. Taking $y = x_t$ in $(*)$ we obtain $< w - x_t - x > \geq 0$. The continuity of F yields now

$< w, x_0 - x > \geq 0$, where $x_0 = (A+B)^{-1}(u+Bx)$. Consequently $x = x_0$.

9. Prove that if $A: X \longrightarrow 2^{X^*}$ is monotone with $0 \in \operatorname{Int} D(A)$, then A is quasi-bounded, i.e. for any M > 0 there is C > 0 such that $(y, v) \in G(A)$, $< v, y > \leq M\|y\|$ and $\|y\| \leq M$, implies $\|v\| \leq C$.

Hint. See Lemma 3.6, Ch. II, Pascali-Sburlan [1].

10. Let $A: X \longrightarrow 2^{X^*}$ be coercive; then A^{-1} is locally bounded.

Hint. From $< u_n, x_n > \geq c(\|x_n\|) . \|x_n\|$ and in X^* we deduce that $c(\|x_n\|) < +\infty$; since $c(t) \longrightarrow \infty$ for $t \longrightarrow \infty$, it follows that $\{x_n\}_n$ is bounded.

11. Let $A: X \longrightarrow 2^{X^*}$ be a maximal monotone mapping and $\varphi(x) = \inf_{u \in Ax} \|u\|$; prove that φ is l.s.c.

Hint. Let $x_n \xrightarrow[n]{} x$ be such that $\varphi(x) > \lim_n \inf \varphi(x_n)$; there are $\varepsilon > 0$ and $u_n \in A x_n$ such that $\|u_n\| < (x) - \varepsilon$. By the local boundedness of A, there is a subsequence $u_{n'} \xrightarrow[n]{w^*} u \in X^*$; then $\|u_n\| \leq \varphi(x) - \varepsilon$.
Since A is maximal monotone, we have $(x, u) \in G(A)$.
Then $\|u\| \geq \varphi(x)$ and this is impossible.

12. Let be $A_1: X \longrightarrow 2^{X^*}$ a monotone mapping and $A_2: X \longrightarrow X^*$ a monotone operator with $D(A_2) = X$; if the sum $A = A_1 + A_2$ is maximal monotone, then A_1 is also maximal monotone.

Hint. See Theorem 2.5. Ch. III, Pascali-Subrlan [1].

13. Let $X = L^2(R_+)$, $Af = -f''$ with $D(A) = \{f \in H^2(R_+), f(0) = 0\}$ and $Bf = -f''$ with $D(B) = \{f \in H^2(R_+), f'(0) = 0\}$; prove that A and B are Maximal monotone but that A + B is not maximal monotone.

Hint. See §3.6. Ch. III, Pascali-Sburlan [1].

14. Let be X reflexive and A, B two maximal monotone mappings; if $\operatorname{Int} D(A) \cap D(B) \neq \phi$, then A + B is maximal monotone.

Hint. See Theorem 1.7., Ch. II, Barbu [2].

15. Let H be a Hilbert space, $A:H \longrightarrow 2^H$ be maximal monotone and $B:H \longrightarrow H$ be monotone with D(B) closed and such that $\|Bx - By\| \leq \alpha\|x - y\|$, $\forall x, y \in D(B)$, for some $\alpha \in (0,1)$; then A + B is maximal monotone.

Hint. We shall prove that $R(A+B+I) = X^*$. Let $u_0 \in H$ and $Cx = (A+I)^{-1}(u_0 - Bx)$; then, since $(A+I)^{-1}$ is contractive (use exercise 6 to prove this property) we get $\|Cx - Cy\| \leq \alpha\|x - y\|$. Then C has fixed point x_0 and $u_0 \in B + (A+I)^{-1} x_0$.

16. Let be X reflexive and $A:X \longrightarrow 2^{X^*}$ monotone; prove that there exists a maximal monotone extension $\tilde{A}$ of A with

$$D(\tilde{A}) \subseteq \overline{\operatorname{conv} D(A)}.$$

Hint. Use Zorn's Lemma to obtain an extension $\tilde{A}$, maximal monotone in the set of the graphs $\subseteq \overline{\operatorname{conv} D(A)} \times X^*$. We can prove that $R(\tilde{A} + J) = X^*$; indeed for $u_0 \in X^*$ we apply Proposition 3.3. to $\tilde{A} - u_0$ and $J|\overline{\operatorname{conv} D(A)}$ to obtain the existence of an $x_0 \in \overline{\operatorname{conv} D(A)}$ with $u_0 \in Ax_0 + Jx_0$.

17. Let be X reflexive, $A:X \longrightarrow 2^{X^*}$ maximal monotone, $u_0 \in X^*$ and $x_\lambda \in (A + \lambda J)^{-1} u_0, \lambda > 0$; then $u_0 \in R(A)$ if and only if $\{x_\lambda\}_{\lambda>0}$ is bounded.

Hint. Suppose $u_0 \in Ax_0$; the monotinicty of A yields $\subseteq < Jx_\lambda, x_0 - x_\lambda > \geq 0$, $\forall \lambda > 0$ and hence $\|x_\lambda\| \leq \|x_0\|$. Conversely, let $x_0 \in X$ be so that $x_{\lambda'} \xrightarrow[\lambda']{w} x_0$ for a subsequence of $\{x_\lambda\}_{\lambda>0}$. Consider the inequality $< u - (u_0 - \lambda' Jx_{\lambda'}), x - x_{\lambda'} > \geq 0$. Letting

$\lambda' \longrightarrow 0$ we get $<u-u_o, x-x_o> \geq 0$, $\forall (x,u) \in G(A)$, i.e. $(x_0, u_0) \in G(A)$.

18. Let be X a reflexive Banach space and $A: X \longrightarrow 2^{X^*}$ a monotone mapping; if A is maximal monotone, resp. cyclically maximal monotone, then the Yosida approximants A_λ are maximal monotone, resp. cyclically maximal monotone.

Hint. Prove that $A_\lambda = (A^{-1} + \lambda^{-1} J)^{-1}$ in order to obtain the maximality of A_λ. We obtain the second assertion if we write:

$$\sum_0^n <A_\lambda x_i, x_i - x_{i+1}> \leq \sum_0^n <A_\lambda x_i, (x_i - J_\lambda x_i) - (x_{i+1} J_\lambda x_{i+1})>$$

$$= \lambda \sum_0^n <A_\lambda x_i, J^{-1} A_\lambda x_i - J^{-1} A_\lambda x_{i+1}> \geq \left(\lambda \sum_0^n \|Ax_i\|^2 - \sum_0^n \|Ax_i\| \|Ax_{i+1}\| \right) \geq 0.$$

19. Let be $A: D(A) = X \longrightarrow 2^{X^*}$ monotone; then the following statements are equivalent:
a) A is maximal monotone
b) A is norm-to-w^*-upper semincontinuous and Ax is a convex w^*-closed subset of X^* for every $x \in X$
c) A is hemicontinuos (i.e. for each $x, y \in X$ and each w^*-open neiborhoud V of $0 \in X^*$, there exists $\delta > 0$ such that $A(x + [0, \delta) y) \subset Ax + V$) and Ax is a convex w^*-closed subset of X^* for every $x \in X$.

Hint. See Cioranescu [1], Ch. 4, §2.

BIBLIOGRAPHICAL COMMENTS:

§1. The continuity results presented in this section are due to Kato [1], [2].

§2. In 1960 Kachurovskii [1] observed that the differential of a convex function is a monotone "operator" - this terminology is due to him. Monotone operators" in Hilbert spaces were studied in the 60-ies by Minty [2], [5], and up to this moment begun an intensive development of the theory of the monotonicity and its applications to nonlinear problems.

The notion of multivalued monotone mappings is due to Browder [4]. Proofs of Theorem 2.4. can be found in Pazy [1] or Rockafellar [6]. The continuity results which we present are due to Browder [15]; we recommend also Browder [2], [4], [12], Fitzpatrick [2] Aubin-Cellina [1]. The applications are taken from Browder [5]. For other properties of the monotone mappings one can consult Kačurovski [2], [3], Minty [1] and the monograph of Pascali-Sburlan [1].

§3. The basic result exposed in Proposition 3.3. in due to Browder [12] [15] and its finite dimensional version is a special case of the early result of Debrunner-Flor [1].

Variational inequalities and their applications to nonlinear functional equations were studied by Browder [14], Lions-Stampacchia [1], Mosco [1], Stampacchia [1], Rockafellar [5].

In Theorem 3.5., due to Browder [5] the reflexivity of the Banach space X is essential for the surjectivity; that this property fails in general Banach spaces was proved by Gossez [4] who constructed a coercive maximal monotone operator A from ℓ^1 into ℓ^∞ with $\overline{R(A)} \neq \ell^\infty$.

Theorem 3.8. is due to Brezis [2] but was previously obtained by Browder [12] in the particular case when X^* is uniformly convex; this theorem was extended to general Banach spaces by Brezis-Browder [1].

Proposition 3.10. and Theorem 3.11. are due to Browder [15] but were obtained in particular cases also by Minty [2] and Moreau [2]. Calvert [3] proved that Theorem 3.11 fails in a non-reflexive Banach space.

Theorem 3.12. is a particular case of a result on the virtual convexity of the domain of a maximal monotone map due to Rackafellar [8]; see also Gossez [2]. The proof given here can be found in Deimling [2].

The properties of the resolvent and the Yosida approximants for a maximal monotone mapping have been stated by Browder [5] in Hilbert spaces and by Brézis-Crandall-Pazy [1] in the general case; see also Pascali-Sburlan [1], Chapter III.

For an introduction in the spectral theory for nonlinear operators one can consult Fucǐk-Nečas-Soucek-Soucek [1] and Zarantonello [1].

§4. The maximal monotonicity of the subdifferential of a convex l.s.c. function was established by Rockafellar [7] to whom is also due its characterization as a cyclically maximal monotone mapping in Theorem 4.4. (see [2], [3]).

CHAPTER VI

ACCRETIVE MAPPINGS AND SEMIGROUPS OF NONLINEAR CONTRACTIONS

In this chapter we first present general result on accretive mappings on a Banach space X in order to tackle then the problem of the genration of semigroups of nonlinear contractions.

§ 1. GENERAL PROPERTIES OF MAXIMAL ACCRETIVE MAPPINGS.

Unless otherwise stated, X will be a real Banach space and J the normalized duality mapping on X.

1.1 DEFINITION. A mapping $A: X \to 2^X$ is said to be

a) accretive if for any (x, u), (y, v) $\in$ G(A) there is $x^* \in J(x - y)$ such that $< x^*, u - v > \geq 0$;

b) maximal accretive it is accretive and the inclusion $A \subseteq B$, with B accretive, implies A = B;

c) hyperacrretive if R(A + I) = X;

d) dissipative (respectively hyperdissipative) if - A is accretive (respectively hyperacretive).

Example 1. In a Hilbert space H, a mapping $A: H \to 2^H$ is accretive if and only if it is monotone,

Example 2. Let $A: X \to X$ be linear; then it is accretive if and only if for every $x \in D(A)$, there is $x^* \in Jx$ with $< x^*, Ax > \geq 0$; in this case - A is dissipative in the sense of Lumer-Phillips.

1.2. REMARK. The maximal accretivity is obviously equivalent to the following property: if $(x, u) \in G(A)$ is such that for each $(y, v) \in G(A)$ there is $x^* \in J(x - y)$ with $< x^*, x - y > \geq$, then $(x, u) \in G(A)$.

1.3. PROPOSITION. A mapping $A: X \to 2^X$ is accretive if and only if $\| x-y+\lambda(u-v) \| \geq \| x-y \|$ for all $\lambda > 0$ and $(x, u), (y, v) \in G(A)$. (1.1)

PROOF. The property is a consequence of Proposition 4.10, Ch. I in which we only have to replace x and y by x - y and u - v, respectively. ∎

1.4 NOTATIONS. For $A: X \to 2^X$ and $\lambda > 0$, we denote

$$J_\lambda = (I+\lambda A)^{-1}, A_\lambda = \lambda^{-1}(I-J_\lambda), D_\lambda = R(I+\lambda A), |Ax| = \inf\{\| u \|;\ u \in Ax\}.$$

1.5. COROLLARY. $A: X \to 2^X$ is accretive if and only if for every $\lambda > 0$, J_λ is single-valued and contravtive on D_λ, i.e.

$$\| J_\lambda u - J_\lambda v \| \leq \| u-v \|,\quad u,v \in D_\lambda.$$

PROOF. Suppose A accretive and that there are $x, y \in (I+\lambda A)^{-1}x_0$ with $x \neq y$ for some $x_0 \in D_\lambda$. Then $x_0 = x+\lambda u = y+\lambda v$ for some $u \in Ax$ and $v \in Ay$ so that by (1.1) x = y. Rewritting now (1.1) we obtain that J_λ is contractive. Conversely, if J_λ is single-valued and contractive then for every $x \in D(A)$ and $u \in Ax$, $J_\lambda(x+\lambda u) = x$; hence $\| x-y+\lambda(u-v) \| \geq \| J_\lambda(x+\lambda u) - J_\lambda(y+\lambda v) \| = \| x-y \|$ for every $(y, v) \in G(A)$. ∎

1.6. PROPOSITION. Let $A: X \to 2^X$ be hyperaccretive; then A is maximal accretive and $R(I+\lambda A) = X$ for every $\lambda > 0$.

PROOF. Let $B: X \to 2^X$ be accretive and such that $A \subseteq B$. Consider $(x, u) \in G(B)$; thin A is hyperaccretive, there is $y \in D(A)$ with $x+u \in (I+A)\,y$; then $x+u = y+v$ for some $y \in Ay$. By proposition 1.3. we obtain that $x = y \in D(A)$ u = v. Hence $(x,u) \in G(A)$, so that A = B. Let further be $\lambda > 0$ and $u_0 \in X$; then $u_0 \in R(I+\lambda A)$ if and only if there is $x_0 \in D(A)$ such that $x_0 = (I+A)^{-1}\left(\frac{u_0}{\lambda} - \frac{1-\lambda}{\lambda}x_0\right)$.

Denote $Bx = (I+A)^{-1}\left(\frac{u_0}{\lambda} - \frac{1-\lambda}{\lambda}x\right)$; then D(B) = X and by (1.1) we have:

$$\| Bx - By \| = \left\| (I+A)^{-1}\left(\frac{u_o}{\lambda} - \frac{1-\lambda}{\lambda}x\right) - (I+A)^{-1}\left(\frac{u_o}{\lambda} - \frac{1-\lambda}{\lambda}y\right)\right\| \le \left|\frac{1-\lambda}{\lambda}\right| \| x-y \|.$$

Suppose $\lambda \ge \frac{1}{2}$; then $\left|\frac{1-\lambda}{\lambda}\right| < 1$ so that we can use the Banach fixed point Theorem to get the existence of a point x_o with $Bx_o = x_o$ and we are done. In order to prove the property also for $0 < \lambda \le \frac{1}{2}$, we select a $\lambda > \frac{1}{2}$ and put $\tilde{A}x = \lambda A$; then by the above considerations $R(I+\gamma\tilde{A}) = X$ for all $\gamma > \frac{1}{2}$, i.e. $R(I+\gamma\lambda A) = X$ for all $\gamma\lambda > \frac{1}{4}$.
Repeating the argument, we obtain that $R(I+\lambda A) = X$ for all $\lambda > \frac{1}{2^n}, n \in N$. ■

The Theorem 3.10 in Ch. V. may be reformulated as follows

1.7. COROLLARY. In a Hilbert space H a mapping $A\colon H \to 2^H$ is maximal accretive if and only if it is hyperaccretive.

1.8. PROPOSITION. Let $A\colon X \to 2^X$ be an accretive mapping; then
i) A_λ is accretive for every $\lambda > 0$, $A_\lambda x \in AJ_\lambda x$, $\forall x \in D_\lambda$,
$\| A_\lambda x - A_\lambda y \| \le 2\lambda^{-1} \| x-y \|$, $\forall x, y \in D_\lambda$ and $\| A_\lambda x \| \le | Ax |$, $\forall x \in D_\lambda \cap D(A)$.
ii) $\lim_{\lambda \to 0_+} J_\lambda x = x$, $\forall x \in D(A) \cap (\cap_{\lambda>0} D_\lambda)$.

PROOF. Let $x, y \in D_\lambda$ and $x^* \in J(x-y)$; we have

$$\begin{aligned} &< x^*, A_\lambda x - A_\lambda y > = \lambda^{-1} < x^*, x-y > - \lambda^{-1} < x^*, J_\lambda x - J_\lambda y > \\ &\ge \lambda^{-1} \| x-y \|^2 - \lambda^{-1} \| x^* \| \, \| J_\lambda x - J_\lambda y \| \ge 0. \end{aligned}$$

Consequently A_λ is accretive. Moreover, as J_λ is contractive, we have that

$$\| A_\lambda x - A_\lambda y \| = \lambda^{-1} \| x - y + (J_\lambda x - J_\lambda y) \| \le 2\lambda^{-1} \| x-y \|.$$

Consider $x \in D_\lambda$ and put $y = J_\lambda x$; then $x = y + \lambda u$ for some $u \in Ay$ and therefore $A_\lambda x = \lambda^{-1}(x - J_\lambda x) = u \in Ay$. Hence $A_\lambda x \in AJ_\lambda x$.
If $x \in D_\lambda \cap D(A)$, then for every $u \in Ax$, $J_\lambda(x+\lambda u) = x$; hence:

$$\| A_\lambda x \| = \lambda^{-1} \| x - J_\lambda x \| = \lambda^{-1} \| J(x+\lambda u) - J_\lambda x \| \le \| u \|$$

and this implies $\| A_\lambda x \| \le | Ax |$, $\forall \lambda > 0$.

ii) Let $x \in D_\lambda \cap D(A)$, $\forall \lambda > 0$; then by i) we have

$$\| J_\lambda x - x \| = \| A_\lambda \| \leq \lambda | Ax | \xrightarrow[\lambda \to 0_+]{} 0. \qquad \blacksquare$$

1.9. DEFINITION. We say that the mapping $A: X \to 2^X$ is closed (respectively demiclosed) if $(x_n, u_n) \in G(A), x_n \xrightarrow[n]{} x_0$ and $u_n \xrightarrow[n]{} u_0$ (respectively $u_n \xrightarrow[n]{w} u_0$) implies $(x_0, u_0) \in G(A)$.

1.10. PROPOSITION. Every maximal accretive mapping $A: X \to 2^X$ is closed; moreover if X^* is uniformly convex, then A is demiclosed.

PROOF. Let $(x_n, u_n) \in G(A)$ be such that $x_n \xrightarrow[n]{} x_0$ and $u_n \xrightarrow[n]{} u_0$; by Proposition 1.3. we have

$$\| x_n - x \| \leq \| x_n - x + \lambda(u_n - u) \|, \quad \forall \lambda > 0, (x,u) \in G(A).$$

Letting $n \to \infty$, we obtain

$$\| x_0 - x \| \leq \| x_0 - x + \lambda(u_0 - u) \|, \quad \forall \lambda > 0, (x,u) \in G(A).$$

Then $G(A) \cup (x_0, u_0)$ defines an accretive mapping and by the maximality of A it follows that $(x_0, u_0) \in G(A)$.

Suppose now X^* uniformly convex; then by Theorem 2.16., Ch. II, J is uniformly continuous on bounded subsets of X. Therefore the inequalities $< J(x_n - x), u_n - u > \geq 0$, $\forall (x,u) \in G(A)$ yield

$$< J(x_0 - x), u_0 - u > \geq 0, \quad \forall (x,u) \in G(A).$$

Since A is maximal, we conclude that $(x_0, u_0) \in G(A)$. $\blacksquare$

1.11. COROLLARY. Let $A: X \to 2^X$ be a hyperaccretive mapping and $\{x_\lambda\}_{\lambda > 0} \subset D(A)$ such that $x_\lambda \xrightarrow[\lambda \to 0_+]{} x_0$; then

i) if $A_\lambda x_\lambda \xrightarrow[\lambda \to 0_+]{} u_0$, then $(x_0, u_0) \in G(A)$.

ii) if X^* is uniformly convex and $\{A_\lambda x_\lambda\}_{\lambda > 0}$ is bounded, then $x_0 \in D(A)$.

Moreover if $A_\lambda x_\lambda \xrightarrow[\lambda \to 0_+]{w}$, then $(x_0, u_0) \in G(A)$.

PROOF. i) Since $A_\lambda x_\lambda \in A_\lambda J_\lambda x_\lambda$, it is sufficient to prove that $J_\lambda x_\lambda \xrightarrow[\lambda \to 0_+]{} x_0$ and to use then closedness of A in order to obtain the desired result. Since $\{A_\lambda x_\lambda\}_{\lambda > 0}$ is bounded, we have

$$J_\lambda x_\lambda - x_0 = (J_\lambda x_\lambda - x_\lambda) + (x_\lambda - x_0) \xrightarrow[\lambda \to 0_+]{} 0.$$

Hence $x_\lambda - J_\lambda x_\lambda = \lambda A_\lambda x_\lambda \xrightarrow[\lambda \to 0]{} 0$.

As $\{A_\lambda x_\lambda\}_{\lambda>0}$ is bounded, it contains a weakly convergent subsequence; hence the result is a consequence of the fact that $J_\lambda x_\lambda \xrightarrow[\lambda\to 0_+]{} x_0$ and that A is demiclosed. ■

1.12. PROPOSITION. Let $A: X \to 2^X$ be a maximal accretive mapping in a smooth Banach space X; then for every $x \in D(A)$, Ax is a closed convex subset of X.

PROOF. Wwe note that J is single-valued; then the maximality of A implies

$$Ax = \bigcap_{(y,v)\in G(A)} \{u \in X; < J(x-y), u-v > \geq 0\}$$

Since each set in the above intersection is closed and convex, Ax has the same properties.

1.13. DEFINITION. The mapping $A°: X \to 2^X$ defined by

$$A°x = \{u \in Ax; \| u \| = | Ax |\}$$

is called minimal selection of A.

1.14. PROPOSITION. If X is reflexive, smooth and strictly convex and $A: X \to 2^X$ is maximal accretive, then $D(A°) = D(A)$ and A° is single-valued.

PROOF. Let $u_n \in Ax$ be so that $\| u_n \| \xrightarrow[n]{} | Ax |$; then passing to a subsequence, we have $u_n \xrightarrow{w} u_0$ for some $u_0 \in X$. Since Ax is closed and convex, $u_0 \in Ax$. From $\| u_0 \| \leq \overline{\lim_n} \| u_u \| \leq | Ax |$, we obtain that $\| u_0 \| = | Ax |$. Suppose that for two elements $u_1 \neq u_2$, we have $\| u_1 \| = \| u_2 \| = | Ax |$; then for $\lambda \in (0,1)$, $\| \lambda u_1 + (1-\lambda) u_2 \| \leq | Ax |$. Hence $\| \lambda u_1 + (1-\lambda) u_2 \| = | Ax |$ which impossible X being strictly convex. ■

1.15. THEOREM. Let X be a uniformly convex Banach space with X^* uniformly convex and $A: X \to 2^X$ be hyperaccretive; then we have:

i) $\| A_\lambda x \| \leq \| A°x \|$, $\forall \lambda > 0$, $x \in D(A)$.

ii) The funcion $R_+ \ni \lambda \to \| A_\lambda x \|$ is nonincreasing, $\forall x \in D(A)$;

iii) $\lim_{\lambda \to 0_+} A_\lambda x = A°x$ and $\lim_{\lambda \to 0_+} A°Jx = A°x$, $\forall x \in D(A)$.

iv) $\overline{D(A)}$ is convex.

PROOF. i) We only need to observe that $D_\lambda = X,\quad \forall \lambda > 0$ and to use Proposition 1.8. i).
ii) Denote $x(\lambda) = J_\lambda x$ and $u(\lambda) = A_\lambda x$; then $(x(\lambda), u(\lambda)) \in G(A)$ and $x = J_\lambda x + \lambda A_\lambda x = x(\lambda) + \lambda u(\lambda),\quad \forall x \in X, \lambda > 0$.
Now for $\lambda,\ \gamma > 0$ we can write

$$\begin{aligned}\| \lambda u(\lambda) - \gamma u(\lambda) \|^2 &= < J(\lambda u(\lambda) - \gamma u(\gamma)),\ \lambda u(\lambda) - \gamma u(\gamma) > \\ &= < J(x(\gamma) - x(\lambda)), \lambda(u(\lambda) - u(\gamma) + (\lambda - \gamma) u(\gamma) > \\ &\le (\lambda - \gamma) < J(x(\gamma) - x(\lambda)), u(\gamma) > \\ &\le |\lambda - \gamma| \cdot \| u(\gamma) \| \| x(\lambda) - x(\gamma) \| \\ &= |\lambda - \gamma| \cdot \| u(\gamma) \| \cdot \| \lambda u(\lambda) - \gamma u(\lambda) \|.\end{aligned}$$

Thus

$$\| \lambda u(\lambda) - \gamma u(\gamma) \| \le |\lambda - \gamma| \cdot \| u(\gamma) \|, \text{ for } \lambda,\ \gamma > 0.$$

Suppose $\lambda > \gamma$; then

$$\begin{aligned}\lambda \| u(\lambda) \| &\le \| \lambda u(\lambda) - \gamma u(\gamma) \| + \| \gamma \cdot u(\gamma) \| \le \\ &\le (\lambda - \gamma) \| u(\gamma) \| + \gamma \| u(\gamma) \| = \lambda \| u(\gamma) \|\end{aligned}$$

Hence $\| u(\lambda) \| \le \| u(\gamma) \|$ as desired.
iii) We show that for every $\lambda_n \xrightarrow[n]{} 0$, there is a subsequence of $\{A_{\lambda_n} x\}_n$ which is convergent to $A^\circ x$ and this proves the first result.
By i) $\{A_{\lambda_n} x\}_n$ is bounded; then there is $\{A_{\lambda_{n'}} x\}_{n'} \subseteq \{A_{\lambda_n} x\}_n$ such that $A_{\lambda_{n'}} x \xrightarrow[n']{w} u$, for some $u \in X$. By Proposition 1.8. , $A_{\lambda_{n'}} x \in AJ_{\lambda_{n'}} x$ and $J_{\lambda_{n'}} x \xrightarrow[n']{} x$; then since A is demiclosed, we have that $(x, u) \in G(A)$. Thus $\| u \| \ge \| A^\circ x \|$.
We further note that the following holds

$$\| u \| \le \liminf_{n'} \| A_{\lambda_{n'}} x \| \le \limsup_{n'} \| A_{\lambda_{n'}} x \| \le \| A^\circ x \|.$$

Hence $\| u \| = \| A^\circ x \|$ and this implies $u = A^\circ x$ and $\| A_{\lambda_{n'}} \| \xrightarrow[n']{} \| A^\circ x \|$.
Then $A_{\lambda_{n'}} \xrightarrow[n']{} A^\circ x$ as desired.
Since $\| A^\circ J_\lambda x \| \le \| A_\lambda x \| \le \| A^\circ x \|$ for all $x \in D(A)$ and $\lambda > 0$, we can use the same argument as above to prove that $A^\circ J_\lambda x \xrightarrow[\lambda \to 0_+]{} A^\circ x$.
iv) We first prove that for any $z = tx + (1-t)y,\ x, y \in D(A),\ t \in (0,1)$ we have $\lim_{\lambda \to 0_+} J_\lambda z = z$.
To do this, let be $x, y \in D(A),\ t \in (0,1)$ and $z = tx + (1-t)y$; fix $u \in Ax$,

$v \in Ay$ and denote $u_\lambda = x + \lambda u$, $v_\lambda = y + \lambda v$. Then $J_\lambda u_\lambda = x$, $J_\lambda v_\lambda = y$ and we have

$$\| J_\lambda z - x \| = \| J_\lambda z - J_\lambda u_\lambda \| \le \| z - u_\lambda \| \le \| z - x \| + \lambda \| u \|. \tag{1.2}$$

Analogously

$$\| J_\lambda z - y \| \le \| z - y \| + \lambda \| v \|. \tag{1.3}$$

In particular $\{J_\lambda z\}_{\lambda > 0}$ is bounded for $\lambda \to 0_+$ hence there is a subsequence $J_{\lambda_n} z \xrightarrow[n]{w} w$. Then using (1.2) and (1.3) we have

$$\| w - x \| \le \liminf_n \| J_{\lambda_n} z - x \| \le \| z - x \| = (1-t) \| x - y \|$$

and

$$\| w - y \| \le \liminf_n \| J_{\lambda_n} z - y \| \le \| z - y \| = t \| x - y \|.$$

It follows that

$$\| w - x \| \ge \| x - y \| - \| w - y \| \ge \| x - y \| - t \| x - y \| = (1-t) \| x - y \|$$

and

$$\| w - y \| \ge \| x - y \| - \| w - x \| \ge \| x - y \| - (1-t) \| x - y \| = t \| x - y \|.$$

Therefore

$$\| w - x \| = \| z - x \| = (1-t) \| x - y \| \text{ and } \| w - y \| = \| z - y \| = t \| x - y \| \tag{1.4}$$

Suppose $w \ne z$; since X is strictly convex, the relation (1.4) yields

$$\| t(w - x) + (1-t)(z - x) \| < (1-t) \| x - y \|$$

and

$$\| t(w - y) + (1-t)(z - y) \| < t \| x - y \|.$$

Consequently

$$\| y - x \| = \| t(w - x) + (1-t)(z - x) - t(w - y) - (1-t)(z - y) \| < \| x - y \|$$

and this is a contradiction. Hence $w = z$ and then

$$J_\lambda z \xrightarrow[\lambda \to 0_+]{w} z. \tag{1.5}$$

Furthermore, form (1.2) we obtain

$$\| z - x \| \le \liminf_{\lambda \to 0_+} \| J_\lambda z - x \| \le \limsup_{\lambda \to 0_+} \| J_\lambda z - x \| \le \| z - x \|$$

hence

$$\lim_{\lambda \to 0_+} \| J_\lambda z - x \| = \| z - x \|. \tag{1.6}$$

Since X is uniformly convex, (1.5) and (1.6) imply that $\lim_{\lambda \to 0_+} J_\lambda z = z$ as desired.

As $J_\lambda z \in D(A)$, it follows that $z \in \overline{D(A)}$. It is an easy matter to check now that for any $x, y \in \overline{D(A)}$ and $t \in (0,1)$, $tx+(1-t)y \in \overline{D(A)}$, i.e. that $\overline{D(A)}$ is convex. ■

The role of the minimal selection is underlined by the following

1.16. THEOREM. Let X and X* be uniformly convex, $A: X \to 2^X$ hyper-accretive and $B: X \to 2^X$ accretive; then we have:

i) if $A^\circ \subseteq B$, then $Bx \subseteq Ax$, $\forall x \in D(A)$;

ii) if B is maximal accretive and $A^\circ \subseteq B$; then $A^\circ \subseteq B^\circ$,

iii) if B is maximal accretive and $A^\circ = B^\circ$, then $A = B$.

PROOF. i) Let be $x_o \in D(A) = D(A^\circ)$ and $u_o \in Bx_o$; then

$$< J(x-x_o), A^\circ x - u_o > \geq 0, \quad \forall x \in D(A). \tag{1.7}$$

We shall prove that $u_o \in Ax_o$. Consider the mapping $\tilde{A}x = Ax - u_o$; then $\tilde{A}$ is hyper-accretive and we have

$$x_o = \tilde{J}_\lambda x_o + \lambda \tilde{A}_\lambda x_o \text{ and } \tilde{A}_\lambda x_o \in \tilde{A}\tilde{J}_\lambda x_o \tag{1.8}$$

Denote $x_\lambda = \tilde{J}_\lambda x_o$ and $u_\lambda = \tilde{A}_\lambda x_o$; then (1.8) shows that

$$x_o = x_\lambda + \lambda(u_\lambda - x_o) \text{ and } u_\lambda \in Ax_\lambda, \ \lambda > 0.$$

Moreover by Theorem 1.15. iii) we have

$$\lim_{\lambda \to 0_+} (u_\lambda - u_o) = \lim_{\lambda \to 0_+} A_\lambda x_o = A^\circ x_o.$$

Hence both $\{u_\lambda\}_{\lambda>0}$ and $\{A^\circ x_\lambda\}_{\lambda>0}$ are bounded; since X is reflexive, there is a subsequence $\{x_{\lambda'}\}_{\lambda'}$ of $\{x_\lambda\}_\lambda$ so that $A^\circ x_{\lambda'} \xrightarrow[\lambda']{w} u_1$, for some $u_1 \in X$. As A is demiclosed, we have that $(x_o, u_1) \in G(A)$, hence $u_1 - u_o \in \tilde{A}x_o$.

On the other hand, by (1.7) we have $J(x_\lambda - x_o), A^\circ x_\lambda - u_o > \geq 0$. Then $< J(-u_\lambda + u_o), A^\circ x_\lambda - u_o > \geq 0$. Letting now $\lambda' \to 0_+$ we obtain

$$< J(-A^\circ x_o), u_1 - u_o > \geq 0.$$

Since $-u_1 + u_o \in -\tilde{A}x_o$, by Porposition 1.11. Ch. II.,

$$\| \tilde{A}^\circ x_o \|^2 \leq < J(-\tilde{A}^\circ x_o), u_o - u_1 > \leq 0.$$

Thus $\tilde{A}^\circ x_o = 0$ i.e. $u_o \in Ax_o$.

ii) By (i) we have $Bx \subseteq Ax$, $\forall x \in D(A)$, hence $\| A^\circ x \| \leq \| u \|$, $\forall u \in Bx$. Then obviously $\| A^\circ x \| \leq \| B^\circ x \|$. But $A^\circ x \in Bx$, hence $\| A^\circ x \| = \| B^\circ x \|$, i.e. $A^\circ x = B^\circ x$.

iii) If $A^\circ = B^\circ$, then $D(A) = D(B)$ and since $A^\circ = B^\circ \subseteq B$, by (i) we conclude that $B \subseteq A$. The maximality of B yields now A = B. ■

We continue this paragraph with discussions on continuity properties of maximal accretive mappings.

1.17. THEOREM. Let X^* be uniformly convex and $A: X \to X$ an accretive operator with D(A) open; then A is demicontinuous if and only if A is hemicontinuos.

PROOF. In order to obtain the result we only have to apply Theorem 1.6, Ch. V, to Y = X^* and F = J; indeed, since X^* is uniformly convex, J is surjective and uniformly continuous on the unit ball of X (see Theorems 3.4 and 2.16. (i) Ch. II); moreover it is obvious that A is "J-monotone". ■

For multivalued mappings we have the

1.18. THEOREM. If X^* is uniformly convex and $A: X \to 2^X$ is locally bounded and maximal accretive, then A is norm-weak*-upper semi-continuous.

PROOF. Suppose that there exists $x_0 \in D(A)$ so that A is not norm-weak*-upper semicontinuous at x_0; then we can find an open nieghbourhood V of Ax_0 in the weak* topology on X, a sequence $\{x_n\}_n \subseteq D(A)$ with $x_n \xrightarrow[n]{} x_0$ and $u_n \in Ax_n$ such that $u_n \notin V$.

Since A is locally bounded, we can assume, passing to a subsequence, that $u_{n'} \xrightarrow[n]{w} u_0$, for some $u_0 \in X \setminus V$. We have

$$< J(x_n - x), u_n - u > \geq 0 \text{ for all } (x,u) \in G(A)$$

therefore

$$< J(x_0 - x), u_0 - u > \geq 0 \text{ for all } (x,u) \in G(A).$$

The maximality of A yields $u_0 \in Ax_0$ and this is a contradiction. ■

1.19. THEOREM. Let X^* be uniformly convex and $A: X \to 2^X$ be accretive , norm-weak*-upper hemicontinuous such that for every x $\in$ X, Ax is a non void closed convex set; then A is maximal accretive.

PROOF. Suppose that there is $x_0 \in X$ and $u_0 \in X \setminus Ax_0$ such that

$$< J(x - x_0), u - u_0 > \geq 0, \quad \forall (x,u) \in G(A). \tag{1.9}$$

Since Ax_0 is closed and convex, there is $y^* \in X^*$ with

$$< y^*, u_0 > > \sup_{u \in Ax_0} < y^*, u >. \tag{1.10}$$

We note that J is surjective, hence there exists $y \in X$ with $Jy = y^*$; then from (1.10) we obtain

$$< Jy, u_0 > > < Jy, u >, \quad \forall u \in Ax_0.$$

Define $V = \{u \in X; < Jy, u - u_0 > < 0\}$; then V is a week* open neighbourhood of Ax_0. Let be $t \in (0, 1)$, $x_t = x_0 + ty$ and $u_t \in Ax_t$; then by the upper-semicontinuity of A, $u_t \in V$, i.e. $< Jy, u_t - u_0 > < 0$.
On the other hand, by (1.9) we have

$$< J(x_t - x_0), u_t - u_0 > = t < Jy, u_t - u_0 > \geq 0$$

and this is a contradiction. ■

1.20 COROLLARY. Let X^* be uniformly convex and $A: X \to X$ be accretive and hemicontinuous with D(A) = X; then A is maximal accretive.

We end our considerations with the notion of T-accretivity in Banach lattices.

1.21. DEFINITION. Let X be a real Banach lattice;
i) $U: X \to X$ is called a T - contraction if

$$\| (Ux - Uy)_+ \| \leq \| (x - y)_+ \|, \quad \forall x, y \in D(U).$$

ii) $A: X \to 2^X$ is called T-accretive if for every (x, u), (y, v) $\in$ G(A) there is $x^* \in J_+(x-y)_+$, such that $< x^*, u - v > \geq 0$.

J_+ was defined in § 5. Ch. I.

1.22. PROPOSITION. Let X be a real Banach lattice and consider a mapping $A: X \to 2^X$; then

i) A is T-accrtive if and only if for every $(x,u),(y,v) \in G(A)$, $\lambda, \gamma \geq 0$ and $z \in (x-y)_+^{\perp}$ we have

$$\|(x-y)_+\| \le \|(1-\gamma)(x-y)_+ + [\lambda(u-v+z)+\gamma(x-y)_+]_+\| \qquad (1.11)$$

ii) If A is T-accretive then J_λ is a T-contraction on $D_\lambda = R(I+\lambda A)$ and A_λ is T-accretive, $\forall \lambda > 0$.

iii) If A is T-accretive and the norm on X has the property *P* i.e.

$$\|x_+\| \le \|y_+\| \text{ and } \|x_-\| \le \|y_-\| \text{ imply } \|x\| \le \|y\|,$$

then J_λ is contractive and A is accretive.

PROOF. i) For (x,u), (y,v) $\in$ G(A) consider $\tilde{x}^* \in J_+(x-y)_+$ given by Proposition 5.5., Ch. I., i.e.

$$<\tilde{x}^*, u-v> = \sup_{x^* \in J_+(x-y)_+} <x^*, u-v> \|(x-y)_+\| \sigma((x-y)_+, u-v).$$

Then it is clear that A is T-accretive iff $<\tilde{x}^*, u-v> \ge 0$. But we have: $<\tilde{x}^*, u-v> \ge 0$ iff $\sigma((x-y)_+, u-v) \ge 0$ or equivalently

$$\|(x-y)_+ + \lambda[(u-v+z)\vee -\alpha(x-y)_+]\| - \|(x-y)_+\| \ge 0, \quad \forall \alpha, \lambda \ge 0, z \in (x-y)_+^\perp.$$

Since

$$(u-v+z)\vee -\alpha(x-y)_+ = (u-v+z+\alpha(x-y))_+ - \alpha(x-y)_+,$$

we have that A is T-accretive iff

$$\|(x-y)_+\| \le \|(1-\lambda\alpha)(x-y)_+ + [\lambda(u-v+z)+\lambda\alpha(x-y)_+]_+\|, \forall \alpha, \lambda \ge 0, z \in (x-y)_+^\perp.$$

Hence (1.11) is proved.

ii) Take in (1.11) $\gamma = 1$ and $z = -\lambda^{-1}(x-y)_-$ to obtain

$$\|(x-y)_+\| \le \|[\lambda(u-v)+(x-y)]_+\|. \qquad (1.12)$$

Suppose $x+\lambda u = y+\lambda v$; then $[x-y+\lambda(u-v)]_+ = [(x-y)+\lambda(u-v)]_- = 0$ and also $(x-y)_- = (y-x)_+ = 0$, hence x = y.

Thus $J_\lambda = (I+\lambda A)^{-1}$ is single-valued and from (1.12) it follows that it is T-contractive.

Let be further $x, y \in D_\lambda$ and $x^* \in J_+(x-y)_+$; we have

$$\begin{aligned}
&<x^*, (I-J_\lambda)x - (I-J_\lambda)y> \\
&= <x^*, (x-y)_+ - (x-y)_- - (J_\lambda x - J_\lambda y)_+ + (J_\lambda x - J_\lambda y)_- > \\
&\ge \|(x-y)_+\|^2 - <x^*, (Jx-Jy)_+> \ge 0.
\end{aligned}$$

(we used the facts that $<x^*, (x-y)_-> = 0$ and $<x^*, (J_\lambda x - J_\lambda y)_-> \ge 0$.)

Hence A_λ is T-accretive.

iii) We have

$$\|(Jx-Jy)_+\| \le \|(x-y)_+\| \text{ and}$$

$$\|(J_\lambda x - J_\lambda y)_-\| = \|(J_\lambda y - J_\lambda x)_+\| \le \|(y-x)_+\| = \|(x-y)_-\|.$$

The property P of the norm implies $\| J_\lambda x - J_\lambda y \| \le \| x-y \|$, $\forall x, y \in D_\lambda$.
From (1.12) we also have $\| (x-y)_- \| \le \| [(x-y)+\lambda(u-v)]_- \|$.
The property P yields again $\| x-y \| \le \| (x-y)+\lambda(u-v) \|$ for every $(x,u),(y,v) \in G(A)$, hence A is accretive. ■

1.23. Remark. On every Banach lattice there exists an equivalent norm which satisfies the condition P.
Indeed, define $|\!\|x\|\!| = \| x_+ \| + \| x_- \|$; then we have $\| x \| \le |\!\|x\|\!| \le 2\| x \|$.
Consider further x, y ∈ X such that: $\| x_+ \| \le \| y_+ \|$ and $\| x_- \| \le \| y_- \|$; then $|\!\|x\|\!| = \| x_+ \| + \| x_- \| \le \| y_+ \| + \| y_- \| = |\!\|y\|\!|$, i.e. $|\!\|\cdot\|\!|$ has the property P.

§ 2. SEMIGROUPS OF NONLINEAR CONTRACTIONS IN UNIFORMLY CONVEX BANACH SPACES

We begin our considerations on semigroups of nonlinear contractions with the following crucial result on the differentiability of absolutely continuous vectorial functions.

2.1. THEOREM. (Komura). Let X be a reflexive Banach space and $f:[0,t_o] \to X$ a stongly absolutely continuous function; then f is a.e. differentiable on $[0,t_o]$, f' is Bochner integrable on $[0,t_o]$.

$$f(t) = \int_0^t f'(s)ds + f(0), \qquad t \in [0,t_o] \tag{2.1}$$

PROOF. As f is continuous, the image set $K=\{f(t);\ t \in [0,t_o]\}$ is compact, hence separable. Passing to the linear closed subspace generated by K, we can suppose without loss of generality that X is separable. Since X is reflexive, it follows that also X^* is separable.
Remark also that f is of bounded variation; for $t \in [0,t_o]$ we denote by V(t) the variation of f on [0, t] and by Var f the variation of f on $[0,t_o]$; then $V(t) \le V(s)$ for $0 \le t \le s \le t_0$, the function $[0,t_o] \ni t \to V(t)$ is a.e. differentiable and

$$\| f(t) - f(s) \| \le V(s) - V(t), 0 \le t \le s \le t_o. \tag{2.2}$$

Denote $f_h(t) = \dfrac{f(t+h)-f(t)}{h}$, $h \in R \setminus \{0\}$. By (2.2) we have

$$\| f(t+h)-f(t) \| \le V(t+h)-V(t) \quad \text{on} \quad [0,t-h'];$$

hence if $h \in (0,t_0)$ then

$$\int_0^{t_0-h} \| f(t+h)-f(t) \| dt \le \int_0^{t_0-h} [V(t+h)-V(t)]dt$$

$$= \int_h^{t_0} V(t)dt - \int_0^{t_0} V(t)dt = \int_{t-h}^{t_0} V(t)dt - \int_0^{h} V(t)dt \le \int_{t_0-h}^{t_0} V(t)dt \le h. \ \text{Varf.}$$

This yields

$$\int_0^{t_0-h} \| f_h(t) \| dt \ \text{Var} f, \ 0<h<t. \tag{2.3}$$

From (2.2) we directly obtain that

$$\limsup_{h \to 0} \| f_h(t) \| < +\infty \quad \text{a.e.} \quad [0,t_0]. \tag{2.4}$$

We denote by $S_0 = \{t \in [0,t_0];\ \limsup_{h \to 0} \| f_h(t) \| = +\infty\}$.

Let $\{x_n^*\}_{n \in N}$ be dense in X^*; for every $n \in N$, the scalar function $[0,t_0] \ni t \to \langle x_n^*, f(t) \rangle$ is absolutely continuous, hence a.e. derivable on $[0,t_0]$. For $n \in N$, denote by S_n the subset of $[0,t_0]$ where the above function is not differentiable and let $S = \bigcup_{n=0}^{\infty} S_n$.

We note that for $t \in [0,t_0] \setminus S$ the $\lim_{h \to 0} \langle x_n^*, f_h(t) \rangle$ exists for every $n \in N$, and that by (2.4) the sequence $\{\| f_h \|\}_{h \to 0}$ is bounded; it follows that the $\lim_{h \to 0} \langle x^*, f_h(t) \rangle$ exists a.e. on $[0,t_0]$ for all $x^* \in X^*$. Since X is weakly complete, we conclude that the sequence $\{f_h(t)\}_{h \to 0}$ weakly converges a.e. on $[0,t_0]$ to a limit which we denote $f'(t)$. Obviously f' is weakly measurable, hence by Pettis's Theorem it is strongly measurable; then f' is Bochner integrable. Indeed $[0,t_0] \ni \to \| f'(t) \|$ is integrable and form (2.3) we get that $\limsup_{h \to 0} \int_0^{t_0} f_h(t)dt < +\infty$; the result is therefore a consequence of Fatou's Lemma.

Let be $g(t) = f(0) + \int_0^t f'(s)ds$, $t \in [0,t_0]$; then is a.e. strongly differentiable on $[0,t_0]$.

Recalling that $t \to \langle x^*, f(t) \rangle$ is absolutely continuous, we may write

$$\langle x^*, f(t) \rangle = \langle x^*, f(0) \rangle + \int_0^t \frac{d}{ds} \langle x^*, f(s) \rangle ds$$

$$=<x^*, f(0)> + \int_0^t <x^*, f'(s)> ds = <x^*, g(t)> \text{ a.e. on } [0,t_o].$$

It follows that $<x^*, f(t)> = <x^*, g(t)>$, a.e. on $[0,t_o]$, $\forall x^* \in X^*$.
Consequently $f(t)=g(t)$, a.e. on $[0,t_o]$. ∎

The fact that the result of this Theorem is not valid in general Banach spaces is illustrated by the example in the Exercise 5.

2.2. DEFINITION. Let X be a Banach space and $C \subseteq X$ closed; a semigroup of nonlinear contractions on C is a function $S: [0,+\infty)\times C \to C$ with the properties:

i) $S_{t+s}\, x = S_t S_s x, \quad \forall t,s \geq 0, \quad x \in C$.

ii) $S_o\, x = x, \quad \forall x \in C$.

iii) $\lim_{t\to 0_+} S_t x = x, \quad \forall x \in C$.

iv) $\| S_t x - S_t y \| \leq \| x-y \|, \quad \forall t \geq 0, \; x,y \in C$.

In what follows $C \subseteq X$ will always be supposed a closed subset of X.

2.3. REMARK. From iii) it follows that for every $x \in X$, the function $R_+ \ni t \to S_t x$ is continuous on $(0,+\infty)$; indeed, for $t > 0$, we have

$$\| S_{t+h}x - S_t x \| = \| S_t S_h x - S_t x\| \leq \| S_h x - x \|, \text{ for } h < 0$$

and

$$\| S_{t-h}x - S_t x \| = \| S_{t-h}x - S_{t-h}S_h x\| \leq \| S_h x - x \|, \; h > 0,\; t-h > 0.$$

2.4. DEFINITION. Let S be a semigroup of nonlinear contractions on $C \subseteq X$ and $A^h x = \dfrac{S_h x - x}{h}, x \in C, h > 0$.

Define $A_s x = \lim_{h\to 0_+} A^h x$ with $D(A_s) = \left\{x \in C; \left\{A^h x\right\}_{h\to 0_+} \text{ is convergent}\right\}$

and

$A_w x = \lim_{h\to 0_+} A^h x$ with $D(A_w) = \left\{x \in C; \left\{A^h x\right\}_{h\to 0_+} \text{ is weakly convergent}\right\}$.

Then the operator A_s, respectively A_w, is called the strong, respectively the weak infinitesimal generator of the semigroup S.

2.5. PROPOSITION. The operators A_s and A_w are dissipative.

PROOF. Let be $x^* \in J(x-y)$ and $x, y \in D(A_w)$; we have

$$<x^*, A^h x - A^h y> = <x^*, \frac{S_h x - S_h y}{h}> - <x^*, \frac{x-y}{h}>$$

$$\leq h^{-1}\left(\| x-y\| \| S_h x - S_h y \| - \| x-y\|^2\right) \leq 0;$$

hence $<x^*, A_w x - A_w y> \leq 0$, i.e. A_w is dissipative. Since $A_s \subseteq A_w$, it follows that also A_s is dissipative. ■

2.6. REMARK. In the case of linear semigroups of class c_0, in particular of linear contractions, it is known that $A_s = A_w$ and that $D(A_s)$ is dense in X (see Pazy [2]). In the nonlinear case, $A_s \neq A_w$. Moreover $\overline{D(A_s)} \neq C$, even in the case C = X, as the following example shows.
Let X = C[0, 1] be the Banach space of the real continuous functions on [0, 1] and

$$F(\xi) = \begin{cases} \xi & \text{for } \xi > 0 \\ 2\xi & \text{for } \xi \leq 0 \end{cases}.$$

Denote $S_t(f)(\xi) = F(t + F^{-1}(f(\xi)))$, $f \in X, \xi[0,1]$; then S is a semigroup of contractions on X and one can prove that $D(A_s)$ consists only of functions with a constant sign. The details are left as Exercise 4.

2.7. LEMMA. Let S be a semigroup of nonlinear contractions on

$C \subseteq X$ and $D = \left\{x \in C; \lim_{h \to 0_+} \inf h^{-1} \| S_h x - x \| = L(x) < +\infty\right\}$; then for every

$x \in D$ we have

$$\| S_t x - S_s x \| \leq L(x) \cdot |t-s|, \quad \forall t, s > 0. \tag{2.5}$$

PROOF. Let $\varepsilon > 0$; there is a nonincreasing sequence $h_k \xrightarrow[k]{} 0$ such that

$$\left\| S_{h_k} x - x \right\| \leq h_k \cdot (L(x) + \varepsilon). \tag{2.6}$$

For t, s ≥ 0, denote $\delta = | t-s |$ and select natural numbers n_k such that:

$$n_k \cdot h_k \leq \delta < (n_k + 1) h_k, \text{ i.e. } \quad 0 \leq \delta - n_k h_k < h_k.$$

Using (2.6) and the inequality $\| S_{nr} x - x \| \leq n \| S_r x - x \|, n \in N, r \in R$ we get

$$\| S_t x - S_s x\| \leq \| S_\delta x - x \| = \left\| S_{\delta - n_k h_k + n_k h_k} x - x \right\|$$

$$\leq \left\| S_{\delta - n_k h_k} x - S_{n_k h_k} x \right\| + \left\| S_{n_k h_k} x - x \right\|$$

$$\leq \left\| S_{\delta-n_k h_k}x - x \right\| + n_k \cdot h_k(L(x)+\varepsilon)$$
$$\leq \left\| S_{\delta-n_k h_k}x - x \right\| + \delta\cdot(L(x)+\varepsilon),\ k\in N.$$

Since $\varepsilon > 0$ was arbitrary, we have the proof of (2.5). ■

2.8. PROPOSITION. Let X be a reflexive Banach space and S a semigroup of nonlinear contraction on $C \subseteq X$; then $\overline{D(A_s)} = \overline{D(A_w)}$.

PROOF. Consider the set D defined in Lemma 2.7.; then $D(A_s) \subseteq D(A_w) \subseteq D$. Let $x \in D$; by Lemma 2.7 the function $t \to S_t x$ is Lipschitz continuous on $[0,+\infty)$. Hence, by Theorem 2.1. it is a.e. differentiable on $[0,+\infty)$ i.e. $S_t x \in D(A_s)$ a.e. on $[0,+\infty)$.
Since $x = \lim_{t\to 0} S_t x$ it follows that $x \in \overline{D(A_s)}$. ■

2.9. THEOREM. Let X be uniformly convex and S a semigroup of nonlinear contractions on $C \subseteq X$; then $A_s = A_w$.

PROOF. Let be $x_o \in D$ and X_o the set of all weak limit points of $\left\{\frac{S_h x_o - x_o}{h}\right\}_{h\to 0}$. Define on $D(A_1) = D(A_s)\cup\{x_o\}$ the mapping A_1 as follows

$$A_1 = A_s \text{ on } D(A_s) \text{ and } A_1 x_o = \overline{\operatorname{conv} X_o}.$$

Then A_1 is dissipative; indeed, let be $x \in D(A_s), u \in X_o$ and $x^* \in J(x - x_o)$; there exists a sequence $h_n \to 0$ so that

$$< x^*, A_s x - u > = \lim_{h_n\to 0} h_n^{-1} < x^*, (S_{h_n}x - x) - (S_{h_n}x_o - x_o) >$$

$$\leq \lim_{h_n\to 0} h_n^{-1}\left(\| x - x_o \| \| S_{h_n}x - S_{h_n}x_o \| - \| x - x_o \|^2\right) \leq 0.$$

Since the function $t \to S_t x_o$ is Lipschitz continuous on $[0,+\infty)$, by Theorem 2.1. it is a.e. differentiable. Hence $S_t x_o \in D(A_s)$ and $\frac{d}{dt}S_t x_o = A_s S_t x_o$, a.e. on $[0,+\infty)$.
Therefore, by the dissipativity of A_1 we see that for same $u \in A_1 x_o$ and $x^* \in J(S_t x_o - x_o)$ we have $< x^*, \frac{d}{dt}(S_t x_o - x_o) > \leq < x^*, u >$, a.e. on $(0,\infty)$.
Using now the Kato's differentiation rule (Proposition 4.9. Ch. I), we get

$$\| S_t x_o - x_o \| \frac{d}{dt} \| S_t x_o - x_o \| = < x^*, \frac{d}{dt}(S_t - x_o - x_o) >$$

$$\leq \| u \| \, \| S_t x_o - x_o \|, \text{ a.e. for } t > 0.$$

Also, then

$$\frac{d}{dt} \| S_t x_o - x_o \| \leq \| u \|, \text{ a.e. for } t > 0, \ u \in A_1 x_o$$

and consequently

$$\| S_t x_o - x_o \| \leq t \cdot | A_1 x_o |, \ t \geq 0. \tag{2.7}$$

Since $A_1 x_o$ is closed and convex and X is reflexive and strictly convex, by Proposition 1.14. there is a unique $u_o \in A_1 x_o$ such that $\| u_o \| = | A_1 x_o |$.

Let be $u \in X_o \subseteq A_1 x_o$; then $\| u \| \leq \liminf_{h_n \to 0} \left\| \frac{S_{h_n} x_o - x_o}{h_n} \right\|$, for some sequence $h_n \to 0$. Now (2.7) yields $\| u \| \leq | A_1 x_o | = \| u_o \|$.

Consequently $u = u_o$, also $X_o = \{u_o\}$. It follows that $x_o \in D(A_w)$ and $A_w x_o = u_o$.

We shall show that $x_o \in D(A_s)$ and this will finish the proof. Again by (2.7) we have

$$\frac{S_h x_o - x_o}{h} \xrightarrow[h]{w} A_w x_o \text{ and } \left\| \frac{S_h x_o - x_o}{h} \right\| \leq \| A_w x_o \|, \ h > 0.$$

Also then

$$\| A_w x_o \| \leq \liminf_{h \to 0} \left\| \frac{S_h x_o - x_o}{h} \right\| \leq \limsup_{h \to 0} \left\| \frac{S_h x_o - x_o}{h} \right\| \leq \| A_w x_o \|.$$

Hence $\lim_{h \to 0} \left\| \frac{S_h x_o - x_o}{h} \right\| = \| A_w x_o \|$. Since X is uniformly convex the existence of the $\lim_{h \to 0} \frac{S_h x_o - x_o}{h} = A_s x_o = A_w x_o$ is a consequence of Proposition 2.8., Ch. II. ∎

2.10. NOTATION. In the case X is uniformly convex, we agree to denote $A = A_s = A_w$.

2.11. THEOREM. Let X^* be uniformly convex and S a semigroup of nonlinear contractions on $C \subseteq X$; then for $x_o \in D(A)$ we have:

i) $S_t x_o \in D(A)$, $\forall t \geq 0$, the function $t \to S_t x_o$ is differentiable at the right of every $t \geq 0$ and $\frac{d^+}{dt} S_t \, x_o = A \, S_t x_o$, $t \geq 0$.

ii) The function $t \to \| A_t S_t x_0 \|$ is nonincreasing on $[0,+\infty)$ and the function $t \to A\, S_t x_0$ is right continuous at every $t \geq 0$;

iii) $\frac{d}{dt} S_t x_0 = A\, S_t x_0$ exists and is continuous on $[0,+\infty)$ except for a countable set of points in $[0,+\infty)$.

PROOF. i) Since $\liminf_{h \to 0_+} \frac{\| S_h S_t x_0 - S_t x_0 \|}{h} \leq \liminf_{h \to 0_+} \frac{\| S_h x_0 - x_0 \|}{h} = \| A x_0 \|$, it follows that $S_t x_0 \in D = D(A)$, $\forall t \geq 0$ and that

$$\lim_{h \to 0_+} \frac{S_h S_t x_0 - S_t x_0}{h} = \frac{d^+}{dt} S_t x_0 = A\, S_t x_0, \; t \geq 0.$$

ii) Recall that the function $t \to S_t x_0$ is absolutely continuous; then $\frac{d}{dt}(S_{t+h} x_0 - S_t x_0) = AS_{t+h} x_0 - AS_t x_0$, a.e. on $[0,+\infty)$. Also, using Kato's differentiation rule we obtain for $x^* = J(S_{t+h} x_0 - S_t x_0)$:

$$\| S_{t+h} x_0 - S_t x_0 \| \frac{d}{dt} \| S_{t+h} x_0 - S_t x_o \| = < x^*, \frac{d}{dt}(S_{t+h} x_o - S_t x_o) >$$

$$= < x^*, AS_{t+h} x_o - AS_t x_o > \leq 0, \text{ a.e. } t \geq 0.$$

Consequently

$$\| S_{t+h} x_0 - S_t x_0 \| \leq \| S_{s+h} x_0 - S_s x_0 \| \text{ for } h > 0 \;\; 0 \leq s \leq t.$$

Using i), we obtain

$$\left\| \frac{d^+}{dt} S_t\, x_0 \right\| = \| AS_t x_0 \| \leq \left\| \frac{d^+}{ds} S_s\, x_0 \right\| = \| AS_s x_o \|, 0 \leq s \leq t.$$

Hence $t \to \| AS_t x_0 \|$ is nonincreasing on $[0,+\infty)$ and therefore also right continuous on $[0,+\infty)$.

Fix $t_0 \geq 0$ and a sequence $t_n \downarrow t_0$; since $\{\| AS_{t_n} x_0 \|\}$ is bounded, there is $y_0 \in X$ and a subsequence $\{t_{n'}\} \subseteq \{t_n\}$ such that $AS_{t_{n'}} x_0 \xrightarrow[n']{w} y_0$.

The right continuity of $t \to \| AS_t x_0 \|$ yields

$$\| y_0 \| \leq \liminf_{n'} \| AS_{t_{n'}} x_0 \| = \lim_{n'} \| AS_{t_{n'}} x_0 \| = \| AS_{t_0} x_0 \| . \quad (2.8)$$

Let $\tilde{A}$ be a maximal dissipative extension of A; then $\tilde{A}$ is demiclosed, so that $y_0 \in \tilde{A} S_{t_0} x_0$.

On the other hand for every $x \in D(A)$ we have

$$<J(S_t x - x), \frac{d}{dt}(S_t x - x) - u> \le 0, \text{ a.e. } t > 0, u \in \tilde{A}\,x$$

and this yields

$$\| S_t x - x \| \frac{d}{dt} \| S_t x - x \| = <J(S_t x - x), \frac{d}{dt}(S_t x - x)>$$

$$\le <J(S_t x - x), u> \le \|u\| \| S_t x - x \|, \text{ a.e. } t > 0, u \in \tilde{A}\,x.$$

Hence:

$$\frac{d}{dt}\| S_t x - x \| \le \|u\|, \text{ a.e. } t > 0, u \in \tilde{A}\,x$$

and therefore

$$\| S_t x - x \| \le t \cdot \|u\|, \; t \ge 0 \quad u \in \tilde{A}\,x.$$

Also, then $\|Ax\| \le \|\tilde{A}^\circ x\|$, i.e. $Ax = \tilde{A}^\circ x, \quad \forall x \in D(A)$.

In particular $AS_{t_0}x_0 = \tilde{A}^\circ S_{t_0}x_0$ and then by (2.8)

$$y_0 = AS_{t_0}x_0 \quad \text{and} \quad \lim_{n'} \| AS_{t_{n'}}x_0 \| = \| y_0 \|.$$

Since X is uniformly convex, by Proposition 2.8., Ch. II we have

$$AS_{t_{n'}}x_0 \xrightarrow[n']{} y_0 = AS_{t_0}x_0.$$

iii) As $\frac{d}{dt}S_t x_0 = AS_t x_0$, a.e. $t > 0$, we have

$$S_{t+h}x_0 - S_t x_0 = \int_t^{t+h} AS_s x_0 ds, \; t, h \ge 0.$$

We shall prove that the set of continuity points of the function $t \to \frac{d}{dt}S_t x_0$ coincides with the set of continuity points of the function $t \to \| AS_t x_0 \|$; this last set being at most countable, this finishes the proof. We note that the set of the continuity points of the function $t \to \frac{d}{dt}S_t x_0$ coincides with the set of the continuity points of the function $t \to AS_t x_0$. Thus we only have to prove that if $t \to \| AS_t x_0 \|$ is continuous at $t_0 > 0$, then aslo $t \to AS_t x_0$ is continuous at t_0. This can be done by the same argument used in the proof of the right continuity of $t \to AS_t x_0$ at t_0. ■

2.12. COROLLARY. In the same conditions as in Theorem 2.11, if $\tilde{A}$ is a maximal dissipative extension of A, then $Ax = \tilde{A}^\circ x, \quad \forall x \in D(A)$.

In order to present some results concearning the boundedness of the $t \to S_t x$ on $[0, +\infty)$ we need the following fixed point theorem

2.13. THEOREM. (Browder-Kirk). Let X be uniformly convex, $C \subseteq X$ be closed, bounded and convex and $T: C \to C$ be a nonlinear contraction; then T has a fixed point. Moreover, the set K of all fixed points of T is convex.

PROOF. We shall first prove that K is convex. Let $x_1, x_2 \in K, r = \|x_1 - x_2\|$ and $x_\lambda = (1-\lambda)x_1 + \lambda x_2$, $\lambda \in (0,1)$; then

$$\|Tx_\lambda - x_1\| = \|Tx_\lambda - Tx_1\| \le \|x_\lambda - x_1\| \le \lambda\|x_1 - x_2\| \qquad (2.9)$$

and analogously

$$\|Tx_\lambda - x_2\| \le (1-\lambda)\|x_1 - x_2\|. \qquad (2.10)$$

Hence

$$\|x_1 - x_2\| \le \|x_1 - Tx_\lambda\| + \|Tx_\lambda - x_2\| \le \|x_1 - x_2\|$$

i.e. $\|x_1 - Tx_\lambda\| + \|Tx_\lambda - x_2\| = \|x_1 - x_2\| = r.$

Then by (2.9) we have $\|x_1 - Tx_\lambda\| = r - \|Tx_\lambda - x_2\| \le \lambda r.$

Hence $\|Tx_\lambda - x_2\| \ge (1-\lambda)r$ and (2.10) implies $\|Tx_\lambda - x_2\| = (1-\lambda)r$. But then we also have $\|x_1 - Tx_\lambda\| = r$. Therefore

$$\left\|\frac{x_1 - Tx_\lambda}{\lambda}\right\| = \left\|\frac{Tx_\lambda - x_2}{1-\lambda}\right\| = \left\|\lambda\frac{x_1 - Tx_\lambda}{\lambda} + (1-\lambda)\cdot\frac{Tx_\lambda - x_2}{1-\lambda}\right\| = r.$$

Since X is strictly convex, it follows that

$$\frac{x_1 - Tx_\lambda}{\lambda} = \frac{Tx_\lambda - x_2}{\lambda}, \text{ i.e. } Tx_\lambda = x_\lambda.$$

In order to prove the existence of a fixed point for T, denote

$\operatorname{diam} C = \sup_{x,y \in C} \|x - y\|$ and $r(C) = \inf\{r > 0; C \subset \overline{S_r}(x), \text{ for some } x \in C\}$.

First we shall note that

$$r(C_o) < \operatorname{diam} C_o, \forall C_o \subset C, C_o \text{ closed, convex, with } \operatorname{diam} C_o > 0. \qquad (2.11)$$

Indeed, by the uniform convexity of X, for every $\varepsilon > 0$ and $r > 0$, there exists $\delta > 0$ such that $\|x + y\| \le 2(r - \delta)$ whenever $\|x - y\| \ge \varepsilon$ and $\|x\|, \|y\| \le r$. (see exercise 5, Ch. II).

For fixed $x_1, x_2 \in C_o$, let $\varepsilon = \|x_1 - x_2\|$ and $r = \operatorname{diam} C_o$; then for $y_o = \frac{1}{2}(x_1 + x_2)$ and every $x \in C_o$, we have $x - y_o = \frac{1}{2}(x - x_1) - \frac{1}{2}(x_2 - x)$.

Since $\|x - x_1\|, \|x - x_2\| \le r$ and $\|(x - x_1) - (x - x_2)\| = \varepsilon$, we get that $\|x - y_o\| \le r - \delta$, for the $\delta > 0$ corresponding to ε and r as above.

Hence $C_o \subset \overline{S_{r-\delta}}(y_o)$, i.e. $r(C_o) \le \operatorname{diam} C_o - \delta$.

Consider now the set $\mathcal{M} = \{M \subset C;\ M \ne \phi, \text{ closed, convex, } T\text{-invariant}\}$, partially ordered by $M_1 \propto M_2$ iff $M_2 \subset M_1$. If $\mathcal{M}_o$ is a totally ordered subset of $\mathcal{M}$ then $\bigcap_{M\in\mathcal{M}_0} M \neq \emptyset$ since C is weakly compact; moreover the above intersection is closed, convex and T-invariant, hence an upper bound for $\mathcal{M}_0$. Then Zorn's Lemma asserts the existence of a maximal element C_0 of $\mathcal{M}$, i.e. a T-invariant, closed, convex subset C_0 of C which is minimal with respect to the inclusion. In particular we have that $\overline{\text{conv}}\, T(C_0) = C_0$.

If we prove that C_0 consists of one point, or equivalently diam $C_0 = 0$, then the proof is finished.

Suppose that diam $C_0 > 0$. Fix $z \in C_0$ and choose $r_n \in (0,1)$ such that $r_n \xrightarrow[n]{} 1$. Let be $T_n = (1-r_n)z + r_n T$; then it is clear that C_0 is T_n invariant and that T_n is a r_n - contraction, $\forall n \in N$; therefore T_n has a unique fixed point $x_n \in C_0$. In order to obtain the assertion we only have to prove that

$$\|x_n - x\| \xrightarrow[n]{} \text{diam } C_0, \quad \forall x \in C_0. \tag{2.12}$$

Indeed, if for some $x \in C_0$ and $r > 0, C_0 \subset \overline{S}_r(x)$, then $\|x_n - x\| \le r$, $\forall n \in N$ and by (2.12), diam $C_0 \le r$; hence diam $C_0 \le r(C_0)$ and this contradicts (2.11).

Suppose that (2.12) is not true; then there is $x_1 \in C_0$ such that $\liminf_n \|x_n - x_1\| = \rho <$ diam C_0.

Select a subsequence $\{x_{n'}\}_{n'} \subseteq \{x_n\}_n$ and x_0 such that $\|x_{n'} - x_1\| \to \rho$ and $x_{n'} \xrightarrow[n']{w} x_0$.

Let be $C_1 = \{x \in C_0;\ \liminf_{n' \to \infty} \|x_{n'} - x\| \le \rho\}$; then $x_1 \in C_1$ and it is immediate that C_1 is convex and closed. Moreover C_1 is also T-invariant; indeed, if $x_1 \in C_1$, then

$$\liminf_{n'} \|x_{n'} - Tx\| \le \liminf_{n'} \|x_{n'} - Tx_{n'}\| + \liminf_{n'} \|Tx_{n'} - Tx\|$$

$$\le \liminf_{n'} (1 - r_{n'}) \|z - Tx_{n'}\| + \liminf_{n'} \|x_{n'} - y\| \le \delta.$$

As C_0 is minimal, $C_1 = C_0$ and therefore $\|x - x_0\| \le \liminf_{n'} \|x - x_{n'}\| \le \rho$ for all $x \in C_0$, i.e. $C_0 \subset \overline{S}_\rho(x_0)$. Finally let be $C_2 = \{x \in C_0;\ C_0 \subset \overline{S}_\rho(x)\}$; then $C_0 \subset \overline{S}_\rho(x_0)$ is convex, closed and T-invariant, in virtue of the property that $\overline{\text{conv}}\, C_0 = C_0$. It follows that $C_2 = C_0$, a contradiction, since diam $C_2 \le \rho <$ diam C_0. ■

2.14. COROLLARY. Let $C \subseteq X$ be a convex, bounded and closed subset of the uniformly convex Banach space X and $\mathcal{T} = \{T_\lambda\}_{\lambda \in \Lambda}$ a family of nonlinear contractions from C into C so that $T_\lambda T_\gamma = T_\gamma T_\lambda$, $\forall \lambda, \gamma \in \Lambda$, then there exists a common fixed point for $\mathcal{T}$.

PROOF. For $T_\lambda \in \mathcal{T}$, let be $K_\lambda = \{x \in C;\ T_\lambda x = x\}$; then K_λ is convex, bounded, closed and T_λ - invariant. We apply the above theorem to $T_\gamma : K_\lambda \to K_\lambda$ to obtain that $K_\lambda \cap K_\gamma \neq \emptyset$. It is an easy matter to see now that the family $\{K_\lambda\}_{\lambda \in \Lambda}$ has the property of the finite intersection. As C is weakly compact and K_λ are weakly closed, it follows that there exists $x_o \in \bigcap_{\lambda \in \Lambda} K_\lambda$. ■

2.15. THEOREM. Let X be uniformly convex, $C \subseteq X$ a convex, closed set and S a semigroup of nonlinear contractions on C; then S is bounded on C (i.e. $\sup_{t \geq 0} \| S_t x \| < +\infty$, $\forall x \in C$) if and only if S has a fixed point $x_o \in C$.

PROOF. Suppose that there is $x_o \in C$ such that $S_t x_o = x_o$, $\forall t \geq 0$; then

$$\|S_t x\| \leq \|S_t x - S_t x_o\| + \|S_t x_o\| \leq \|x - x_o\| + \|x_o\|, \quad \forall x \in C, t \geq 0$$

i.e. S is bounded.

Conversely, let S be bounded and $x_o \in C$ fixed; denote

$$M = \sup_{t \geq 0} \|S_t x_o\| \quad \text{and} \quad C(x) = \{y \in C; \|y - x\| \leq \|x_o\| + M\}, x \in X.$$

Write further $C_s = \bigcap_{t \geq s} C(S_t x_o)$, $s \geq 0$. It is clear that C_s is a closed convex subset of C and since $x_o \in C(S_t x_o)$, $\forall t \geq 0$, it is also nonvoid. Moreover it is evident that $C_s \subseteq \overline{S}_r(0)$, where $r = \|x_o\| + 2M$. Let be $C^* = \bigcup_{s \geq 0} C_s$; then C^* is a convex, bounded subset of C, invariant for every $S_t, t \geq 0$. Indeed if $x \in C_{s_o}$, then $\|x - S_s x_o\| \leq \|x_o\| + M, \forall s \geq s_o$ and this yields $\left\|S_t - x - S_{t+s} x_o\right\| \leq \|x_o\| + M$, $\forall s \geq s_o$, i.e. $S_t x \in C_{t+s_o}$. Hence $S_t(C^*) \subset C^*$. Then we also have: $S_t(\overline{C}^*) \subseteq \overline{C}^*$, $\forall t \geq 0$. Finally by the Corollary 2.14. there exists $x \in \overline{C}^* \subset C$, with $S_t x = x$, $\forall t \geq 0$. ■

2.16. COROLLARY. Let X be unformly convex, $C \subseteq X$ be a convex, closed subset and S a semigroup of nonlinear contractions on C; then S is bounded on C if and only if $0 \in R(A)$.

PROOF. If S is bounded, then it has a fixed point x_o; hence $A^h x_o = \frac{S_h x_o - x_o}{h} = 0$, $\forall h \geq 0$; so that $Ax_o = 0$.
Conversely, if $0 \in R(A)$, then there is $x_o \in D(A)$ such that $Ax_o = 0$. By Lemma 2.7. we obtain that $\|S_t x_o - x_o\| \leq t \cdot \|Ax_o\| = 0$, $\forall t \geq 0$. Hence $x_o = S_t x_o$, $\forall t \geq 0$ and the statement is a consequence of above theorem. ■

§ 3. THE EXPONENTIAL FORMULA OF CRANDALL-LIGGETT

We start our discussion with the following computational result:

3.1. LEMMA i) For m, n $\in$ N with $m \leq n$ and $\alpha, \beta > 0$ let $a_{k,j} \in R$, $0 \leq k \leq m$, $0 \leq j \leq n$, be such that

$$a_{k,j} \leq \alpha \cdot a_{k-1,j-1} + \beta \cdot a_{k,j-1}; \tag{3.1}$$

then

$$a_{m,n} \leq \sum_{j=n}^{m-1} \alpha^j \beta^{n-j} \binom{n}{j} a_{m-j,0} + \sum_{j=m}^{n} \alpha^m \cdot \beta^{j-m} \binom{j-1}{m-1} a_{o,n-j}. \tag{3.2}$$

ii) For $0 < m \leq n$ and $0 < \alpha \leq 1$, we have

$$\sum_{j=0}^{m} \binom{n}{j} \alpha^j (1-\alpha)^{n-j} (m-j) \leq \left[(n\alpha - m)^2 + n\alpha(1-\alpha)\right]^{1/2}$$

$$\sum_{j=m}^{n} \binom{j-1}{m-1} \alpha^m (1-\alpha)^{j-m} (n-j) \leq \left[\frac{m(1-\alpha)}{2} + \left(\frac{m(1-\alpha)}{\alpha} + m - n\right)^2\right]^{1/2}$$

PROOF. i) Let m = 1; we shall prove by induction that

$$a_{1,n} \leq \beta^n a_{1,0} + \sum_{i=1}^{n} \alpha \cdot \beta^{j-1} a_o, n - j, \qquad n \in N. \tag{3.3}$$

For n = 1, (3.2) gives $a_{1,1} \leq \beta \cdot a_{1,0} + \alpha \cdot a_{o,o}$; this holds by (3.1).
Suppose that (3.3) is true for n; (3.1) yields us $a_{1,n+1} \leq \alpha \cdot a_{o,n} + \beta \cdot a_{1,n}$, hence

$$\begin{aligned} a_{1,n+1} &\leq \alpha \cdot a_{o,n} + \beta^{n+1} a_{1,0} + \sum_{j=1}^{n} \alpha \cdot \beta^j a_{o,n-1} \\ &= \beta^{n+1} \cdot a_{1,0} + \sum_{j=1}^{n+1} \alpha \beta^{j-1} a_{o,n+1-j}. \end{aligned}$$

Hence (3.3) is valid also for n + 1.

Suppose further that (3.2) is true for $m \le n$ and let us prove it for $n + 1$. Then we have

$$a_{m,n+1} \le \alpha \cdot a_{m-1,n} + \beta a_{m,n}$$

so that the induction hypothesis yields

$$a_{m,n+1} \le \sum_{j=0}^{m-2} \alpha^j \beta^{n-j} \binom{n}{j} a_{m-j,0} + \alpha \sum_{j=m-1}^{n} \alpha^{m-1} \beta^{j+1-m} \binom{j-1}{m-2} a_{0,n-j}$$

$$+ \beta \sum_{j=0}^{m-1} \alpha^j \beta^{n-j} \binom{n}{j} a_{m-j,0} + \beta \sum_{j}^{n} \alpha^m \alpha^{j-m} \binom{j-1}{m-1} a_{0,n-j}$$

$$+ \sum_{j=0}^{m-1} \alpha^j \beta^{n+1-j} \binom{n}{j} a_{m-j,0} + \sum_{j=m}^{n} \alpha^m \beta^{j+1-m} \binom{j-1}{m-2} a_{0,n-j}$$

$$= \sum_{j=0}^{m-1} \alpha^j \beta^{n+1-j} \binom{n+1}{j} a_{m-j,0} + \sum_{j=m-1}^{n} \alpha^m \beta^{j+1-m} \binom{j}{m-1} a_{0,n-j}$$

$$= \sum_{j=0}^{m-1} \alpha^j \beta^{n+1-j} \binom{n+1}{j} a_{m-j,0} + \sum_{j=m}^{n+1} \alpha^m \beta^{j-m} \binom{j-1}{m-1} a_{0,n+1-j}$$

ii) Consider the identity

$$\sum_{j=0}^{n} \binom{n}{j} t^j \alpha^j \beta^{n-j} = (\alpha t + \beta)^n, \quad \forall t \in R.$$

Differentiating twice with respect to t and taking $t = 1$, $\beta = 1 - \alpha$; we obtain:

$$\sum_{j=0}^{n} \binom{n}{j} \alpha^j (1-\alpha)^{n-j} = 1, \sum_{j=0}^{n} j \binom{n}{j} \alpha^j (1-\alpha)^{n-j} = \alpha \cdot n$$

and

$$\sum_{j=0}^{n} j^2 \binom{n}{j} \alpha^j (1-\alpha)^{n-j} = \alpha^2 \cdot n(n-1) + \alpha n.$$

Then, by Schwartz' inequality, we have

$$\sum_{j=0}^{n} \binom{n}{j} \alpha^j (1-\alpha)^{n-j} (m-j)$$

$$= \sum_{j=0}^{n} \left[\binom{n}{j} \alpha^j (1-\alpha)^{n-j} \right]^{1/2} \left[\binom{n}{j} \alpha^j (1-\alpha)^{n-j} (m-j)^2 \right]^{1/2}$$

$$\le \left[\sum_{j=0}^{n} \binom{n}{j} \alpha^j (1-\alpha)^{n-j} \right]^{1/2} \left[\sum_{j=0}^{n} \binom{n}{j} \alpha^j (1-\alpha)^{n-j} (m-j)^2 \right]^{1/2}$$

$$= \left[m^2 - 2mn\alpha + \alpha^2 n(n-1) + \alpha n \right]^{1/2} = \left[(n\alpha - m)^2 + n\alpha(1-\alpha) \right]^{1/2}.$$

In order to prove the last inequality in ii), we recall the formula:

$$\sum_{j=m}^{\infty} \binom{j-1}{m-1} t^{j-m} = (1-t)^{-m}, \quad \text{for } |t| < 1.$$

Then, differentiating twice with respect to t the identity

$$\sum_{j=m}^{\infty}\binom{j-1}{m-1}t^{j-m}(1-t)^m=1 \tag{3.4}$$

we obtain

$$\sum_{j=m}^{\infty} j\binom{j-1}{m-1}t^{j-m}(1-t)^m={}^m\!/_{1-t} \quad \text{and} \quad \sum_{j=m}^{\infty} j^2\binom{j-1}{m-1}t^{j-m}(1-t)^m$$

$$=m^2+m\cdot\frac{1-t}{t}+2\frac{m^2t^2}{1-t}+\frac{m^2\cdot t^2}{2}+m\frac{t^2}{(1-t)^2}. \tag{3.5}$$

We use now again Schwartz' inequality together with the identities (3.4) and (3.5) to obtain

$$\sum_{j=m}^{n}\binom{j-1}{m-1}\alpha^m\cdot(1-\alpha)^{j-m}(n-j)$$

$$\le\left[\sum_{j=m}^{\infty}\binom{j-1}{m-1}\alpha^m(1-\alpha)^{j-m}\right]^{1/2}\left[\sum_{j=m}^{\infty}\binom{j-1}{m-1}\alpha^m(1-\alpha)^{j-m}(n-j)^2\right]^{1/2}$$

$$\le\left[m\frac{1-\alpha}{\alpha^2}+\left(m\frac{1-\alpha}{\alpha}+m-n\right)^2\right]^{1/2}. \qquad \blacksquare$$

With the notations from § 1 we have the

3.2. PROPOSITION. Let $A:\to 2^X$ be accretive in the Banach space X; then

i) $\|J_\lambda^n x-x\|\le n\cdot\lambda\cdot|Ax|, \quad \forall x\in D(J_\lambda^n)\cap D(A), n\in N, \lambda>0.$

ii) If $x\in D, \lambda>0$, then for every $\gamma>0, \gamma/\lambda x+(1-\gamma/\lambda)J_\lambda x\in D_\gamma$ and $J_\lambda x=J_\gamma(\gamma/\lambda x+(1-\gamma/\lambda)J_\lambda x)$.

iii) For every $x\in D(J_\lambda^m)\cap D(J_\gamma^n)\cap D(A), 0<\gamma\le\lambda$ and $m\le n$, m, n $\in$ N

$$\left\{\left[(m\lambda-\gamma n)^2+n\gamma(\lambda-\gamma)\right]^{1/2}+\left[(m\lambda-n\gamma)^2+m\lambda(\lambda-\gamma)\right]^{1/2}\right\}|Ax|.$$

PROOF. i) From the formula $J_\lambda^n x-x=\sum_{i=0}^{n-1}\left(J_\lambda^{n-i}x-J_\lambda^{n-i-1}x\right)$ we obtain that

$$\|J_\lambda^n x-x\|\le\sum_{i=0}^{n-1}\|J_\lambda^{n-i-1}x-J_\lambda^{n-i-2}x\|$$

$$\le\sum_{i=0}^{n-1}\|J_\lambda x-x\|=n\|J_\lambda x-x\|=n\lambda\|A_\lambda x\|.$$

The result is now a consequence of Proposition 1.8.

ii) Let $x\in D_\lambda=R(I+\lambda A)$; then there is $(x_o,u_o)\in G(A)$ such that $x=x_o+\lambda u_o$. It follows that $J_\lambda x=x_o$ and

$$\gamma/\lambda x+(1-\gamma/\lambda)J_\lambda x=\gamma/\lambda(x_o+\lambda u_o)+(1-\gamma/\lambda)x_o=x_o+\gamma u_o.$$

Hence

$$\gamma/\lambda x+(1-\gamma/\lambda)J_\lambda x\in R(I+\gamma A)=D_\gamma$$

and

$$J_\lambda(\gamma/\lambda x+(1-\gamma/\lambda)J_\lambda x)=J_\gamma(x_o+\gamma u_o)=x_o=J_\lambda x.$$

iii) Let $m,n\in N$, $m\le n$; for $0\le k\le m$, $0\le j\le n$ and $x\in D\left(J_\lambda^m\right)\cap D\left(J_\gamma^n\right)\cap D(A)$, with $0<\gamma\le\lambda$, we denote $a_{k,j}=\left\|J_\gamma^j x-J_\lambda^k x\right\|$.

Since $J_\lambda^{k-1}x\in D_\lambda$, ii) yields

$$\gamma/\lambda J_\lambda^{k-1}x+(1-\gamma/\lambda)J_\lambda^k x\in D_\gamma \quad\text{and}\quad J_\lambda^k x=J_\gamma\left[\gamma/\lambda J_\lambda^{k-1}x+(1-\gamma/\lambda)J_\lambda^k x\right].$$

Hence we write:

$$\begin{aligned}a_{k,j}&=\left\|J_\gamma^j x-J_\gamma\left[\gamma/\lambda J_\lambda^{k-1}x+(1-\gamma/\lambda)J_\lambda^k x\right]\right\|\\&\le\left\|J_\gamma^{j-1}x-\left[\gamma/\lambda J_\lambda^{k-1}x+(1-\gamma/\lambda)J_\lambda^k x\right]\right\|\\&\le\gamma/\lambda\left\|J_\gamma^{j-1}x-J_\lambda^{k-1}x\right\|(1-\gamma/\lambda)\left\|J_\gamma^{j-1}x-J_\lambda^k x\right\|\\&=\alpha\cdot a_{k-1,j-1}+(1-\alpha)\cdot a_{k,j-1},\quad\text{with } \alpha=\gamma/\lambda.\end{aligned}$$

We use now Lemma 3.1 and i) to obtain

$$\begin{aligned}&\left\|J_\gamma^n x-J_\lambda^m x\right\|=a_{m,n}\\&\le\left[\lambda\sum_{i=0}^{m-1}\alpha^j(1-\alpha)^{n-j}\binom{n}{j}(m-j)+\gamma\sum_{j=m}^{n}\alpha^m(1-\alpha)^{j-m}\binom{j-1}{m-1}(n-j)\right]|Ax|\\&\le\left\{\lambda\left[(n\alpha-m)^2+n\alpha(1-\alpha)\right]^{1/2}+\gamma\left[\frac{m(1-\alpha)}{\alpha^2}+\left(\frac{m(1-\alpha)}{\alpha}+m-n\right)^2\right]^{1/2}\right\}|Ax|\\&=\left\{\left[(n\gamma-\lambda m)^2+n\gamma(\lambda-\gamma)\right]^{1/2}+\left[m\lambda(\lambda-\gamma)+(m\lambda-\gamma n)^2\right]^{1/2}\right\}\cdot|Ax|.\qquad\blacksquare\end{aligned}$$

3.3. DEFINITION. We say that a mapping $A: X\to 2^X$ on Banach space X satisfies the condition (R) if it is accretive and $\overline{D(A)}\subseteq\bigcap_{\lambda>0}R(I+\lambda A)$.

We can now give the main result from this section

3.4. THEOREM. Let $A: X\to 2^X$ be a mapping which satisfies the condition (R); then the following limit

$$\lim_{n\to\infty}(I+{}^t\!/\!_n A)^{-n}x=S_t x \tag{3.6}$$

exists for every $x\in\overline{D(A)}$, uniformly for t in bounded subintervals of $[0,+\infty]$; moreover (3.6) defines a semigroup of nonlinear contractions on $\overline{D(A)}$.

PROOF. Since A satisfies the condition (R) it follows that $D(A)\subseteq D\left(J_\lambda^n\right)$, $\forall n\in N, \lambda>0$. Let be $x\in D(A), m, n\in N, m\le n$; take in Proposition 3.2., iii), $\gamma=\dfrac{t}{n}$ and $\lambda=\dfrac{t}{m}$; we obtain

$$\left\|J^n_{t/n}x-J^m_{t/m}x\right\|\le 2t(1/m-1/n)^{1/2}|Ax|. \tag{3.7}$$

Therefore $\lim_{n\to\infty}J^n_{t/n}x=S_tx$ exists uniformly for t in compact subsets of $[0,+\infty]$. Moreover, since

$$\left\|J^n_{t/n}x-J^n_{t/n}y\right\|\le\|x-y\|,\ x,y\in D(A), t\ge 0,$$

then

$$\|S_tx-S_ty\|\le\|x-y\|,\ x,y\in D(A).$$

Consequently we can extend S_t by continuity to $\overline{D(A)}$. As $J^n_{t/n}x\in D(A)$ for all $x\in D(A)$, $n\in N$ and $t\ge 0$, then $S_tx\in\overline{D(A)}$, for each $x\in\overline{D(A)}$ and $t\ge 0$. We shall show that

$$\|S_tx-S_sx\|\le 2|Ax|\cdot|t-s|,\quad \forall x\in D(A), s,t\ge 0. \tag{3.8}$$

Let be $0\le s\le t$ and $n\in N$, taking in the inequality iii) Proposition 3.2., $\gamma=\dfrac{s}{n}$, $\lambda=\dfrac{t}{n}$, we have

$$\left\|J^n_{t/n}x-J^n_{s/n}x\right\|\le|Ax|\cdot\left\{\left[(s-t)^2+\frac{s(t-s)}{n}\right]^{1/2}+\left[(s-t)^2+\frac{t(t-s)}{n}\right]^{1/2}\right\}.$$

Letting $n\to\infty$, we obtain (3.8.). In particular it follows that $t\to S_tx$ is continuous on $[0,+\infty)$ for every $x\in D(A)$, hence also for $x\in\overline{D(A)}$. To end the proof we only need to verify the semigroup property for S.
It is not difficult to see that for $x\in D(A)$, $S_t^2x=\lim_{n\to\infty}J^{2n}_{t/n}x$ and inductively, that $S_t^mx=\lim_{n\to\infty}J^{mn}_{t/n}x$. Then

$$S_{mt}x=\lim_{n\to\infty}J^n_{mt/n}x=\lim_{k\to\infty}J^{mk}_{mt/mk}x=\lim_{k\to\infty}J^{mk}_{t/k}x=S_t^mx.$$

Consider t, s rational, i.e. t = p/q s = r/v; then

$$S_{t+s}x = S_{(pv+rq)/qv}x = S^{pv+rq}_{(qv)^{-1}}x = S^{pv}_{(qv)^{-1}}\left[S^{rq}_{(qv)^{-1}}x\right] = S_tS_sx.$$

It is now clear that the semigroup property is true also for all real positive t, s. ■

3.5. REMARK. By (3.8) we have that for every $x \in D(A)$ the function $t \to S_tx$ is absolutely continuous on bounded subintervals of R_+.

3.6. COROLLARY. In the conditions of the Theorem 3.3., for every $x \in \overline{D(A)}$ we have

$$\lim_{\lambda \to 0_+} J_\lambda^{[t/\lambda]} x = \lim_{n \to \infty} J^n_{t/n} x = S_tx.$$

PROOF. We use Proposition 3.2. iii) to prove that $\left\{J_\lambda^{[t/\lambda]}x\right\}_\lambda$ is a Cauchy sequence and that $\left\{J_\lambda^{[t/\lambda]}x - J^n_{t/n}x\right\}_\lambda$ is a zero-sequence. ■
The details are left as Exercise 8.

3.7. COROLLARY. Let X be a Banach lattice and $A: X \to 2^X$ T-accretive satisfying the condition (R); then S given by the formula (3.6) is a semigroup of T-contractions on $\overline{D(A)}$.

PROOF. Consider on X an equivalent norm having the property (P) (see Remark 1.24); then by Proposition 1.22 the mapping A is accretive for this norm. We apply Theorem 3.4. to obtain a semigroup S on $\overline{D(A)}$. Since all $J^n_{t/n}$ are T-contractions and the map $x \to x_+$ is continuous, we have

$$\|(S_tx - S_ty)_+\| = \lim_{n \to \infty}\left\|\left(J^n_{t/n}x - J^n_{t/n}y\right)_+\right\| \le \|(x-y)_+\|, \quad x \in \overline{D(A)}. \ ■$$

3.8. REMARK. If $A: D(A) \to X$ is linear with $\overline{D(A)} = X$ and satisfies the condition (R), then by Lummer-Phillips' Theorem (see Pazy [2] (-A) is the infinitesimal generator of a c_0-semigroup of contractions on X for which the exponential formula (3.6) is true.
In the nonlinear case, without suplimentary conditions on the space X, the coincidence between the generator of the semigroup obtaind by the formula (3.6) and (-A) may not occur. The particular case when X and X^* are uniformly convex will be studied in the next section.
On the other hand, the converse problem, i.e. if for a given semigroup of nonlinear contraction, there exists an accretive mapping such that

the formula (3.6) is true, is unsolved, exept in the case of Hilbert spaces, where the answer is affirmative.

We finally remark that if X^* is not strictly convex, even in the finite dimensional case we have no unicity of the accretive mapping A in the representation formula (3.6), as the following example due to Crandall and Liggett [1] ilustrates:

3.9. EXAMPLE. Consider the space $X = R^2$ endowed with the norm $\|(a,b)\| = \max\{|a|,|b|\}$ and let be $g:[-1,1]\to[-1,1]$ continuous a function, strictly decreasing with g(-1) = 1 and g(1) = -1. We define the mapping $A: R^2 \to 2^{R^2}$ by

$$D(A) = R^2 \text{ and } A(a,b) = \begin{cases} (-1,1) & \text{if } b > a \\ \{(t,g(t)), t \in [-1,1]\} & \text{if } b = a. \\ (1,-1) & \text{if } b < a \end{cases}$$

We shall verify that A is accretive, i.e. that for every $\lambda > 0$ and $(c,d) \in A(a,b)$, $(c'd') \in A(a',b')$ we have

$$\|(a,b)-(a',b')+\lambda[(c,d)-(c',d')]\| \geq \|(a,b)-(a',b')\|. \tag{3.9}$$

Suppose b > a and b' > a' or b < a and b' < a'; then (3.9) is an identity. Suppose b > a and b' < a'; then b-b' > a-a' and

$$\|(a,b)-(a',b')+\lambda[(c,d)-(c',d')]\| = \max\{|a-a'-2\lambda|,|b-b'+2\lambda|\}.$$

Hence

$$\|(a,b)-(a',b')\| = \max\{|a-a'|,|b-b'|\} = \begin{cases} |b-b'| & \text{if } |b-b'|>|a-a'| \\ |a-a'| & \text{if } |b-b'|<|a-a'| \end{cases}$$

$$\leq \begin{cases} |b-b'+2\lambda| & \text{if } |b-b'|>|a-a'| \\ |a-a'-2\lambda| & \text{if } |b-b'|<|a-a'| \end{cases} = \max\{|a-a'-2\lambda|,|b-b'+2\lambda|\}.$$

The cases $b \gtrless a$ and b' = a', or b = a and $b' \gtrless a'$ can be treated analogously.

Consider finally a = b and a' = b'; then, then $(c-c')(g(c)-g(c')) \leq 0$

$$\|(a,b)-(a',b')+\lambda[(c,d)-(c',d')]\| = \max\{|a-a'-\lambda(c-c')|;\ |a-a'+\lambda(g(c)-g(c'))|\}.$$
$$\geq \max|a-a'| = \|(a,b)-(a',b')\|.$$

We shall prove now that $R(I+\lambda A) = R^2, \forall\lambda > 0$; to this purpose we remark that for a fixed $\lambda > 0$ we have

$$(a,b) = \begin{cases} (a+\lambda,\ b-\lambda)+\lambda(-1,1) & \text{if } 2\lambda < b-a \\ (a-\lambda,\ b+\lambda)+\lambda(1,-1) & \text{if } 2\lambda < a-b \\ (a-\lambda t_\lambda,\ b-\lambda g(t_\lambda))+\lambda(t_\lambda,g(t_\lambda)) & \text{if } |a-b| \leq 2\lambda \end{cases} \text{ with } t_\lambda \text{ s.t.}$$

$g(t_\lambda)-t_\lambda=\dfrac{b-a}{\lambda}.$

Since

$$(-1,1)=A(a+\lambda,b-\lambda),(1,-1)=A(a-\lambda,b+\lambda),(t_\lambda,g(t_\lambda))\in A(a-\lambda t_\lambda,b-\lambda t_\lambda)$$

it follows that $(a,b)\in R(I+\lambda A)$. It is now easy to check that

$$\left(I+\frac{t}{n}A\right)^{-1}(a,b)=\begin{cases}\left(a\pm\frac{t}{n},b\mp\frac{t}{n}\right) & \text{if } \pm(b-a)>2t/n\\ \left(a-\frac{t}{n}t_0,a-\frac{t}{n}t_0\right) & \text{if } a=b,t>0 \text{ and } g(t_0)=t_0\end{cases}$$

It follows that

$$\left(I+\frac{t}{n}A\right)^{-n}(a,b)=\begin{cases}(a\pm t,b\mp t) & \text{if } 2t\le\pm(b-a)\\ (a-t\cdot t_0,a-t\cdot t_0) & \text{if } a=b,t>0\end{cases}\cdot \tag{3.10}$$

We note that since $g(\pm1)=\pm1$, the fixed point t_0 of g exists and is unique. Making use of the formula (3.6), we can deduce from (3.10), that:

$$S_t(a,b)=\begin{cases}(a\pm t,b\mp t) & \text{if } 0<t<\pm\dfrac{(b-a)}{2}\\ \left[\dfrac{a+b}{2}-\left(t-\dfrac{|b-a|}{2}\right)t_0,\dfrac{a+b}{2}-\left(t-\dfrac{|b-a|}{2}\right)t_0\right] & \text{if } t\ge\dfrac{|b-a|}{2}\end{cases} \tag{3.11}$$

We shall verify only the last equality. Suppose for example that b - a > 0; then we have

$$S_t(a,b)=S_{t-\frac{b-a}{2}}\cdot S_{\frac{b-a}{2}}(a,b)=S_{\frac{b-a}{2}}\left(a+\frac{b-a}{2},b-\frac{b-a}{2}\right)$$

$$=S_{t-\frac{b-a}{2}}\left(\frac{a+b}{2},\frac{a+b}{2}\right)=\left[\frac{a+b}{2}-\left(t-\frac{b-a}{2}\right)t_0,\frac{a+b}{2}-\left(t-\frac{b-a}{2}\right)t_0\right].$$

The formula (3.11) shows that S_t depends only of the fixed point t_0 of g and not on g itself. Consider the functions $g_1(t)=-t$ and $g_2(t)=-t^3$; it is clear that g_1 and g_2 have the common fixed point $t_0 = 0$, hence they "generate" by formula (3.6) the same semigroup although the corresponding mappings A_{g_1} and A_{g_2} are distinct.

§ 4. THE ABSTRACT CAUCHY PROBLEM FOR ACCRETIVE MAPPINGS

Let X be a real Banach space and $A: X \to 2^X$ an accretive mapping. In this section we shall present results on the Cauchy Problem for the following evolution equation

$$0 \in \frac{du}{dt} + Au, \quad u(0) = x, x \in D(A) \tag{4.1}$$

where the notion of a solution of the Problem 4.1. is understood in the sense of the following

4.1. DEFINITION. A function $u: R_+ \to X$ is a (strong) solution of the Problem (4.1) if it is absolutely continuous on bounded subintervals of R_+, a.e. differentiable on R_+, u(0) = x satisfies the equation (4.1) a.e. on R_+.

4.2. PROPOSITION. The Problem (4.1) has at most a solution.

PROOF. Suppose that we have two solutions u and v; using the differentiation rule of Kato, we have that for every $x^* \in J(u(t)-v(t))$

$$\|u(t)-v(t)\| \frac{d}{dt}\|u(t)-v(t)\| = <x^*, \frac{du(t)}{dt} - \frac{dv(t)}{dt}> \text{ a.e. on } R_+.$$

Since $\frac{du}{dt}(t) - \frac{dv}{dt}(t) \in -Au(t) + Av(t)$, a.e., it follows that

$$\frac{d}{dt}\|u(t)-v(t)\|^2 \le 0 \text{ a.e. on } [0,+\infty].$$

Hence:

$$\|u(t)-v(t)\|^2 \le \|u(0)-v(0)\|^2 = 0, \quad \forall t \ge 0 \text{ i.e. } u = v \text{ on } R_+. \qquad ■$$

The next result is our further considerations

4.3. LEMMA. Let $A: X \to 2^X$ be a mapping satisfying the condition (R) and S the semigroup given by the exponential formula (3.6); then for every $x \in D(A)$ and $(x_o, u_o) \in G(A)$, we have:

$$\sup_{x^* \in J(x-x_o)} \limsup_{t \to 0_+} <x^*, \frac{S_t x - x}{t}> \le \sup_{x^* \in J(x_o - x)} <x^*, u_o>.$$

PROOF. Let be $\lambda > 0$ and $n \in N$; since $A_\lambda J_\lambda^{n-1}x \in AJ_\lambda^n x$ and $u_o \in Ax_o$, the accretivity of A provides the existence of $x^* \in J\left(x_o - J_\lambda^n x\right)$ with

$$x^*, u_o - A_\lambda J_\lambda^{n-1} x > \ge 0. \tag{4.2}$$

We note that

$$<x^*,A_\lambda J_\lambda^{n-1}x,> = \lambda^{-1} <x^*,J_\lambda^{n-1}x - J_\lambda^n x>$$
$$= \lambda^{-1} <x^*,x_o - J_\lambda^n x> + \lambda^{-1} <x^*,J_\lambda^{n-1}x - x_o>$$
$$> \lambda^{-1}\left(\left\|x_o - J_\lambda^n x\right\|^2 - \left\|x_0 - J_\lambda^{n-1}x\right\| + \left\|x_0 - J_\lambda^n x\right\|\right)$$
$$\geq (2\lambda)^{-1}\left(\left\|x_0 - J_\lambda^n x\right\|^2 - \left\|x_0 - J_\lambda^{n-1}x\right\|^2\right).$$

Hence, using (4.2) we obtain

$$\left\|J_\lambda^n x - x_0\right\|^2 - \left\|J_\lambda^{n-1}x - x_0\right\|^2 \leq 2\lambda \cdot <x^*,u_o>. \tag{4.3}$$

Denote:

$$f(x,y) = \sup_{x^* \in Jx} <x^*, y>; \tag{4.4}$$

then (4.3) can be written as

$$\left\|J_\lambda^n x - x_0\right\|^2 - \left\|J_\lambda^{n-1}x - x_0\right\|^2 \leq 2\int_\lambda^{(n+1)\lambda} f\left(x_0 - J_\lambda^{[s/\lambda]}x, u_0\right)ds. \tag{4.5}$$

Let be $t > 0$ and consider now $\lambda \leq t$; suming the inequalities (4.5) corresponding to $n = 1, 2, \ldots\ldots, \left[\frac{t}{\lambda}\right]$, we have

$$\left\|J_\lambda^{[t/\lambda]}x - x_0\right\|^2 - \left\|x - x_0\right\|^2 \leq 2\int_\lambda^{([t/\lambda]+1)\lambda} f\left(x_0 - J_\lambda^{[s/\lambda]}x, u_0\right)ds. \tag{4.6}$$

We shall prove that the function $f: XxX \to R$ defined by (4.4) is upper semicontinuous. Indeed, since Jx is w^*-compact, there is $x_0^* \in Jx$ so that $f(x,y) = <x_o^*,y>$. Let $x_n \xrightarrow[n]{} x$ and $y_n \xrightarrow[n]{} y$; it is enough to prove that $\limsup_n f(x_n,y_n) \leq f(x,y)$.

Since $|f(x,y)| \leq \|x\| \cdot \|y\|$, it follows that $\limsup_n f(x_n,y_n) < +\infty$; thus, passing eventually to a subsequence, we can suppose that $\{f(x_n,y_n)\}_n$ is convergent. On the other hand, there are $x_n^* \in Jx_n$ so that $f(x_n,y_n) = <x_n^*,y_n>, n \in N$. Let x^* be a limit point in the w^*-topology of the sequence $\{x_n^*\}_n$; then we have

$$<x^*,x> = \lim_n <x_n^*,x_n> = \lim_{n\to\infty} \|x_n\|^2 = \|x\|^2 .$$

Since $\left\|x^*\right\| \leq \liminf_n \left\|x_n^*\right\| = \|x\|$ it follows that $x^* \in Jx$. Hence

$$\lim_{n\to\infty} f(x_n,y_n) = \lim_{n\to\infty} <x_n^*,y_n> = <x^*,y> \leq f(x,y).$$

Using now Corollary 3.6 we can pass with $\lambda \to 0_+$ in (4.6) to obtain

$$\|S_t x - x_0\|^2 - \|x - x_0\|^2 \le 2\int_0^t f(x_0 - S_s x, u_0)ds. \tag{4.7}$$

Finally we observe that for every $x^* \in J(x - x_0) = \partial\left(\frac{1}{2}\|x - x_0\|^2\right)$ we have

$$\frac{1}{2}\|S_t x - x_0\|^2 - \frac{1}{2}\|x - x_0\|^2 \ge \langle x^*, S_t x - x \rangle$$

so that (4.7) yields

$$\langle x^*, S_t x - x \rangle \le \int_0^1 f(x_0 - S_{t\tau} x, u_0)d\tau, \quad \forall x^* \in J(x - x_0).$$

Now, the continuity of $t \to S_t x$ and the upper semicontinuity of f yield

$$\lim_{t \to 0_+} \langle x^*, \frac{S_t x - x}{t} \rangle \le f(x_0 - x, u_0), \ \forall x^* \in J(x - x_0).$$

This completes the proof. ■

We can now give our main existence result for the Cauchy Problem (4.1)

4.4. THEOREM. Let X be a Banach space, $A: X \to 2^X$ a closed mapping satisfying the condition (R) and S the semigroup generated by the exponential formula (3.6). If for $x \in D(A)$, the function $R_+ \ni t \to S_t x$ is a.e. differentiable on R_+, then $u(t) = S_t x$ is the solution of the Cauchy Problem (4.1).

PROOF. By Remark 3.5., the function $t \to S_t x$ is absolutely continuous on bounded subsets of R_+.

Consider a point $t_0 > 0$ at which $t \to S_t x$ is differentiable and denote by $S'_{t_0} x$ this derivative. We shall prove that $\left(S_{t_0} x, -S'_{t_0} x\right) \in G(A)$.

By hypothesis, we have

$S_{t_0 - \lambda} x = S_{t_0} x - S'_{t_0} x + \lambda \cdot r(\lambda, t_0)$, $0 < \lambda < t_0$ where $r(\lambda, t_0) \xrightarrow[\lambda \to 0]{} 0$.

Since $S_{t_0 - \lambda} x \in \overline{D(A)} \subset R(I + \lambda A)$, there is $(x_\lambda, u_\lambda) \in G(A)$ so that:

$$x_\lambda + \lambda u_\lambda = S_{t_0 - \lambda} x = S_{t_0} x - \lambda S'_{t_0} x + \lambda r(\lambda, t_0). \tag{4.8}$$

Applying Lemma 4.1. to $S_{t_0} x \in \overline{D(A)}$ and $(x_\lambda, u_\lambda) \in G(A)$, we have

$$\sup_{x^* \in J(S_{t_0} x - x_\lambda)} \langle x^*, S'_o x \rangle \le \sup_{x^* \in J(x_\lambda - S_{t_0} x)} \langle x^*, u_\lambda \rangle. \tag{4.9}$$

As the set $J(x_\lambda - S_t x)$ is w^*-compact in X^*, the supremum in the right side of (4.9) it assumed at a point $x^* \in J(x_\lambda - S_{t_0}x)$. Then (4.9) yields

$$<x^*, S'_{t_o}x> \le <x^*_\lambda, u_\lambda>, \quad \forall x^* \in J(S_{t_0}x - x_\lambda);$$

in particular

$$<x^*_\lambda, u_\lambda + S'_{t_o}x> \ge 0.$$

Using now (4.8) we obtain

$$<x^*_\lambda, \frac{S_{t_o}x - x_\lambda}{\lambda} + r(\lambda, t_o)> \ge 0, \quad 0 < \lambda < t_0.$$

But since $x^*_\lambda \in J(x_\lambda - S_{t_0}x)$, we have

$$\lambda^{-1}\left\|S_{t_0}x - x\right\|^2 \le <x^*_\lambda, r(\lambda, t_o)> \le \left\|S_{t_0}x - x_\lambda\right\| \cdot \left\|r(\lambda, t_0)\right\|,$$

i.e.
$$\frac{\left\|S_{t_0}x - x_\lambda\right\|}{\lambda} \le \left\|r(\lambda, t_0\right\|.$$

Hence $\lim\limits_{\lambda \to 0_+} \dfrac{S_{t_0}x - x_\lambda}{\lambda} = 0$, in particular $\lim\limits_{\lambda \to 0_+} x_\lambda = S_{t_0}x$.

We note finally that by (4.8)

$$\lim_{\lambda \to 0_+} u_\lambda = \lim_{\lambda \to 0_+} \left(\frac{S_{t_0}x - x_\lambda}{\lambda} - S'_{t_0}x + r(\lambda, t_0)\right) = -S'_{t_0}x.$$

Since A is closed, we conclude that $(S_{t_0}x, -S'_{x_0}x) \in G(A)$ and this ends the proof. ■

4.5. COROLLARY. i) If X is reflexive and $A: X \to 2^X$ is hyperaccretive, then for every $x \in D(A)$ the Cauchy Problem (4.1) has a unique solution given by

$$S_t x = \lim_{n \to \infty} (I + t/n\, A)^{-n} x, \quad t \ge 0.$$

ii) Moreover, if X and X^* are uniformly convex, then the generator of the semigroup S coincides on D(A) with $-A^o$.

PROOF. i) Since A is hyperaccretive, Proposition 1.2. shows that the condition (R) is satisfied. By Proposition 1.4. A beeing maximal monotone, is also closed. We note that by Komura's Theorem 2.1. the function $t \to S_t x$ is a.e. differentiable on R_+ (by Remark 3.5 it is absolutely continuous on bounded subsets of R_+). Now the statement is a consequence of the previous theorem.

ii) By i), $\frac{d}{dt}(S_t x - x) \in -AS_t x$, a.e. on R_+ for $x \in D(A)$ so that using Kato's differentiation rule we have

$$\|S_t x - x\| \frac{d}{dt} \|S_t x - x\| \le \langle x^*, -u \rangle, \quad \text{a.e. on } R_+, \quad \forall u \in Ax, x^* \in J(S_t x - x).$$

Therefore

$$\|S_t x - x\| \le t \|A^\circ x\|, \quad \forall t \ge 0. \tag{4.10}$$

Hence $x \in D$, where D is defined in Lemma 2.7., and therefore $x \in D(A_s) = D(A_w) = D$. But then by Theorem 2.11., we have

$$\frac{d^+}{dt} S_t x = A_s S_t x, \quad \forall x \in D(A), \quad t \ge 0$$

so that $A_s S_t x \in -AS_t x$, a.e. on R_+.

Since $t \to A_s S_t x$ is right continuous on R_+ and A is closed, we can select a sequence $t_n \downarrow 0_+$ so that $A_s S_{t_n} x \in -AS_{t_n} x$ and

$$A_s x = \lim_{t_n \downarrow 0_+} A_s S_{t_n} x \in -Ax, \quad \forall x \in D(A).$$

By (4.10) $\|A_s x\| \le \|A^\circ x\|$, so that $A_s x = -A^\circ x$; $\forall x \in D(A)$. ■

We complete our considerations with the following "unicity" result:

4.6. THEOREM. Let X be uniformly convex, with X^* uniformly convex and $A, B: X \to 2^X$ two hyperaccretive mappings; let S and T be the semigroups generated by A, respectively B, on $\overline{D(A)}$ and $\overline{D(B)}$; if $S = T$ then $A = B$.

PROOF. We shall first show that $D(A) = D(B)$. It is clear that $\overline{D(A)} = \overline{D(B)}$. Consider $x \in D(B)$; by Lemma 4.3., we have:

$$\limsup_{t \to 0_+} \left\langle J(x - x_0), \frac{S_t x - x}{t} + u_0 \right\rangle \le 0, \quad \forall (x_0, u_0) \in G(A). \tag{4.11}$$

But $\left\| \frac{S_t x - x}{t} \right\| \le \|B^\circ x\|$, $\forall t \ge 0$ (see 4.10.) so that there exist $u \in x$ and a sequence so that $\frac{S_{t_n} x - x}{t_n} \xrightarrow[n]{w} u$. Now (4.11) yields

$$\langle J(x - x_0), u + u_0 \rangle \le 0, \quad \forall (x_0, u_0) \in G(A).$$

Since A is maximal accretive $(x, -u) \in G(A)$ and therefore $D(B) \subseteq D(A)$. We can prove analogously that $D(A) \subseteq D(B)$. Then by Corollary 4.4. (ii), $A^\circ = B^\circ$ and in view of Theorem 1.16 (iii), $A = B$. ■

We end this section with another aproximation result for the semigroup S generated by an hyperaccretive mapping.

4.7. LEMMA. Let $C \subseteq X$ be convex and closed and $T: C \to C$ be a contractiveoperator; then for every $x \in C$, the equation

$$\frac{du}{dt} + (I-T)u = 0 \quad u(0) = x \tag{4.12}$$

has a unique solution $u \in C^1((0,+\infty);X)$ such that $u(t) \in C$, $\forall t \geq 0$; moreover:

$$\left\| u(n) - T^n x \right\| \leq \sqrt{n}\, \| x - Tx \|, \; x \in N. \tag{4.13}$$

PROOF. It is clear that the Problem (4.12) is equivalent to the integral equation:

$$u(t) = e^{-t}x + \int_0^t e^{s-t} Tu(s) ds. \tag{4.14}$$

Let be $t_0 > 0$ and $K = \left\{ u \in C\left([0,t_0];\ X\right) = Y;\ u(t) \in C, 0 \leq t \leq t_0 \right\}$; then K is convex and closed. Denote by Q the operator on K defined by $(Qu)(t) = e^{-t}x + \int_0^t e^{s-t} Tu(s) ds$. Then $Qu \in K$; indeed, we may write

$e^{-t}x = \left(1 - \int_0^t e^{s-t} ds \right) x$ and note that

$$\sum_{i=1}^{k} e^{s_i - t}(s_{i+1} - s_i)\, Tu(s_i) + \left(1 - \sum_{i=1}^{k} e^{s_i - t}(s_{i+1} - s_i) x \right) \in C.$$

Moreover, for $u, v \in K$ we have

$$\| Qu - Qv \|_Y = \sup_{t \in [0,t_0]} \left\| \int_0^t e^{s-t}(Tu(s) - Tv(s) ds \right\|$$

$$\leq \sup \int_0^t e^{s-t} \| u(s) - v(s) \| ds \leq (1 - e^{-t_0}) \| u - v \|_Y.$$

Then by Banach's fixed point Theorem it follows that the equation (4.14) has a unique solution $u \in K$; since $t_0 > 0$ was arbitrary, the first part of the Proposition is proved.

We note further that I - T is (global) accretive; indeed, for $\lambda > 0$, $x, y \in C$ and $x^* \in J(x-y)$, we have:

$$< x^*, (I-T)x - (I-T)y > = \| x - y \|^2 - < x^*, Tx - Ty >$$
$$\geq \| x - y \|^2 - \| x - y \| \, \| Tx - Ty \| \geq 0.$$

Using Kato's differentiation rule we obtain

$$\| u(t) - x \| \frac{d}{dt} \| u(t) - x \| = < x^*, (T-I)u(t) >$$
$$\leq < x^*, (T-I)x >, \; \forall t \geq 0, x^* \in J(u(t) - x).$$

Hence $\frac{d}{dt}\|u(t)-x\| \le \|Tx-x\|$, $\forall t \ge 0$, and this yields

$$\|u(t)-x\| \le t\|Tx-x\|, \quad \forall t \ge 0. \tag{4.15}$$

From (4.14) we easily obtain

$$u(t)-T^n x = e^{-t}(x-T^n x)+\int_0^t e^{s-t}(Tu(s)-T^n x)ds. \tag{4.16}$$

Since we also have $\|x-T^n x\| \le n\|x-Tx\|$, then (4.16) yields

$$\|u(t)-T^n x\| \le ne^{-t}\|x-Tx\|+\int_0^t e^{s-t}\|u(s)-T^{n-1}x\|ds. \tag{4.17}$$

Denote $\varphi_n(t)=\|u(t)-T^n x\|, n \in N \cup \{0\}$; then (4.15) and (4.17) yield

$$\varphi_n(t) \le ne^{-t}\varphi_1(0)+\int_0^t e^{s-t}\varphi_{n-1}(s)ds, \ \varphi_o(t) \le \varphi_1(0)\cdot t, \quad \forall t \ge 0. \tag{4.18}$$

We shall show by induction that

$$\varphi_n(t) \le \varphi_1(0)\cdot\left[t+(n-t)^2\right]^{1/2}, \ t \ge 0, \ n \in N \cup \{0\} \tag{4.19}$$

and complete the proof taking t = n in (4.19).

For $n=0$, $\varphi_o(t) \le \varphi_1(0)\cdot t \le \varphi_1(0)\cdot(t+t^2)^{1/2}, t \ge 0$; hence (4.19) is true. Supose it is satisfied for n - 1. Then we may write:

$$\varphi_n(t) \le ne^{-t}\cdot\varphi_1(0)+\int_0^t e^{s-t}[s+(n-1-s)]^{1/2}\varphi_1(0)ds$$

$$=\varphi_1(0)\cdot e^{-t}\cdot\psi_n(t), \ t \ge 0,$$

where $\psi_n(t)=n+\int_0^t e^s[s+(n-1-s)]^{1/2}ds$. To obtain (4.19) we only have to show that

$$\psi_n(t) \le e^t\left[t+(n-t)^2\right]^{1/2}, \ t \ge 0.$$

But this is a consequence of the inequality on R_+

$$\psi_n'(t) \le e^t[t+(n-1-t)]^{1/2} \le \frac{d}{dt}\left[e^t[t+(n-t)^2\right]^{1/2}$$

$$=e^t\left[t+(n-t)^2\right]^{1/2}+e^t\left(\frac{1}{2}-n+t\right)\left[t+(n-t)^2\right]^{-1/2}$$

which can be verified by easy computations. ■

4.8. COROLLARY. In the conditions of the Lemma 4.7., let u be the solution of the Problem (4.12); then for every $h>0, u_h(t)=u(h^{-1}t)$ is the solution of the equation

$$\frac{du_h}{dt}+\frac{I-T}{h}u_h(t)=0,\quad u_h(0)=h,\quad t\geq 0$$

and we have:

$$\left\|u_h(nh)-T^n x\right\|\leq\sqrt{n}\,\|x-Tx\|,\quad \forall n\in N\cdot$$

4.9. PROPOSITION. If $A: X\to 2^X$ is hyperaccretive, then for every $x\in D(A)$ and $\lambda>0$, the Problem

$$\frac{du_\lambda}{dt}+A_\lambda u_\lambda=0,\quad u_\lambda(0)=x \tag{4.20}$$

has a unique solution $u_\lambda\in C^1((0,+\infty);\ X)$ and

$$\lim_{\lambda\to 0}u_\lambda(t)=S_t x=\lim_{n\to\infty}J^n_{t/n}x,$$

uniformly on compact subsets of R_+.

PROOF. We apply the above Corollary to $T=J_\lambda$ to we obtain the first part of the result. Moreover, for any $n\in N$, we have

$$\left\|u_\lambda(n\lambda)-J^n_\lambda x\right\|\leq\sqrt{n}\,\|I-J_\lambda x\|=\lambda\sqrt{n}\,\|A_\lambda x\|\leq\sqrt{n}|Ax|. \tag{4.21}$$

We may now use the differentiation rule of Kato and the (global) accretivity of A_λ to obtain that:

$$\|u_\lambda(t)-x\|\frac{d}{dt}\|u_\lambda(t)-x\|=\langle x^*,-A_\lambda u_\lambda(t)\rangle\leq\langle x^*,-A_\lambda x\rangle$$

for some $x^*\in J(u_\lambda(t)-x)$

$$\|u_\lambda(t+h)-u_\lambda(t)\|\frac{d}{dt}\|u_\lambda(t+h)-u_\lambda(t)\|=\langle x^*,A_\lambda u_\lambda(t)-Au(t+h)\rangle\leq 0$$

for some $x^*\in J(u_\lambda(t+h)-u_\lambda)(t))$

Consequently

$$\left\|u_\lambda(t)-x\right\|\leq t\cdot\|A_\lambda x\|\leq t|Ax|$$

and

$$\left\|u_\lambda(t+h)-u_\lambda(t)\right\|\leq\|u_\lambda(h)-x\|\leq h|Ax|. \tag{4.22}$$

For t > 0, let $n\in N$ be so that $t=n\lambda+\gamma$, with $0\leq\gamma<\lambda$. Then by (4.22) we have:

$$\left\|u_\lambda(t)-u_\lambda(n\lambda)\right\|\leq\gamma|Ax|. \tag{4.23}$$

On the other hand, by (3.8)

$$\|S_t x-S_{n\lambda}x\|\leq 2\gamma\cdot|Ax|. \tag{4.24}$$

and by (3.7)

$$\left\|S_{n\lambda}x-J^n_\lambda x\right\|\leq\frac{2n\lambda}{\sqrt{n}}|Ax|=2\lambda\sqrt{n}\ |Ax|. \tag{4.25}$$

Finally, from (4.21), (4.23), (4.24) and (4.25) we obtain

$$\left\|u_\lambda(t)-S_t x\right\|$$

$$\leq \|u_\lambda(t)-u_\lambda(n\lambda)\|+\|u_\lambda(n\lambda)-J_\lambda^n x\|+\|J_\lambda^n x-S_{n\lambda}x\|+\|S_{n\lambda}x-S_t x\|$$
$$\leq (3\gamma+3\lambda\sqrt{n})\cdot|Ax|\leq 3(\gamma+\sqrt{\lambda t})\cdot|Ax|$$

and this yields the result. ■

4.10. REMARK. The Proposition is also true for operators $A: X \to 2^X$ satisfying the following generalized condition $\overline{\text{conv}D(A)} \subseteq \bigcap_{\lambda>0} R(I+\lambda A)$. Indeed, denote $C = \overline{\text{conv}D(A)}$; then $J_\lambda: C \to C$ and we can apply the Proposition 4.8 to J_λ and C. All of our last considerations underline the role of the hyperaccretive mappings in the abstract Cauchy Problem. In what follows we shall present an important hyperaccretivity criterion. We shall need the following result of Peano type:

4.11. PROPOSITION. Let X be a real Banach space and $A: X \to X$ a continuous accretive operator with $D(A) = X$; then for every $x \in X$, there exists a unique solution $u \in C^1((0,+\infty); X)$ of the Problem:

$$\frac{du(t)}{dt} + Au(t) = 0, \ u(0) = x, \ \ t \geq 0. \tag{4.26}$$

PROOF. The unicity is assured by Proposition 4.8. For the existence, we fix $x \in X$ and consider a neighbourhood of x, $V = \{y \in X; \|x-y\| \leq r\}$ such that $\|Ay\| \leq M$, $\forall y \in V$ for some given $M > 0$. Let be $t_o > 0$ so that $M \cdot t_o < r$ and $\varepsilon_n \xrightarrow[n\to\infty]{} 0_+$. We shall show that for every $n \in N$ we can construct a partition $\{t_i^n\}_{o \leq i \leq k_n}$ of the interval $[0, t_o]$, with $t_o^n = 0, t_{k_n}^n = t_o$ and a function u_n defined on the set $\{t_i^n\}_{o\leq i\leq k_n}$ by the recursive formula

$$u_n(t_i^n) = u_n(t_{i-1}^n) - \ell_i^n A u_n(t_{i-1}^n), \ u_n(0) = x \tag{4.27}$$

where $\ell_i^n = t_i^n - t_{i-1}^n > 0$, $i = 1, 2, \ldots, k_n$, such that

$$\|y - u_n(t_{i-1}^n)\| \leq M \cdot \ell_i^n \Rightarrow \|Ay - Au_n(t_{i-1}^n)\| \leq \varepsilon_n \tag{4.28}$$

Indeed, define $t_o^n = 0$ and $u_n(0) = x$; the continuity of A provides the existence of a $\delta_o^n > 0$ such that

$$\|Ay - Au_n(t_o^n)\| \leq \varepsilon_n \text{ whenever } \|y - u_n(t_o^n)\| \leq \delta_o^n.$$

Define $\ell_1^n = \delta_o^n/_M$, $t_1^n = t_o^n + \ell_1^n$, $u_n(t_1^n) = u_n(t_o^n) - \ell_1^n A u_n(t_o^n)$.

Then there is $\delta_1^n > 0$ such that

$$\left\|y-u_n(t_1^n)\right\| \le \delta_1^n \Rightarrow \left\|Ay-Au_n(t_1^n)\right\| \le \varepsilon_n.$$

Put now $\ell_2^n = \delta_{1/M}^n, t_2^n = t_1^n + \ell_2^n, u_n(t_2^n) = u_n(t_1^n) - \ell_2^n Au_n(t_1^n)$ and continue the procedure. We can suppose that at each step the number ℓ_i^n is maximal relatively to the property (4.28). We shall prove that after a finite number of steps the point t_0 is attained.

Indeed, suppose the contrary; then $\lim_{i\to+\infty} t_i^n = t_1 \in (0,t_0]$. We note that $u_n(t_1^n) = x - \ell_1^n \cdot Ax$, hence $u_n(t_1^n) \in V$.

Further, since $u_n(t_2^n) = x - \ell_1^n Ax - \ell_2^n Au_n(t_1^n)$, then

$$\left\|u_n(t_2^n) - x\right\| \le (\ell_1^n + \ell_2^n)M < r, \text{i.e.}\ u_n(t_2^n) \in V.$$

Finally, if we suppose $u_n(t_{i-1}^n) \in V$, then the relation

$$u_n(t_i^n) = x - \sum_{j=1}^{i} \ell_j^n Au_n(t_{j-1}^n)$$

yields

$$\left\|u_n(t_i^n) - x\right\| \le \left(\sum_{j=1}^{i} \ell_j^n\right) \cdot M \le t_0 \cdot M < r.$$

Hence $u_n(t_i^n) \in V$, $i \in N$. Therefore from (4.27) we have

$$\left\|u_n(t_i^n) - u_n(t_{i-1}^n)\right\| \le M(t_i^n - t_{i-1}^n),\ i = 1,2,\dots \tag{4.29}$$

Consequently the limit $u_0 = \lim_{i\to\infty} u_n(t_i^n)$ exists.

We note now that by the maximality property of the numbers ℓ_i^n, $i \in N$, there exist $y_i \in X$, $i \in N$ such that $M\ell_i^n \le \left\|y_i - u_n(t_{i-1}^n)\right\| \le M\ell_i^n + i^{-1}$ and

$$\left\|Ay_i - Au_n(t_{i-1}^n)\right\| \ge \varepsilon_n, i \in N. \tag{4.30}$$

But then $\lim_{i\to\infty} y_i = \lim_{i\to\infty} u_n(t_i^n) = u_0$ exist, and by the continuity of A,

$\lim_{i\to+\infty} Ay_i = \lim_{i\to+\infty} Au_n(t_i^n) = Au_0$. This is in contradiction with (4.30).

Define the function u_n on $[0,t_0)$ by the formula

$$u_n(t) = u_n(t_{i-1}^n) - (t - t_{i-1}^n)Au_n(t_{i-1}^n),\ \text{for}\ t_{i-1}^n \le t \le t_i^n. \tag{4.31}$$

Then u_n is continuous on $[0,t_0)$ and $\frac{du_n(t)}{dt} = -Au_n(t_{i-1}^n)$, for $t \in (t_{i-1}^n, t_i^n)$. Since $\left\|u_n(t) - u_n(t_{i-1}^n)\right\| \le M\ell_i^n$, from (4.28) we obtain

$$\frac{du_n(t)}{dt} = -Au_n(t) + g_n(t), \quad t \in \left(t_{i-1}^n, t_i^n\right), \text{ with } \|g_n(t)\| \le \varepsilon_n. \tag{4.32}$$

Analogously, for every m, n $\in$ N and $t \in \Delta_{ij} = \left(t_{i-1}^n, t_i^n\right) \cap \left(t_{j-1}^m, t_j^m\right)$ we have

$$\frac{d}{dt}(u_n(t) - u_m(t)) = Au_m\left(t_{j-1}^m\right) - Au_n\left(t_{i-1}^n\right).$$

Since $\left\|u_n(t) - u_n\left(t_{i-1}^n\right)\right\| \le M\ell_i^n, \left\|u_m(t) - u_n\left(t_{j-1}^m\right)\right\| \le M\ell_j^m, \ \forall t \in \Delta_{ij}$, then 4.28) yields

$$\frac{d}{dt}(u_n(t) - u_m(t)) = Au_m(t) - Au_n(t) + g_{m,n}(t), \quad \forall t \in \Delta_{ij}$$

with $\left\|g_{m,n}(t)\right\| \le \varepsilon_m + \varepsilon_n$.

Then, by Kato's differentiation rule we have

$$\|u_n(t) - u_m(t)\| \frac{d}{dt} \|u_n(t) - u_m(t)\|$$

$= \langle x^*, Au_m(t) - Au_n(t) \rangle + \langle x^*, g_{m,n}(t) \rangle$, for $x^* \in J(u_n(t) - u_m(t))$.

But as $\langle x^*, Au_n(t) - Au_m(t) \rangle \ge 0$ then

$$\|u_n(t) - u_m(t)\| \frac{d}{dt} \|u_n(t) - u_m(t)\| \le \|g_{m,n}\| \|u_n(t) - u_m(t)\|, t \in \Delta_{ij}.$$

Since $(0, t_o) = \bigcup_{i,j} \Delta_{i,j}$, integrating we obtain

$$\|u_n(t) - u_m(t)\| \le (\varepsilon_m + \varepsilon_n) t_o, \quad \forall t \in [0, t_o].$$

Therefore $u(t) = \lim_{n \to \infty} u_n(t)$ exists uniformly on $[0, t_o]$. We shall prove that u yields the solution we are looking for. Indeed, from (4.32) we have

$$u_n(t) = x - \int_0^t Au_n(s)ds + \int_0^t g_n(t), \quad \forall t \in (0, t_o).$$

Letting $n \to \infty$ we obtain $u(t) = x - \int_0^t Au(s)ds, \quad t \in (0, t_o)$; hence A satifies (4.26) on $[0, t_o)$.

To end the proof we only have to show that u can be extended to the whole $[0, +\infty)$. Let $[0, t_{max})$, be the largest interval of existence of u and suppose that $t_{max} < +\infty$. Since u is a solution of the equation (4.26) on $[0, t_o)$ we have,

$$\|u(t) - u(s)\| \le |t - s| \ \|Ax\|, \quad s, t \in [0, t_o).$$

Then the $\lim_{t \to t_{max}} u(t) = u(t_{max})$ exists and by the continuity of A we also have

$$\lim_{t \to t_{max}} \frac{d}{dt} u(t) = - \lim_{t \to t_{max}} Au(t) = -Au(t_{max}).$$

We consider now the following Problem

$$\frac{d}{dt} u(t + t_{max}) + Au(t + t_{max}) = 0, \quad u(t_{max}) = x.$$

By the first part of this proof, there exists $t_o' > 0$ such that the above equation has a solution on $[0, t_o')$, i.e. u can be extended at the right of t_{max}. This contractiets the hypothesis, hence the proof is complete.

■

4.12. THEOREM. Let A be a continuous and accretive operator on the real Banach space X with D(A) = X; then A is hyperaccretive.

PROOF. We shall prove that for every $y \in X$ the equation x + Ax = y has at least a solution.
Let y, $x_o \in X$ be fixed and apply the above existence result to the operator $\overline{A}x = (A + I)x - y$ to obtain the existence of a unique function $u \in C^1((0, +\infty); X)$ such that

$$\frac{du}{dt}(t) + (A + I)u(t) - y = 0, \quad u(0) = x_o. \tag{4.33}$$

We have

$$\frac{du}{dt}(u(t+h) - u(t)) = -Au(t+h) + Au(t) - u(t+h) + u(t), \quad t \geq 0;$$

hence by a method now standard, we obtain

$$\|u(t+h) - u(t)\| \frac{d}{dt} \|u(t+h)\| \leq -\|u(t+h) - u(t)\|^2, \quad t \geq 0.$$

Consequently

$$\|u(t+h) - u(t)\| \leq e^{-t} \|u(h) - x_o\|, \quad t, h \geq 0. \tag{4.34}$$

Analogously, from $\frac{d}{dt}(u(t) - x_o) = -Au(t) - u(t) + y$, it follows

$$\|u(t) - x_o\| \frac{d}{dt} \|u(t) - x_o\| \leq -\|u(t) - x_o\|^2 + \|u(t) - x_o\| (\|Ax_o\| + \|x_o\| + \|y\|).$$

Thus

$$\|u(t) - x_o\| \leq (1 + e^{-t}) (\|Ax_o\| + \|x_o\| + \|y\|), \quad t \geq 0. \tag{4.35}$$

Then (4.35) and (4.34) show that $\lim_{t \to \infty} u(t) = x$ exists and that $\lim_{t \to +\infty} \frac{du}{dt}(t) = 0$.

Letting $t \to +\infty$ in (4.33) we obtain (A + I) x= y. ■

§ 5. SEMIGROUPS OF NON LINEAR CONTRACTIONS IN HILBERT SPACES

In this section we shall give the nonlinear version of the generation theorem of Hille-Yosida-Lumer-Phillips.

For a semigroup $S: C \to C$ on a closed subset of the Hilbert space H, we denote by A_0 its infinitesimal generator and by $A^h x = \frac{S_h x - x}{h}$, $x \in C$.

It is clear that A^h is dissipative and therefore $(I - \lambda A^h)^{-1}$ exists for every $\lambda > 0$ and $h > 0$. Moreover we have the

5.1. LEMMA. If C is closed and convex then
$$C \subseteq D((I - \lambda A^h)^{-1}), \text{ for all } \lambda, h > 0.$$

PROOF. Let be $x \in C$ and λ, $h > 0$; the equation $(I - \lambda A^h) y = x$ is equivalent to

$$y = \frac{h}{h+\lambda} x + \frac{\lambda}{h+\lambda} S_h y. \tag{5.1}$$

Define $T y = \frac{h}{h+\lambda} x + \frac{\lambda}{h+\lambda} S_h y$; then C isT-invariant and
$$\|Ty - Ty'\| \le \frac{\lambda}{h+\lambda} \|y - y'\|, \quad \forall y, y' \in C.$$
By the Banach fixed point theorem, there exists a solution $y \in C$ of the equation (5.1), hence $(I - \lambda A^h)\, y = x$, i.e. $C \subseteq D((I - \lambda A^h)^{-1})$. ■

The following result is fundamental in our considerations.

5.2. THEOREM. If S is a semigroup of nonlinear contractions on a closed convex set $C \subseteq H$, then $\overline{D(A_0)} = C$.

PROOF. First step. For $x \in C$ and λ, $h > 0$ denote by $x_{\lambda,h} = (I - \lambda A^h)^{-1} x$. Let $\varepsilon > 0$ and $0 < \delta \le 1$ be such that $\|x - S_h x\| \le \varepsilon$, $\forall h \in (0, \delta)$; we shall prove that for every $\lambda > 0$ we have:

$$\|x_{\lambda,h} - x_{\lambda,nh}\|^2 \le 2\varepsilon \|x_{\lambda,h} - x\| \text{ for } h > 0,\ n \in N \text{ with } nh \in (0, \delta) \tag{5.2}$$

and

$$\|x_{\lambda,h} - x\| \le 2\varepsilon(1+4\lambda/\delta),\ \forall h \in (0,\delta). \tag{5.3}$$

Indeed, we can suppose without restriction of the generality that $x = 0$. Then we have

$$S_h x_{\lambda,h} = (1+h/\lambda)x_{\lambda,h},\quad \forall h>0,\ \lambda>0, \tag{5.4}$$

so that for $1 \le k \le n$ we may write

$$\begin{aligned}\|x_{\lambda,h} - S_{(k-1)}(x_{\lambda,h})\|^2 &\ge \|S_h x_\lambda - S_{kh}x_{\lambda,nh})\|^2\\ &= \|(1+h/\lambda)x_{\lambda,h} - S_{kh}x_{\lambda,nh}\|\\ &\ge \|x_{\lambda,h} - S_{kh}x_{\lambda,nh})\|^2 + 2h/\lambda < x_{\lambda,h}, x_{\lambda,h} - S_{kh}x_{\lambda,nh} >.\end{aligned}$$

Further we note that we also have

$$\begin{aligned}&\|x_{\lambda,h} - S_{nh}x_{\lambda,nh})\|^2 = \|x_{\lambda,h} - (1+nh/\lambda)x_{\lambda,nh}\|^2\\ &\|x_{\lambda,h} - x_{\lambda,nh}\|^2 + 2\,nh/\lambda < x_{\lambda,nh}, x_{\lambda,nh} - x_{\lambda,h} >.\end{aligned}$$

Summing these $n + 1$ inequalities, we obtain

$$\begin{aligned}&0 \ge 2nh/\lambda\|x_{\lambda,h}\|^2 + 2nh/\lambda\|x_{\lambda,nh}\|^2 - 2h/\lambda \sum_{k=1}^{n} < x_{\lambda,h}, S_{kh}x_{\lambda,nh} >\\ &-2nh/\lambda < x_{\lambda,nh}, x_{\lambda,h} >\end{aligned}$$

Hence

$$\|x_{\lambda,h}\|^2 + \|x_{\lambda,nh}\|^2 \le \frac{1}{n}\sum_{k=1}^{n} < x_{\lambda,h}, S_{kh}x_{\lambda,nh} > + < x_{\lambda,nh}, x_{\lambda,h} >. \tag{5.5}$$

Since $kh \in (0, \delta)$, we have

$$\|S_{kh}x_{\lambda,nh}\| = \|S_{kh}x_{\lambda,nh} - S_{kh}(0)\| + \|S_{kh}(0)\| \le \|x_{\lambda,nh}\| + \varepsilon,\ 1 \le k \le n.$$

Then from (5.5) we deduce that

$$\|x_{\lambda,h}\|^2 + \|x_{\lambda,nh}\|^2 \le \|x_{\lambda,h}\|(\|x_{\lambda,nh}\| + \varepsilon) + < x_{\lambda,nh}, x_{\lambda,h} >.$$

Consequently

$$\begin{aligned}&\|x_{\lambda,h} - x_{\lambda,nh}\|^2 = 2\left(\|x_{\lambda,h}\|^2 + \|x_{\lambda,nh}\|^2\right) - \|x_{\lambda,h} + x_{\lambda,nh}\|^2\\ &\le 2\|x_{\lambda,h}\|(\|x_{\lambda,nh}\| + \varepsilon) + 2 < x_{\lambda,nh}, x_{\lambda,h} > - \|x_{\lambda,nh}\|^2 - 2 < x_{\lambda,h}, x_{\lambda,nh} > - \|x_{\lambda,h}\|^2\\ &= 2\,\varepsilon\|x_{\lambda,h}\| - (\|x_{\lambda,h}\| - \|x_{\lambda,nh}\|)^2 \le 2\,\varepsilon\|x_{\lambda,h}\|.\end{aligned}$$

In order to prove (5.3), consider first $h \in (\delta/2, \delta)$; then from (5.4) we obtain that

$$\begin{aligned}\|x_{\lambda,h}\| &= \frac{\lambda}{h+\lambda}\|S_h x_{\lambda,h}\| = \frac{\lambda}{h+\lambda}(\|S_h x_{\lambda,h} - S_h(0)\| + \|S_h(0)\|)\\ &\le \frac{\lambda}{h+\lambda}(\|x_{\lambda,h}\| + \varepsilon),\end{aligned}$$

hence

$$\|x_{\lambda,h}\| \le \frac{\lambda\varepsilon}{h} \le 2\frac{\lambda\varepsilon}{\delta}. \tag{5.6}$$

If $h \in (0, \delta/2)$, there is $n \in N$ such that $nh \in (0, \delta/2)$. We can use (5.2) and (5.6) to obtain

$$\|x_{\lambda,h} - x_{\lambda,nh}\| \le 2\varepsilon \|x_{\lambda,h}\| \le 2\varepsilon \|x_{\lambda,h} - x_{\lambda,nh}\| + 2\varepsilon \|x_{\lambda,nh}\|$$
$$\le 2\varepsilon \|x_{\lambda,h} - x_{\lambda,nh}\| + \frac{4\lambda\varepsilon^2}{\delta}.$$

Now a simple computation yields

$$\|x_{\lambda,h} - x_{\lambda},nh\| \le \varepsilon\left(1 + \left(1 + \frac{4\lambda}{\delta}\right)^{1/2}\right) \le 2\varepsilon\left(1 + \frac{2\lambda}{\delta}\right),$$

hence, by (5.6)

$$\|x_{\lambda,h}\| \le \|x_{\lambda,h} - x_{\lambda,nh}\| + \|x_{\lambda,nh}\| \le 2\varepsilon\left(1 + \frac{2\lambda}{\delta}\right) + \frac{2\lambda\varepsilon}{\delta} \le 2\varepsilon\left(1 + \frac{4\lambda}{\delta}\right).$$

Second step. We shall show that for every $\lambda > 0$, the limit $x_\lambda = \lim_{\substack{h \to 0 \\ h \in Q}} x_{\lambda,h}$ exists and moreover $x_\lambda \in D(A_o)$. Indeed, for $\lambda > 0$ fixed, by (5.3) there is $M > 0$ such that $\|x_{\lambda,h} - x\| \le M$ for $h \in (0,\delta)$. Then (5.2) yields

$$\|x_{\lambda,h} - x_{\lambda,nh}\| \le 2(2\varepsilon M)^{1/2} \quad \text{for } n \in N,\ h > 0 \text{ with } nh \in (0, \delta).$$

This last inequality implies

$$\|x_{\lambda,h} - x_{\lambda,t}\| \le (2\varepsilon M)^{1/2} \quad \text{for } h,t \in (0,\delta) \text{ with } t/h \in Q. \tag{5.7}$$

Indeed if $h \in (0,\delta)$ and $t = \frac{n}{k} h \in (0,\delta)$, then for $h' = h/k$ we have:

$$\|x_{\lambda,h} - x_{\lambda,t}\| \le \|x_{\lambda,h} - x_{\lambda,h'}\| + \|x_{\lambda,h'} - x_{\lambda,t}\|$$
$$= \|x_{\lambda,kh'} - x_{\lambda,h'}\| + \|x_{\lambda,h'} - x_{\lambda,nh'}\| \le 2(2\varepsilon M)^{1/2}.$$

In particular (5.7) is true for every $h,t \in (0,\delta) \cap Q$, consequently $\lim_{\substack{h \to 0 \\ h \in Q}} x_{\lambda,h} = x_\lambda$ exists for every $\lambda > 0$.

In order to prove that $x_\lambda \in D(A_o)$, it is sufficient to show that $x_\lambda \in D = \left\{x \in C;\ \lim_{t \to 0_+} \inf \frac{\|S_t x - x\|}{t} < +\infty\right\}$. Consider $t \in Q$; for every $n \in N$, we may write $t = nh$, $h \in Q$; then

$$\|S_t x_{\lambda,h} - x_{\lambda,h}\| \le \sum_{k=1}^{n} \|S_{(k-1)h} x_{\lambda,h} - S_{kh} x_{\lambda,h}\|$$
$$\le n \|x_{\lambda,h} - S_h x_{\lambda,h}\| = t/\lambda \|x - x_{\lambda,h}\|.$$

Letting now $h \to 0_+$ we obtain

$$\|S_t x_\lambda - x_\lambda\| \le t/\lambda \|x - x_\lambda\|, \ \forall\lambda > 0. \tag{5.8}$$

It is now clear that (5.8) is true for every $t \ge 0$, hence $x \in D$.

Third step. We shall fnally prove that for every $x \in C$, the sequence $\{x_\lambda\}_{\lambda>0}$ obtained in the second step converges to x. To this purpose, we note that letting $h \to 0_+$, $h \in \mathbb{Q}$ in (5.3), we obtain

$$\|x_\lambda - x\| \le 2\varepsilon(1 + 4\lambda/\delta), \lambda > 0.$$

In particular, if $\lambda < \delta/4$, then $\|x_\lambda - x\| < 4\varepsilon$, i.e. $x_\lambda \xrightarrow[\lambda \to 0_+]{} x$. Since $x_\lambda \in D(A_o)$, it follows that $x \in \overline{D(A_o)}$. ■

We can give now the non linear version of the generation Theorem of Hille-Yosida-Lumer-Phillips.

5.3. THEOREM. Let $S: C \to C$ be a semigroup of nonlinear contractions on the convex, closed subset C of the Hilbert space H; then there exists a unique maximal accretive set A which generates S.
Conversely, every maximal accretive mapping $A: H \to 2^H$ generates an unique semigroup of contractions on $\overline{D(A)}$.
In both cases we have $A_o = -A^\circ$.

PROOF. If A_o is the generator of S, then by the Theorem 5.2. we have that $\overline{D(A_o)} = C$. We assert that there exists a maximal accretive extension A of $-A_o$ such that $D(A) \subseteq C$. Indeed, with Zorn's Lemma we can obtain in an extension A, which is maximal accretive in the set of graphs $\subseteq C\text{x}H$. Moreover $R(A+I) = H$; indeed for $u_o \in H$ we apply Proposition 3.3., Ch. V. to $A - u_o$ and to the restriction of I to C to get $x_o \in C$, with $u_o \in Ax_o + x_o$ (see also Exercise 16, Ch. V). We shall prove that S is generated by A (in the sense of the formula 3.6).
Since A is hyperaccretive, by Corollary 4.5. there is a semigroup T defined on $\overline{D(A)} = C$ such that $T_t x = \lim_{n \to \infty} (I + t/n\, A)^{-1} x, x \in D(A), t > 0$ and $\frac{dT_t x}{dt} = -A^\circ T_t x \in -AT_t x$, a.e. $t > 0$. But we also have

$$\frac{d}{dt} S_t x = A_o S_t x \in -AS_t x, \text{ a.e. } t > 0, \ \forall x \in D(A_o) \subset D(A),$$

Then the unicity result in Proposition 4.2. yields

$$S_t x = T_t x, \quad \forall x \in \overline{D(A_o)} = \overline{D(A)} = C, \ t \ge 0.$$

Hence S is generated by A and $A_o = -A^\circ$ (see Corollary 4.5. ii).
A is unique by Theorem 4.6. Finally we note that the remained part of the theorem was proved in the more general context of the Corollary 4.5. ■

§ 6. THE INHOMOGENEOUS CASE

Consider the Problem

$$\frac{du}{dt} + Au \ni f, \text{a.e. } t \in [0,T], u(0) = x_0, f \in L^1((0,T);X),\ 0 < T < \infty. \quad (6.1)$$

6.1. Definition. A function $u:[0,T] \to X$ is called an integral solution of the Cauchy-Problem (6.1) if it is continuous, $u(0) = x_0$ and satisfies:

$$\frac{1}{2}\|u(t)-x\|^2 \le \frac{1}{2}\|u(s)-x\|^2 + \int_s^t \sup_{x^* \in J(u(\tau)-x)} < x^*, f(\tau)-u > d\tau.$$

for all $(x,u) \in G(A), 0 \le s \le t \le T$

6.2. REMARK. Let A be accretive and satisfying the condition (R); then for any $x_0 \in \overline{D(A)}$, $u(t) = S_t x_0 = \lim_{n \to \infty} J^n_{t/n} x_0$ is an integral solution of the homogeneous Problem (6.1) (i.e. $f \equiv 0$). Indeed, we only need to rerwrite the inequality (4.7) in the proof of Lemma 4.3 replacing

$$x \to S_s x_0 = u(s), t \to t-s, (x_0, u_0) \to (x,u);$$

we obtain

$$\frac{1}{2}\|u(t)-x\|^2 - \frac{1}{2}\|u(s)-x\|^2$$

$$\le \int_0^{t-s} \sup_{x^* \in J(x-u(\tau-s))} < x^*, u > d\tau = \int_s^t \sup_{x^* \in J(u(\tau)-x)} < x^*, -u > d\tau.$$

6.3. REMARK. Let A be hyperaccretive and $f \in L^1((0,T);X)$; then for $\lambda > 0$, the Problem

$$\frac{du}{dt} + A_\lambda u_\lambda = f, u_\lambda(0) = x_0, t \in [0,T] \quad (6.2)$$

has a unique solution of class $C^1([0,T]; X)$, for every $x \in X$.
This can be proved using the Banach's Fixed Point Theorem as is the proof of the Lemma 4.7.

The main result in this section is the following:

6.4. THEOREM. Let $A: X \to 2^X$ be hyperaccretive; then for every $f \in L^1((0,T);X)$ and $x_0 \in \overline{D(A)}$, there is an unique integral solution of the Problem (6.1) such that $u(t) \in \overline{D(A)}$, $\forall t \in [0,T]$.
Moreover, if u, v are two integral solutions of the equations

$$\frac{du}{dt} + Au \ni f, \frac{dv}{dt} + Av \ni g,\ u(0) = v(0) = x_0,\ f, g \in L^1((0,T);X),$$

then for $0 \le s \le t \le T$.

$$\frac{1}{2}\|u(t)-v(t)\|^2 \le \frac{1}{2}\|u(s)-v(s)\|^2 + \int_s^t \sup_{x^* \in J(u(\tau)-v(\tau))} \langle x^*, f(\tau)-g(\tau)\rangle \, d\tau \qquad (6.3)$$

PROOF. **First step.** We shall prove the existence of an integral solution u.

i) We assume first that f(t) = y_0 on [0, T]. The mapping $A_0 = A - y_0$ is hyperaccretive, hence it generates by formula (3.6) a semigroup S_0 of nonlinear contractions on $\overline{D(A)}$. By the Remark 6.2., $u_0(t) = S_0(t)x_0$ is an integral solution of the equation

$$\frac{du}{dt} + Au(t) \ni y_0, \; u(0) = x_0.$$

We note that the Yosida approximants of A_0 are $A_{0\lambda}x = A_\lambda(x+\lambda y_0) - y_0$, $x \in X$. Let $x_{0\lambda}$ be the solution of the Problem:

$$\frac{du_{0\lambda}}{dt} = -A_{0\lambda}u_{0\lambda} = -A_\lambda(u_{0\lambda}+\lambda y_0) + y_0, \; u_{0\lambda}(0) = x_0 \qquad (6.4).$$

By Propositon 4.9., the $\lim_{\lambda \to 0_+} u_{0\lambda}(t) = u_0(t)$, exists uniformly for $t \in [0, T]$. Consider also the solution u_λ given by the Remark 6.3., of equation

$$\frac{du_\lambda}{dt} = -A_\lambda u_\lambda + y_0, \; u_\lambda(0) = x_0. \qquad (6.5)$$

Then from (6.4) and (6.5) we obtain

$$\frac{d}{dt}\left(u_\lambda(t) - u_{0\lambda}(t)\right) = -A_\lambda u_\lambda + A_\lambda\left(u_{0\lambda}+\lambda y_0\right);$$

consequently for some $x^* \in J(u_\lambda(t) - v_\lambda(t) - \lambda y_0)$ we have

$$\left\|u_\lambda(t) - u_{0\lambda}(t) - \lambda y_0\right\| \frac{d}{dt}\left\|u_\lambda(t) - u_{0\lambda}(t) - \lambda y_o\right\| =$$

$$\langle x^*, -A_\lambda u_\lambda(t) + A_\lambda\left(u_{o\lambda}(t)+\lambda y_o\right)\rangle \le 0.$$

Hence $\frac{d}{dt}\left\|u_\lambda(t) - u_{0\lambda}(t) - \lambda y_0\right\| \le 0$, a.e. on [0, T].

Integrating we see that

$$\left\|u_\lambda(t) - u_{0\lambda}(t) - \lambda y_0\right\| \le \lambda\left\|y_0\right\|, \; t \in [0, T],$$

so that the limit $\lim_{\lambda \to 0_+} u_\lambda(t) = \lim_{\lambda \to 0_+} u_{0\lambda}(t) = u_0(t)$ exists uniformly on [O, T].

ii) Suppose now that f is a step function on [0, T], namely that f(t) = y_i for $t_{i-1} \le t < t_i, i = 1,2...n, t_0 = 0, t_n = T$.

Then the solutions u_λ of the equation (6.2) are given by

$$u_\lambda(t) = u_\lambda^i(t - t_{i-1}) \text{ for } t_{i-1} \le t \le t_i,$$

where $u_\lambda^i, 1 \le i \le n$, are the solutions of the problems

$$\frac{du_\lambda}{dt} = -A_\lambda u_\lambda^i + y_i, 0 \le t \le t_i - t_{i-1}, u_\lambda^i(0) = u_\lambda(t_{i-1}) = u_\lambda^{i-1}(t_i - t_{i-1}).$$

Thus, according to i) $\lim_{\lambda \to 0_+} u_\lambda(t) = u(t)$ exists uniformly on [0, T]. Since u(t) is an integral solution on every interval $[t_{i-1}, t_i]$, one can verify that u is an integral solution of (6.1) on [0, T].

iii) We shall consider finally the general case $f \in L^1((0,T); X)$. There exists a sequence of step functions $\{f_n\}_n$ such that $f_n \xrightarrow[n]{L^1} f$. Let u_λ^n be the solutions of the Cauchy Problem

$$\frac{du_\lambda^n}{dt} = -A_\lambda u_\lambda^n + f_n, \ t \in [0, T], \ u_\lambda^n(0) = x_0 \tag{6.6}$$

and let u_λ be the solution of (6.2). From (6.2) and (6.5) we have

$$\frac{d}{dt}(u_\lambda - u_\lambda^n) = -A_\lambda u_\lambda + f + A_\lambda u_\lambda^n - f_n.$$

Then for some $x^* \in J(u_\lambda - u_\lambda^n)$,

$$\|u_\lambda - u_\lambda^n\| \frac{d}{dt} \|u_\lambda - u_\lambda^n\| = < x^*, -A_\lambda u_\lambda + A_\lambda u_\lambda^n > + < x^*, f - f_n >$$
$$\le \|u_\lambda - u_\lambda^n\| \cdot \|f - f_n\|.$$

This implies

$$\frac{d}{dt}\|u_\lambda(t) - u_\lambda^n(t)\| \le \|f(t) - f_n(t)\|, \text{ a.e. on } [0, T].$$

Hence:

$$\|u_\lambda(t) - u_\lambda^n(t)\| \le \int_0^T \|f(t) - f_n(t)\| dt, \text{ on } [0, T].$$

Let $\lambda, \gamma > 0$; then

$$\|u_\lambda(t) - u_\gamma(t)\| \le \|u_\lambda(t) - u_\lambda^n(t)\| + \|u_\lambda^n(t) - u_\gamma^n(t)\| + \|u_\gamma^n(t) - u_\gamma^n(t)\|. \tag{6.7}$$

We note nowe that the first and last terms in the right side of (6.7) converge to 0 when $n \to \infty$. According to the preceeding point of the proof, for a fixed n, $u^n(t) = \lim_{\lambda \to 0} u_\lambda^n(t)$ exists uniformly on $t \in [0, T]$; thus by (6.7) $\{u_\lambda\}$ is a Cauchy sequence. We set $u(t) = \lim_{\lambda \to 0_+} u_\lambda(t)$. We shall prove now that u is an integral solution. To this end we remark that each u^n is an integral solution of (6.1); hence:

$$\|u^n(t) - x\|^2 \le \|u^n(s) - x\|^2 + 2\int_s^t \sup_{x^* \in J(u^n(\tau) - x)} < x^*, f_n(\tau) - u > d\tau \tag{6.8}$$

for all $(x, u) \in G(A), 0 \le s \le t \le T$.

It follows from (6.6) that $\lim_{n\to\infty} u^n(t) = u(t)$, uniformly on $[0, T]$.
Since $u^n_\lambda(t) \in \overline{D(A)}$ for $t \in [0, T]$, then $u^n(t) \in \overline{D(A)}$ and consequently $u(t) \in \overline{D(A)}$ for $t \in [0, T]$. Since the integrand in (6.8) is uppersemicontinuous, letting $n \to \infty$ we obtain that u is an integral solution.

Second step. Let us prove the relation (6.3). Suppose that u and v are as in the theorem; we have:

$$\|v(t)-x\|^2 \le \|v(s)-x\|^2 + 2\int_s^t \sup_{x^* \in J(v(\tau)-x)} \langle x^*, g(\tau)-u\rangle d\tau \qquad (6.9)$$

for $(x,u) \in G(A)$, $0 \le s \le t \le T$.
Consider the solutions u_λ of (6.2) and denote by $w_\lambda(t) = J_\lambda u_\lambda(t)$; then $\lim_{\lambda\to 0_+} w_\lambda(t) = u(t)$, uniformly on $[0, T]$. Take in (6.9) $x = w_\lambda(\sigma)$ and $u = f(\sigma) - \frac{du_\lambda}{d\sigma}(\sigma) = A_\lambda u_\lambda(\sigma) \in AJ_\lambda u_\lambda(\sigma) = Aw_\lambda(\sigma)$; we have

$$\|v(t)-w_\lambda(\sigma)\|^2 \le \|v(s)-w_\lambda(\sigma)\|^2 +$$

$$2\int_s^t \sup_{x^* \in J(v(\tau)-w_\lambda(\sigma))} \langle x^*, g(\tau)-f(\tau)+\frac{du_\lambda(\sigma)}{d\sigma}\rangle d\tau. \qquad (6.10)$$

On the other hand, using the fact that the integrand in (6.10) is uppersemicontinuous and the identity

$$\frac{d}{d\sigma}\|u_\lambda(\sigma)-v(\tau)\| = \sup_{x^* \in J(u_\lambda(\sigma)-v(\tau))} \langle x^*, \frac{du_\lambda(\sigma)}{d\sigma}\rangle$$

we obtain

$$\limsup_{\lambda\to 0} \int_\alpha^\beta\int_s^t \sup_{x^* \in J(v(\tau)-w(\sigma))} \langle x^*, g(\tau)-f(\sigma)+\frac{du_\lambda(\sigma)}{d\sigma}\rangle d\tau\, d\sigma \ge \qquad (6.11)$$

$$\lim_{\lambda\to 0} \int_\alpha^\beta\int_s^t \sup_{x^* \in J(v(\tau)-w(\tau))} \langle x^*. g(\tau)-f(\sigma) - \frac{1}{2}\int_s^t \left(\|u_\lambda(\beta)-v(\tau)\|^2 - \|u_\lambda(\alpha)-v(\tau)\|^2\right) d\tau$$

For $0 \le \alpha \le \beta \le T$.
For $\sigma, \tau \in [0, T]$, put

$$\varphi(\sigma,\tau) = \frac{1}{2}\|u(\sigma)-v(\tau)\|^2 \text{ and } \psi(\sigma,\tau) = \sup_{x^* \in J(u(\sigma)-v(\tau))} \langle x^*, f(\sigma)-g(\tau)\rangle.$$

Then from (6.10) and (6.11) we obtain:

$$\int_\alpha^\beta (\varphi(\sigma,t)-\varphi(\sigma,s))d\sigma \le \int_\alpha^\beta\int_s^t \psi(\sigma,\tau)d\tau d\sigma + \int_s^t (\varphi(\beta,\tau)-\varphi(\alpha,\tau))d\tau. \qquad (6.12)$$

Let φ_n, ψ_n be the regularisations of the functions φ and ψ; it can easily be proved that (6.12) is satisfied by φ_n and ψ_n, if $\frac{1}{n} \le \alpha \le \beta \le T$ and $\frac{1}{n} \le s \le t \le T$. Hence:

$$\frac{\partial \varphi_n}{\partial \tau}(\sigma,\tau) + \frac{\partial \varphi_n}{\partial \sigma}(\sigma,\tau) \le \psi_n(\sigma,\tau) \text{ for } \frac{1}{n} \le \sigma,\tau \le T$$

and this implies

$$\varphi_n(t,t) \le \varphi_n(s,s) + \int_s^t \psi_n(\tau,\tau)d\tau, \frac{1}{n} \le s \le t \le T.$$

Letting $n \to \infty$, we obtain

$$\varphi(t,t) \le \varphi(s,s) + \int_s^t \psi(\tau,\tau)d\tau, \text{ for } 0 \le s \le t \le T$$

and hence (6.3) is proved.

Third step. Let us prove the unicity of the solution; suppose that there are two integral solutions u and v of (6.1) corresponding to the same initial value x_o; then from (6.3) it follows that the function $t \to \|u(t) - v(t)\|$ is nonincreasing on [0, T]; since u(0) = v(0), then $\forall t \in [0, t]$. ■

6.5. THEOREM. Let X be reflexive and $A: X \to 2^X$ hyperaccretive; then for every $x_o \in D(A)$ and $f \in H^{1,1}((0,T),X)$ the Problem (6.1) has a unique solution $u \in H^{1,\infty}((0,T);X)$.

PROOF. By Theorem 6.4. there is a unique continuous integral solution of (6.1); hence

$$\frac{1}{2}\|u(t) - x\|^2 \le \frac{1}{2}\|u(s) - x\|^2 + \int_s^t \sup_{x^* \in J(u(\tau)-x)} <x^*, f(\tau) - u> d\tau$$

$\forall (x,u) \in G(A), 0 \le s \le t \le T$. In particular, for s = 0, x = x_0 and $u \in Ax_0$ arbitrary, we have:

$$\frac{1}{2}\|u(h) - x_0\|^2 \le \int_0^h \|f(\tau) - u\| \|u(\tau) - x_0\| d\tau, \ 0 < h < T. \qquad (6.12)$$

Replace in (6.3), f by f(t + h), g by f(t), u by u(t + h), v by u and take s = 0; we obtain:

$$\|u(t+h - u(t)\|^2 \le \|u(h) - u(0)\|^2 +$$
$$+2\int_0^t \|u(\tau+h) - u(\tau)\| \|f(\tau+h) - f(\tau)\| d\tau. \qquad (6.13)$$

We shall solve the integral inequalities (6.12) and (6.13). From (6.12) we have

$$\frac{d}{dh}\|u(h) - x_0\| \le \|f(h) - u\|, \text{ a.e. on } [0, T], \ \forall u \in Ax_0,$$

hence:

$$\|u(h)-x_0\| \le \int_0^h (\|f(\tau)\|+|Ax_0|)d\tau. \tag{6.14}$$

Analogously, (6.13) yields

$$\frac{d}{dt}\|u(t+h)-u(t)\| \le \|f(t+h)-f(t)\| \text{ a.e. on } [0,T]$$

hence

$$\|u(t+h)-u(t)\| \le \|u(h)-x_0\| + \int_0^t \|f(\tau+h)-f(\tau)\|d\tau \tag{6.15}$$

$$\le \int_0^h (\|f(\tau)\|+|Ax_0|)d\tau + \int_0^t \|f(\tau+h)-f(\tau)\|d\tau.$$

Since u is absolutely continuous and X is reflexive, du/dt exists a.e. on (0, T) so that from (6.14) and (6.15) we obtain

$$\|du/_{dt}(t)\| \le \|f(0)\| + |Ax_0| + \int_0^T \|df/_{d\tau}(\tau)\|d\tau, \text{ a.e. on } (0,T).$$

Hence $du/_{dt} \in L^\infty(0,T);X)$, i.e. $u \in H^{1,1}((0,T);X)$.

To end the proof we only have to verify that u is a solution of the Problem (6.1). Let t_0 be a differentiability point of u and $(x,u_0) \in G(A)$; we have:

$$\tfrac{1}{2}\left(\|u(t)-x\|^2 - \|u(t_0)-x\|^2\right) \le \int_{t_0}^t \sup_{x^* \in J(u(\tau)-x)} \langle x^*, f(\tau)-u_o \rangle d\tau.$$

Then, for each $x^* \in \partial\left(\tfrac{1}{2}\|u(t_0)-x\|^2\right) = J(u(t_0)-x)$ it follows that

$$\langle x^*, u(t)-u(t_o) \rangle \le \int_{t_o}^t \sup_{x^* \in J(u(\tau)-x)} \langle x^*, f(\tau)-u_o \rangle d\tau.$$

Consequently

$$\sup_{x^* \in J(u(t_0)-x)} \langle x^*, \frac{u(t)-u(t_o)}{t-t_o} \rangle = \frac{1}{t-t_0}\int_{t_0}^t \sup_{x^* \in J(u(\tau)-x)} \langle x^*, f(\tau)-u_o \rangle d\tau.$$

Letting $t \to t_0$, we obtain:

$$\sup_{x^* \in J(u(t_0)-x)} \langle x^*, du/_{dt}(t_o) \rangle \le \sup_{x^* \in J(u(t_0)-x)} \langle x^*, f(t_o)-u_o \rangle.$$

Then there exists $x^* \in J(u(t_0)-x)$ such that

$$\langle x^*, \frac{du}{dt}(t_o) - f(t_o) + u_o \rangle \le 0. \tag{6.16}$$

Since A is hyperaccretive, we can find $(x,u_0) \in G(A)$ such that

$$x + u_0 = -\frac{du}{dt}(t_0) + f(t_0) + u(t_0).$$

Taking in (6.11) (x,u_0) as above, we obtain:

$$\langle x^*, u(t_o)-x \rangle = \|u(t_o)-x\|^2 \le 0.$$

Hence $u(t_o)=x\in D(A)$ and $-\frac{du}{dt}(t_o)+f(t)=u_o\in Ax=Au(t_o)$. ■

6.6. EXAMPLE. We shall present the following typical initial boundary valued Problem.

Consider the operator $A: X=H^{1,2}(\Omega)\to H^{1,2}(\Omega)^*\approx H^{1,2}(\Omega)$ from Example 4, 2.8, Ch. V, with adquate condition on the functions a_i, $0\le i\le n$, in order that A be maximal monotone then A is hyperaccretive and we cna apply the above Theorem 6.5 to obtain

Let be $u_o\in H^{1,2}(\Omega)=X$ and $f\in H^{1,1}((0,T);\ X)$; then the problem:

$$\begin{cases}\frac{\partial u}{\partial t}+a_o(x,u,\ \text{grad}u)-\sum_i D_i a_i(x,u,\ \text{grad}u\ u)=f,\ \text{a.e. } t\in(0,T)\\ u(0,x)=u_o(x)\end{cases}.$$

has a unique solution $u\in H^{1,\infty}(0,T);X)$.

EXERCISES

1. Let $A: X\to 2^X$ be an accretive mapping and $Y=L^p([0,1];X)$; define $\mathcal{A}: Y\to 2^Y$ by $G(\mathcal{A})=\{(x,u)\in Y\times Y;\ (x(t),u(t))\in G(A),\ \text{a.e.}\}$
Prove that $\mathcal{A}$ is accretive.

2. Let f: R→ R be non decreasing; show that the operator $\mathcal{A}: L^p\to L^p$, $1<p<+\infty$, defined by
$G(\mathcal{A})=\{(x,u)\in L^p[0,1]\times L^p[0,1];\ (x(t),u(t))\in G(f)\ \text{a.e.}\}$ is T - accretive:

   Hint. Prove that an operator $A: L^p[0,1]\to L^p[0,1], 1<p<+\infty$ is accretive if and only if and only if $\forall(x,u),(y,v)\in G(A)$
$\int_{E_{x,y}}[u(t)-v(t)][x(t)-y(t)]^{p-1}dt\ge 0$, where $E_{x,y}=\{t;x(t)\ge y(t)\}$.

3. Let X be such that X* is uniformly convex, and
i) let A: $X\to X$ be hemicontinuous and f-maximal accretive (i.e. A has no single valued accretive extension) with D(A) open or D(A) convex and dense in X; then A is maximal accretive.
ii) Let A: $X\to X$ be linear, accretive and such that it has no linear accretive extension; then A is f-maximal accretive and if $\overline{D(A)}=X$, then A is maximal accretive.

Solution. i) Consider $B: X \to 2^X$, B accretive and $B \supseteq A$; then $D(B) = D(A)$. Let $u_0 \in Bx_0$; then for $x_t = x_0 + t(u_0 - Ax_0) \in D(A)$ (t small enough) in the case D(A) is open, respectively for $x_t = tx + (1-t)x_0 \in D(A) (t \in (0,1), x \in D(A))$ in the case D(A) is convex and dense, we have:

$$< J(x_t - x_0), Ax_t - u_0 > \geq 0.$$

Now the result is a consequence of the hemicontinuity of A.
ii) Suppose that A is not f-maximal accretive; then there exist $x_0 \notin D(A)$ and $u_0 \in X$ so that $< J(x_0 - x)m\ u_0 - Ax > \geq 0,\ \forall x \in D(A)$. We extend A to the linear subspace X_0 generated by D(A) and $\{x_0\}$ by $\tilde{A}(x + tx_0) = Ax + tu_0, x \in D(A), t \in R$. Then $\tilde{A}$ is accretive and this is a contradiction. The last part of ii) is a direct consequence of i).

4. Let X = C [0, 1] and $F(\xi) = \begin{cases} \xi, & \xi > 0 \\ 2\xi & \xi \leq 0 \end{cases}$; then $(S_t f)(\xi) = F(t + F^{-1}(f(\xi))), \xi \in [0,1], f \in X$, defines a semigroup of non-linear contractions on X with $\overline{D(A_s)} \neq X$.

Hint. For h > 0 sufficiently small we have

$$\frac{(S_h f)(\xi) - f(\xi)}{h} = \begin{cases} 1 & \text{if} \quad f(\xi) > 0 \\ 2 & \text{if} \quad f(\xi) \leq 0 \end{cases}.$$

5. Let be $X = L^1[0,1]$ and $f = \chi[0,t], t \in [0,1]$, where $\chi[0, 1]$ is the characteristic function of the interval [0, 1]; then f is absolutely continuous but is not differentiable at any point of (0, 1).

Hint. Suppose that f is differentiable at $t_0 \in (0,1)$; then for every $g \in L^\infty[0,1]$, the function $[0,t] \ni t \to \int_0^t g(s)ds$ is differentiable at t_0; in the case

$$g(s) = \begin{cases} 1 & \text{if} \quad s > t_0 \\ -1 & \text{if} \quad s < t_0 \end{cases},$$

this yields a contradiction.

6. Let X be a real Banach lattice and S: C → C a semigroup of T-contractions on C; prove that A_w and A_s are T-accretive.

Hint. See the proof of ii) in Proposition 1.23.

7. Let X be uniformly convex with an uniformly convex dual, $C \subseteq X$ closed and S: C→ C be a semigroup of contractions on C with the property:

$$x_n \in D(A), \|x_n\| \xrightarrow[n]{} +\infty \Rightarrow \|Ax_n\| \xrightarrow[n]{} +\infty$$

Prove that S is bounded on D(A).

Solution. Suppose that there $x_o \in D(A)$ such that $\|S_{t_n} x_o\| \geq n$, for a sequence $\{t_n\} \subseteq R_+$. Since $R_+ \ni t \to S_t x_o$ is continuous, then $t_n \xrightarrow[n]{} +\infty$. The hypothesis implies that $\|AS_{t_n} x\| \xrightarrow[n]{} +\infty$; since by ii) in Theorem 2.11, $t \to \|AS_{t_n} x\|$ is non increasing, this is a impossible.

8. Let $\varepsilon_n > 0, \varepsilon_n \xrightarrow[n]{} 0$ and $k_n \in N$, with $\varepsilon_n \cdot k_n \xrightarrow[n]{} t$; let $A: X \to 2^X$ be a mapping which satisfies the condition (R); then

$$\lim_{n \to \infty} J_{\varepsilon_n}^{k_n} x = \lim_{n \to \infty} J_{t/n}^{n} x, \ \forall x \in D(A).$$

Hint. Use twice Proposition 3.2 (iii) to prove that the sequences $\left\{J_{\varepsilon_n}^{k_n} x\right\}_n$ and $\left\{J_{\varepsilon_n}^{k_n} x - J_{t/n}^{n} x\right\}_n$ are Cauchy, respectively zero-sequences.

9. Let H be a Hilbert space and $T: C \subseteq H \to H$ be a nonlinear contraction; there exists a contractive extension of T to all H.

Hint. Use Kirzbaum's theorem (see Schoenberg [1]; also Cioranescu [1]).

10. Let be $C = \{f \in C[0,2]; 0 \leq f(s) \leq s, s \in [0,1]\}$ and define $(S_t f)(s) = [t + f(s)] \wedge s, t \in R_+$. Prove that the semigroup S is generated (in the sense of the exponential formula) by the mapping

$$G(A) = \bigcup_{f \in C, \lambda > 0} \left\{(\lambda + f) \wedge id, \lambda^{-1}[f - (\lambda + f) \wedge id]\right\}$$

(here id (s) = s, $\forall s \in [0,1]$).

Solution. See Crandall - Liggett [1]

11. Let $A:\ H \to H$ be linear and with $\overline{D(A)} = H$; the following conditions on A are equivalent:
i) A is the generator indinitesimal of a semigroup of linear contractions on H.
ii) A is dissipative and has no linear dissipative extention.
iii) A is disipative and $R(I - A) = H$.

Hint. Use Lummer-Phillips' results [1], to prove that i) $\Rightarrow$ ii) and exercise 3, ii) to prove ii) $\Rightarrow$ iii).

12. Let X be reflexive, with the property (I) and an OP-scheme $\{X_n, P_n\}$; let $T:\ X \to X$ be a demicontinuous and accretive mapping with the property that $< Jx, Tx > \geq \alpha(\|x\|)\ \|Jx\|,\ x \in X$, where J is a duality mapping on X and $\alpha:\ R_+ \to R_+$ is such that $\lim_{t \to \infty} \alpha(t) = \infty$; then $T(X) = X$.

Solution. (see Browder De Figueiredo [1]). We need some preparative results.
i) If X_o is a finite dimensional Banach space and $T:\ X_o \to X_o$ is continuous so that

$$< Jx, Tx > \geq 0 \text{ for } \|x\| = r, \qquad (*)$$

then there exists $x_o \in S_r$ so that $Tx_o = 0$.
Indeed, we can prove that $U = I - T$ has fixed point applying the Brouwer's fixed point theorem to te map $V: S_r \to S_r$ defined as, $Vx = Ux$ if $Ux \in S_r$ and $Vx = r\ \ \|Ux\|$ if $Ux \notin S_r$.
ii) Let X be as in the exercise 12 and $T:\ X \to X$ accretive, demicontinuous and such that (*) is satisfied; then $T_n = P_n T$ has a zero in $X_n \cap S_r$. Indeed, using the Proposition 5.17 in Ch. II we can prove that $J_n = P_n^* J$ is a duality mapping on X_n; moreover, for $x \in X_n, \|x\| = r$ we have

$$< J_n x, T_n x > = < Jx, P_n Tx > = < Jx, Tx > \geq 0.$$

Now the result is a consequence of i).
iii) If the conditions in the statement ii) are satisfied, then there exists $x_o \in S_r$ such that $Tx_o = 0$. Indeed, consider $x_n \in S_r \cap X_n$ such that $T_n x_n = 0$; put $G_n = \{x_n, x_{n+1}, \ldots\}$ and let be $x_o \in \bigcap_{n \in N} \overline{G_n}^w$.
We shall prove that $Tx_o = 0$. For n fixed, $y \in X_n$ and $k \geq n$ we have $< J(y - x_k), Tx_k > = < P_k^* J(y - x_k), Tx_k > = < J(y - x_k), T_k x_k > = 0$.
Since $< J(y - x_k), Ty - Tx_k > \geq 0$, it follows that $< J(y - x_k), Ty > \geq 0$.
Let be $f(x) = < J(y - x), Ty >,\ x \in S_r$; then f is weakly continuous

and nonnegative on G_n; hence $f(x_0) = < J(y - x_0), Ty > \geq 0$, $\forall y \in X_n, n \geq 1$. This yields

$$< J(P_n x - x_0), T P_n x > \geq 0, \ \forall x \in X, n \geq 1,$$

and finally

$$< J(x - x_0), T_x > \geq 0, \ \forall x \in X. \qquad (**)$$

Let $x_t = x_0 + tx, t > 0$; from (**) we obtain $< Jx, Tx_t \geq 0$ and letting $t \to 0$, it follows that $< Jx, Tx_0 > \geq 0, \ \forall x \in X$. Hence $< Jx, Tx_0 > = 0$, $\forall x \in X$ and since J is surjeetive it follows that $Tx_0 = 0$.

iv) We can now give the proof of the exercise 12.

For $y \in X$ fixed, consider $T_y x = Tx - y, x \in X$ and r so that $\alpha(r) \geq \|y\|$; then for $x \in X$ with $\|x\| = r$ we have

$$< Jx, T_y x > = < Jx, Tx > - < Jx, y >$$
$$\geq \alpha(r)\|Jx\| - \|y\| \|Jx\| = (\alpha(r) - \|y\|)\|Jx\| \geq 0.$$

We apply iii) to obtain that T_y has a zero in S_r, i.e. there exists x_0 with $Tx_0 = y$.

13. Let be X a Banach space with a projection scheme of type $(\pi)_1$ and with X^* uniformly convex and let $T: X \to X$ be demicontinuous and strongly accretive, i.e. $T - \lambda I$ is accretive for some $\lambda > 0$. Then T is A-prper and surjective.

Solution. See J. R. L Webb [1].

14. The following assertions are equivalent for any Banach space X:
i) X is smooth;
ii) for any maximal accretive operator $A: X \to 2^X$, the set Ax is convex, $\forall x \in D(A)$.

Solution. See L. Vesely [1].

15. The following assertions are equivalent for any Banach space X
i) $\dim Jx < \infty$ for any $x \in X$, J beeing the normalited duality mapping;
ii) for any maximal accretive operator $A: X \to 2^X$, the set Ax is weak closed, $\forall x \in D(A)$.

Solution. See L. Vesely [1].
We note that it would be interesting to characterize Banach space with the property that for any maximal accretive $A: X \to 2^X$ the set Ax is convex (resp. weak closed) for $x \in \text{int } D(A)$. Does a general Banach space satisfy this property?

BIBLIOGRAPHICAL COMMENTS:

§ 1. The study of nonlinear accretive mappings developed in connection with the theory of semigroups of nonlinear contractions in works of Browder [6], [8] - [11], Browder - De Figueiredo [1] (who called them J-monotone mappings) and Kato [4] (who used the term of m-accretivity for hyper-accretivity). Minty [2] proved that in Hilbert spaces the hyperaccretivity coincides with the maximal accretivity. The convexity of D(A) in Theorem 1.5. (iv) was proved by Minty [1] in finite dimensional case and by Komura [1] in the Hilbert case.
The results in this section can be found in Crandall-Pazy [1], [2], Kato [4] and in the monograph of Barbu [2]. Concerning the local boundedness of a maximal accretive map, Kato-Hess-Fitzpatrick [1] gave an affirmative answer in the case X^* is uniformly convex and Prüss [1] when X is uniformly convex. Various other properties of the accretive mappings were obtained by Cernes [1], Corduneanu [1], Da Prato [1], [2], Deimling [1], Webb, J.R.L. [1].
The last part concearning the T-accretivity is due to Calvert [1] [2] (see also Sato [1], [2]).

§ 2. The basic result on the differentiability of absolutely continuous X-valued function presented in Theorem 2.1. is due to Komura [1].
The study of the semigroups of nonlinear contractions begun with works of Neuberger [1] and Komura [1], [2] and was continued by Brezis-Pazy [1] [2], Browder [9], [10], [15] Crandall-Liggett [1], Crandall-Pazy [1], [2], [3] Crandall [1], Dorroh [1] to [4], Ianelli [1], [2], Kato [3], [4], [5], Mermin [1], [2], Miyadera [1] to [4], Oharu [1]. [2], Segal [1], Watanabe [1], [2], Webb [1], [2], [4].
The results in the first part of this section are taken from Crandall-Pazy [1]. Theorem 2.13. is due to Browder [3] and Kirk [1]; more results on fixed point theorems for nonexpansive maps can be found in Deimling [2].

§ 3. The generation Theorem 3.4. was stated by Crandall-Liggett [1]; the Corollary 3.7. can be found in Picard [1]; for more results see also Amann [1], Konishi [1], [2], Oharu [1], [2], Segal [1], Watanabe, J. [1], [2]. We mention that other approximation results for nonlinear semigroups, such as of Trotter-Chernoff type were obtained by Brezis-Pazy [3], Miyadera [4], [5], Miyadera-Oharu [1].

§ 4. Evolution equations of the form (4.1) were studied by Barbu [1], Brezis-Pazy [1], Browder [1], [10], [15], [16], Crandall-Pazy [1], [3], Haraux [1], Kato [3], [4], Martin [1], [2], Watanabe [1], Webb [3], [5].
Theorem 4.7. is due Crandall-Pazy [2]; we note that this result is true only under the hypothesis that X^* is uniformly convex (see Brezis [1]).

The existence result of Peano type in Proposition 4.12. is due to Martin [1]; for particular cases see Browder [15], Webb [3]. The hyperaccretivity result in Theorem 4.13. was obtained by Kato [4]; we mention that Barbu-Cellina [1] generalized it.

§ 5. Theorem 5.2. is due to Komura [2] but the proof given here is from Kato [5]. Theorem 5.3. was independently obtained by Dorroh [4] and Crandall-Pazy [3]. For an extended study of the semigoups of nonlinear contractions in Hilbert spaces we sent to Brezis [3].

§ 6. Theorem 6.4 is due to Benilan [1], [2], [3], to whom the notion of integral solution is due. For a complete treatment of the inhomgeneous Problem (6.1) we recommend Barbu's book [2]; see also Pavel [1] and Zeidler [1], [2]

Finally for those who are interested in the theory of linear semigroups of operators we recommend the monographs of Pazy [1] and Goldstine [1].

REFERENCES

Adams, R. A.

1. Sobolev spaces, Academic Press, New York, (1975).

Altman, M.

1. A unified theory of nonlinear operators and evolution equation, Dekker, New-Yorks (1986).

Amann, H.

1. Nonlinear operators in ordered Banach spaces and some applications to nonlinear boundary-value problems, Lectures Notes Math, 543, Springer Verlag, (1976), 1-55.

2. Semigroups and nonlinear evolution equations, Linear Algebra Appl. 84, (1986), 3-32.

Amann, H. - Weiss, S.

1. On the uniqueness of the topological degree, Math. Z. 130 (1973), 39-54.

Amir, D. - Lindenstrauss, J.

1. The structure of weakly compact sets in Banach spaces, Ann. of Math. (2) 88 (1968), 35-46.

Asplund, E.

1. Averaged norms, Israel J. Math. 5 (1967), 227-233.
2. Positivity of duality mappings, Bull. Amer. Math. Soc. 73 (1967).
3. Fréchet differentiability of convex functions, Acta Math. 121 (1968), 31-47.

Asplund, E. - Rockafellar, R. T.

1. Gradients of convex functions, Trans. Amer. Math. Soc., 139, (1969), 443-467.

Aubin, J. P.

1. Approximation des espaces des distribution set des operateurs differentiels, Bull. Soc. Math. France, Memoire,12, (1967), 139.

Aubin, J. P. - Cellina, A.

1. Differential Inclusions: Set-Valued Maps and Viability Theory, Springer-Verlag, Berlin, Meidelberg, New York, Toky, (1986)

Barbu, V.

1. Dissipative sets and nonlinear pertubated equations in Banach spaces. Ann. Sc. Norm. Pisa, XXVI (1972), 365-390.

2. Nonlinear Semi-groups and differential equations, Ed. Acad. Rom.-Noordhoff Inter. Pub. (1976).

Barbu, V. - Cellina, A.
1. On the surjectivity of multivalued dissipative mappings. Boll. U.M.I. (3) (1970), 817-826.

Barbu, V. - Precupanu, I.
1. Convexity and Optimizations in Banach spaces, Ed. Acad. Rom.-Sijthoff & Noordhoff Intern. Publ. (1978).

Beauzamy, B.
1. Introduction to Banach Spaces and their Geometry, North Holland, Notas de Mat. (1982).

Benilan Ph.
1. Solutions faibles d'équations d'évolution dans un espace réflexif, Séminaire Deny sur les semigroupes nonlinéares, Orsay (1970-1971).
2. Equations d'évolution dans un espace de Banach quelconque et applications, These, Orsay (1972).
3. Solution integrales d'équations d'évolutions, C. R. Acad. Sci. Paris, 274 (1972), 47-50.

Berger, M.S.
1. Nonlinearity and Functional Analysis, Acad. Press. (1977).

Beurling, A-Livingston, A.E.
1. A theorem on duality mappings in Banch spaces, Ark. Math., 4 (1962), 405-411.

Bishop, E. - Phelps, R.
1. The support functionals of a convex set. Proc. Symp. on Pure Math., 7 (1963), 27-35.

Bonic, R. - Reis, Fl.
1. A caracterization of Hilbert spaces, Annais da Acad. Brasil de Ci, 383, (1966), p 239-241.

Borisovish, Yu. G. - Gel'man, B.D. - Nyshkis, A.D. - Dbukhovskü, V.V.
1. Multivalued mappings - J. of Soviet Math., Vol. 24, No. 6, (1984), 719-739.

Bourbaki, N.,
1. Fonctions d'une variable réele. Chap. 1, 2, 3, IX, livre IV, Hermann, Paris.

Brezis, H.
1. On a Problem of Kato, Comm. Pure Appl. Math., 24 (1971), 1-6.
2. Nonlinear perturbations of monotone operators, Techn. Report,25 Univ. of Kansas (1972).
3. Opérateurs maximaux monotone et semigroupes de contractions dans les espaces de Hilbert, Math. Studies, 5, North Holland, (1973).
4. Quelques proprietés des opérateurs monotones et les semigroupes non linéaires, Lect. Notes Math, 543, SpringerVer., (1976), 56-82.

Brezis, H. - Browder F.
1. Nonlinear integral equations and systems of Hammerstein type, Bull. A.M.S., 81, (1975), 73-78.

Brezis, H. - Crandall M. - Pazy A.
1. Perturbations of nonlinear maximal monotone sets. Comm. Pure Appl. Math., 23 (1970), 367-383.

Brezis, H. - Pazy A.,
1. Accretive sets and differential equations in Banach spaces, Israel J. of Math., 8 (1970), 367-383.
2. Semigroups of nonlinear contractions on convex sets, J. Funct. Anal., 6, 2, (1970), 237-280.
3. Convergence and approximation of nonlinear semigroups, J. Funct. Anal. 1. (1972)

Brøndsted A.,
1. Conjugate convex functions in topological vector spaces, Math. Phys. Medd. Dansk Vid. Selsk, 34 (1964), p. 2-27.

Brøndsted, A. - Rockafeller R. T.
1. On the subdifferentiability of convex functions, Proc. Amer. Math Soc., 16 (1965), 605-611.

Browder F.
1. Nonlinear equations of evolution, Ann. of Math., 80 (1964), 485-523.
2. Continuity properties of monotone nonlinear operators in Banach spaces, Bull. A.M.S. 70, (1964), 551-553.
3. Nonexpansive non linear operators in Banach spaces, Proc. Mat. Acad. Sci. USA, 54, (1965), 1041-1066.
4. Multivalued and monotone non-linear mappings and duality mapping in Banach spaces, Trans. Amer., Math. Soc., 118, (1965), 338-361.
5. Problemes non-linéaires, Univ. of Montreal, Press (1966).
6. Fixed points theorems for non-linear semicontractive mappings in Banach spaces, Arch. Mech. and Anal., 21 (1966), 259-270.

7. On a theorem of Beurling and Livingston, Canadian J. Math., 17 (1967), 367-372.
8. Nonlinear accretive operators in Banach spaces, Bull. Amer. Math. Soc. 73 (1967), 470-476.
9. Nonlinear mappings of non-expansive and accretive type in Banach spaces, Bull. Amer. Math. Soc. 73, (1967), 875-882.
10. Nonlinear equations of evolution and nonlinear accretive operators in Banach spaces, Bull. Amer. Math. Soc., 73 (1967), 867-874.
11. Nonlinear monotone and accretive operators in Banach spaces, Proc. Nat Acad. Sci. USA, 61, (1968), 388-393.
12. Nonlinear maximal monotone operators in Banach spaces, Math. Ann., 175, (1968), 89-113.
13. Nonlinear eigenvalue problems and Galerkin approximations, Bull. Amer. Math. Soc., 74, (1968), 651-656.
14. Nonlinear variational inequalities and maximal monotone mappings in Banach spaces, Math. Ann., 73 (1969), 213-232.
15. Nonlinear operators and nonlinear equations of evolution in Banach spaces, Proc. Symp. Nonlin. Funct. Anal. Chicago, Amer. Math. Soc., 18, II, (1972).
16. Nonlinear operators and nonlinear equations of evolution in Banach spaces, Proc. Symp. Pure Math. Vol. 18/2, AMS, (1976).

Browder, F. - De Figuieredo D.G.
1. Monotone nonlinear operators in Banach spaces, Koukl. Nederl. Akad. Wetersch, 69 (1966), 412-420.

Browder, F. - Petryshyn, W.
1. The topological degree and Galerkin approximation for non-compact operators in Banach spaces, Bull. Amer. Math. Soc., 74 (1968), 641-646.
2. Approximation methods and the generalised topological degree for mappings in Banach spaces, J. Func. Anal., 3 (1969), 217-245.

Bru, B. - Heinich, H.
1. Applications de dualité dans les espaces de Köthe, Studia Mat. t. XCIII (1989), 47-69.

Calvert, B.
1. Nonlinear evolution equations in Banach lattices, Bull. Amer. Math. Soc., 76 (1970), 845-950.
2. Nonlinear equations of evolution, Pacific J. Math., 39 (1971), 293-350.
3. Maximal accretive is not m-accretive. Boll. U.M.I., 6 (1970), 1042-1044.

Cernes, A.
1. Ensembles maximaux accrétifs et m-accrétifs, Israel J. Math. 19 (1974), 335-348.

Cioranescu, I.
1. Aplicatii de dualitate în alaliza functională neliniară, Ed. Acad. Rom. (1974).

Clarkson, J. A.
1. Uniformly convex spaces, Trans. Amer. Math. Soc., 40, (1936), 396-414.

Corduneanu, A.
1. Some remarks on the sum of two m-dissipative mappings, Rev. Roum. Math., 20 (1975), 411-415.

Crandall, M.
1. Semigroups of nonlinear transformations in Banach spaces, Contributions to Nonlinear Funct. Anal., Ed. Zarantonello, Acad. Press, 1971.

Crandall, M. - Liggett T.
1. Generation of semigroups of nonlinear transformations on genral Banach spaces, Amer. J. Math. 93 (1971), 265-298.

Crandall, M. - Pazy, A.
1. Non linear semigroups of contractions and dissipative sets, J. Funct. Anal., 3 (1969), 376-418.
2. On accretive sets in Banach spaces, J. Funct. Anal., 5 (1970), 204-217.
3. Nonlinear evolutions in Banach spaces, Israel J. Math, 11 (1972), 57-94.

Cristescu, R.
1. Toplogical vector spaces, Ed. Acad. Rom.-Noordhoff Intern. Publ. (1977).

Cudia, D.F.
1. On the localisation and direction of lisation of uniform convexity, Bull. Amer. Math. Soc., 69 (1963).
2. The geometry of Banach spaces, Smoothness, Amer. Math. Soc., 110 (1964), 284-314.

Da Prato, G.
1. Somme d'applications nonlinéaires, Sump. Math. Acad. Press (1971), 233-268.
2. Applications accrétives dans des espaces d'opérateurs hermitiens. Atti Lincei, 4 LII (1972), 485-498.

Day, M.M.
1. Reflexive Banach spaces not isomorphic to uniformly convex spaces, Bull. Amer. Math. Soc., 47 (1941), 313-317.
2. Uniform convexity, III, Bull. Amer. Math. Soc., 49 (1943), 745-750.
3. Uniform convexity in factor and conjugate spaces, Ann. Math. 2, 45 (1944).
4. Strict convexity and smoothness of normed spaces, Trans. Amer. Math. Soc., 78 (1955), 516-528.

Debrunner, H. - Flor, P.
1. Ein Erweiterungssatz für monotone Mengen, Arch. Math. 15, (1964), 445-447.

De Figueiredo, G.
1. Topic in nonlinear functional Analysis, Univ. of Maryland, (1967).
2. The Ekland variational principle with applications and detours, Tata Inst. for fundamental research, Springer Verlag, (1989).

Deimling, K.
1. Zeros of accretive operators, Manuscripta Math. 13, (1974), 365-374.
2. Nonlinear functional analysis, Springer Verlag, (1985).

Diestel, J.
1. Geometry of Banach spaces-Selected topics, Lect. Notes in Math. 485, Springer Verlag (1975).

Ditzian, Z. and Totik, V.
1. Moduli of smoothness, Springer-Verlag, New York, (1987).

Dorroh, I. R.
1. Semigroups of nonlinear transformations, Michigan Math. J. 12 (1965), 317-320.
2. Semigroups of nonlinear transformations, Notices Amer. Math. Soc., 15 (1968), 128.
3. Some classes of semigroups of nonlinear transformations and their generators, J. Math. Soc. Japa, 20, 3 (1968), 437-455.
4. A non linear Hille-Yosida-Phillips Theorem, J. Func. Anal., 5 (1969), 345-353.

Dunford, N. - Schwartz, J.
1. Linear operators, Part I, New York, 1958.

Ekeland, I. - Temam, R.
1. Convex Analysis and Variational Problems, North Holland. (1958).

Enflo, P.
1. Banach spaces which can be given an equivalent uniformily convex norm, Israel J. of Math., 13, 3-4 (1972), 281-289.

Fan, K. - Glicksberg I.
1. Some geometric properties of spheres in a normed linear space, Duke Math. J., 25 (1958), 553-568.

Fígiel, T.
1. On the moduli of convexity and smoothness, Studia. Math. 26, (1976), 121-155.

Fitzpatrick, P. M.
1. Surjectivity results for nonlinear mappings from a Banach space to its dual, Math. Am., 204, (1973), 177-188.
2. Continuity of nonlinear monotone operators, Proc. A.M.S. 62 (1977), 111-116.

Fucik, S. - Necas, J. - Soucek, J. - Soucek. S.
1. Spectral analysis of nonlinear operators, Lect. Notes Math., 346, (1973), Springer Verlag.

Giles, J.R.
1. Convex Analysis with application in differentiation of convex functions, Research Notes Math 58, (1982), Pitmann-Boston-London.

Godini G.
1. Contributii la studiul geometriei spatiilor Banach (thesis), St. Cerc., Mat., 24, 2 (1972), 193-238.

Goldstein, J.A.
1. Semigroups of Linear Operators and applications, Oxford University Press, 1985.

Gossez, J.P.
1. A note on multivalued monotone operators. Mich. Math. J. 17, 4 (1970).
2. Ensembles virtuellement convexes et opérateurs monotones, Bull. Sci. Math. Paris, 2e série, 94 (1970), 73- 80.
3. Opérateurs monotones non-linéaires dans les espaces de Banach non-réflexifs, J. Math. Anal. and Appl. 34, 2 (1971), 371-395.
4. On the range of a coercive maximal monotone operator in a nonreflexive Banach space, Proc. Amer. Math. Soc. 35, (1972), 88-92.

Gossez, J.P. - Lami Dozo, E.
1. Some geometric properties related to the fixed point theory for nonexpansive mappins, Pacific J. Math., 40 (1972). 565-573.

Gromes, W.
1. Ein einfacher Beweis des Satzes von Borsuk, Math, Z., 178 (1981), 339-400.

Haraux, A.
1. Nonlinear evolution equations, Lecture Notes in Math., 841, Springer Verlag, (1981).

Heinz E.
1. An elementary analytic theory of the degree of mappings in n-dimensional spaces, J. Math. Mech., 8 (1959).

Hille, E. - Phillips, R.
1. Functional analysis and semigroups, Amer. Math. Soc. Coll. Publ. 39 (1957).

Hirzebruch, F. - Scharlau W.
1. Einführung in die Functional analysis, Hochschultaschenbücher Band 296, (1971).

Holmes R.
1. A course on optimisation, Springer Verlag, 257, (1972).
2. Geometric functional Analysis and its applications, G.M.T. 24, Springer Verlag (1975).

Ianelli, M.
1. Nonlinear semigroups on cones on a nonreflexive Banach space, Boll. Un. Mat. Ital. 3 (1970), 412-419.
2. Opérateurs derivables et semi-groupes non-linéaires non-contractifs, J. Math. Anal. App, 46, (1974), 700-724.

Istratescu, V.
1. Inner Product Structures, Kluwer Acad. Publ. (1987).

James, R.
1. Reflexivity and the supremum of linear functionals, Ann. of Math. 66. I (1957), 159-169.
2. Characterisations of reflexivity. Studia Mat. XXIII (1964), 205-216.
3. Reflexivity and the sup of linear functionals. Israel J. of Math., 13, 3-4 (1972), 289-301.

John, K. - Zizler, V.

1. A renorming of dual spaces, Israel J. Math., 12 (1972), 331-336.
2. A note on renorming of dual spaces, Bull. Acad. Polon. Sci. 21, (1973), 49-50.

Kacurovski, R. I.

1. On monotone operators and convex functionals, Uspehi Mat. Nauk, 15, 4 (1960), 213-215.
2. Monotone operators in Banach spaces, Dokl. Akad. Nauk, S.S. S.R., 163 (1965), 559-562.
3. Nonlinear monotone operators in Banach spaces, Uspehi Mat. Nauk, 23. 2 (1968), 121-168.

Kadec, M.I.

1. Spaces isomorphic to a locally uniformly convex space, Izv. Vyss. Ucebn. Zaved. Mat. 13 (1959), p. 51-57.

Kaplanski, I.

1. Set Theory and metric spaces, Allyn and Bacon, Inc. Boston-London- Sydney.

Kato T.

1. Demicontinuity, hemicontinuity and monotonicity, I, Bull. Amer. Soc. 70 (1964), 548-550.
2. Demicontinuity, hemicontinuity and monotonicity, II, Bull. Amer. Soc. 73 (1967), 886-889.
3. Nonlinear semigroups and evolution equation, J. Math. Soc. Japan, 19. 4 (1967), 508-519.
4. Accretive operators and nonlinear evolution equation in Banach spaces, Proc. Symp. Nonlin. Funct. Anal. 18, Chicago. Amer. Math. Soc. (1970).
5. Differentiability of nonlinear semigroups, Global Anal. Proc. Sump, Pure Math. Amer. Math. Soc., 1970.

Kato, T. - Hess, P. - Fitzpatrick, P.M.

1. Local boundedness of monotone type operators, Proc. Japan. Acad. Ser. A. 48, (1972), 275-277.

Kirk, W.A.

1. A fixed point theorem for mappings which do not increase distance, Amer. Math. Monthly, 72 (1965), 1004-1006.

Klee, V.L.

1. Convex bodies and periodic homeomorphisms in Hilbert space, Trans. A.M.S. 74 (1953), 10-43.
2. Some new results on smoothness and rotundity in normed linear spaces, Math. Ann., 139 (1959), 51-63.

3. Some characterizations of reflexivity, Rev. Ci. (Lima), 52 (1950), 15-23.

Kolomy, Josef
1. A note on duality mappings and its applications. Bull Math. Soc. Sci. Math. Rép. Soc. Roum., Série. 30, 78, (1986), 231-237.

Komura, J.
1. Nonlinear semigroups in Hilbert space, J. Math. Soc. Japan 19, 4 (1967) 493-507.
2. Differentiability of nonlinear semigroups, J. Math. Soc. Japan, 21 (1969), 375-402.

König, H.
1. On basic concepts in convex analysis, In Korte (Ed), (1982), 107-144.

Konishi, Y.
1. Nonlinear semi-groups in Banach lattices, Proc. Japan Acad. 47, (1971), 24-28.
2. Some examples of nonlinear semi-groups in Banach lattices J. Fac. Sci. Univ. Jokyo, 18 (1972), 537-543.

Köthe, G.
1. Topological vector spaces, Springer Verlag (1969).

Krasnoselskü, M.A. - Zabreiko, P.P.
1. Seometrical methods of nonlinear analysis, Springer Verlag (1984).

Krein, M.G.
1. Sur quelques questions de la géométrie des ensembles convexes situés dans un espace linéaire normé et complet, Dokladi Acad. Nauk. S.S.S.R. 14 (1937), p. 5-7.

Ladas, G.E. - Laksmikantham, V.
1. Differential Equations in abstract spaces, Acad. Press, New-York-London. (1976).

Leray, J. - Schauder J.
1. Toplogie et équations fonctionelles, Ann. Sci. Ecole Norm Sup. 51, (1934) 45-78.

Lindenstrauss, J.
1. On the moduls of smoothness and divergent series in Banach space, Mich. Math J. 10 (1963), 241-252.

2. On nonseparable reflexive Banach spaces, Bull Amer. Math. Soc. 72 (1966).
3. Weakly compact sets, Ann. Math. Studies, 69, (1972), 235-273.

Lindenstrauss, J. - Tzafriri, L.
1. Classical Banach Spaces I, II, Springer Verlag (1977), (1978).

Lions, J.L.
1. Quelques méthodes de résolution des problemes aux limites nonlineaires, Dunod, et. Gauthier Villars Ed. Paris, (1969).

Lions, J.L. - Stampachia G.
1. Variational inequalities, Comm. Pures Math., 20 (1967), 493-519.

Lovaglia, A.R.
1. Locally uniformly convex Banach spaces, Trans. Amer. Math. Soc., 78 (1955), 225-238.

Lumer, G. - Phillips, R.
1. Dissipative operators in a Banach space, Pacific J. Math. 11 (1961), 679-698.

Martin, R.N.
1. A global existence theorem for autonomous differential equations in a Banach space, Proc. Amer. Math. Soc., 26, (1970), 307-314.
2. Nonlinear operators and differential equations in Banach spaces, Willey & Sons, (1976).

Mermin, J.
1. Accretive operators and nonlinear semigroups, Thesis, Univ. of California, Berkeley (1968).
2. On exponential limit formula for nonlinear semi-groups, Trans. Amer. Math. Soc. 150 (1970), 469-476.

Milman, D.P.
1. On some criteria for the regularity of spaces of the type (B), Dokl. Acad. Nauk. S.S.S.R. N.S., 20 (1938), 243-246.

Milman, V.D.
1. The geometric theory of Banach spaces. Part I, Usp, Math. Nauk, 25 (1970). English Trans. Math Surweys 25 (1970), 111-170. Part II, Usp, Math Nauk, 26 (1971), 73-149. English Trans. Math. Surweys 26 (1971), 79-163.

Minty, G.
1. On the maximal domain of a monotone function, Mich. Math. J., 8 (1961), 135-137.
2. Monotone nonlinear operators in Hilbert spaces, Duke Math. J., 29 (1962), 341-346.
3. On a monotonicity method for the solution of nonlinear equations in Banach spaces, Proc. Nat. Acad. Sci. U.S.A. 50 (1963), 1038-1041.
4. On the monotonicity of the gradient of a convex function, Pacific J. Math. 14 (1964), 243-247.
5. A theorem on maximal monotone sets in Hilbert space, J. Math. Appl. 11 (1965), 434-439.
6. Monotone operators and certain systems of nonlinear ordinary differential equations, Proc. Symp. on Systems, Theor. Polytechnic Instit. of Brooklyn, (1965), 39-55.

Miyadera, J.
1. Note on nonlinear contraction semigroups, Proc. Amer. Math. Soc., 21 (1969), 219-225.
2. On the convergence of nonlinear semigroups, Tohoku, J., 21 (1969), 221-236.
3. On the convergence of nonlinear semigroups II, J. Math. Soc. Japan., 21, 3 (1969), 203-412.
4. Some remarks on semigroups of nonlinear transformations, Tohoku Math. J., 23 (1971), 254-258.

Miyadera, J., Oharu, S.
1. Approximation of semigroups of nonlinear operators, Tohoku Math. J., 2.22 (1970), 24-47.

Moreau, J.J.
1. Fontionnelles sous-différentiables, C.R. Acad. Sci. Paris, 257 (1963), 4117-4119.
2. Proximité et dualité dans un space de Hilbert, Bull. Sco. Math. France, 93 (1965), 273-299.
3. Fontionelles convexes, Sém. Eq. Der. Part. College de France, (1967).

Morosanu, G.
1. Nonlinear Evolution equations and applications, D. Reidel Publ. Comp. (1988).

Mosco, U.
1. Convergence of convex sets and of solutions of variational inequalities, Arch. Math. 3,4 (1969), 510-585.

2. On the continuity of the Young - Fenchel transform, J. of Math. Anal. and Appl., 35, 3(1971), 518-535.

Nagumo, M.
1. A theory of degree of mappings based on infinitesimal analysis, Amer. J. Math. 73 (1951), 485-496.

Neuberger, J.W.
1. An exponential formula for one parmeter semigroups of nonlinear transformations, J. Math. Soc. Japan 18 (1966), 154-157.

Nirenberg, L.
1. Topics in Nonlinear Functional Analysis, Courant Trust, New York, (1974).

Nördlander G.
1. The modulus of convexity in normed linear spaces, Arkiv Mat. 4, (1960) 15-17.

Oharu, S.
1. Note on the representation of nonlinear operators, Proc. Japan. Acad., 42 (1966), 1149-1154.
2. On the generation of semigroups of nonlinear contractions, J. Math. Soc. Japan., 22, 4 (1970), 526-550.

Pascali, D. - Sburlan, S.
1. Nonlinear mappings of monotone type, Ed. Acad., Buc. Rom.-Sijthoff & Noordhoff Inter. Publ. (1978).

Pavel, N. H.
1. Nonlinear Evolution Operators and Semigroups, Lectures Note 1280, Springer Verlag (1987).

Pazy, A.,
1. Semigroups of nonlinear contractions in Hilbert spaces, Problems of Nonlinear Analysis, Cremonese, Roma, (1971), 343-430.
2. Semigroups of linear operators and applications to partial differential equations, Springer Verlag, (1983).

Petryshyn, W.V.
1. On a fixed point theorem for nonlinear P-compact operators in Banach spaces, Bull. Amer. Math. Soc., 72 (1966), 329-334.
2. On nonlinear P-compact operators in Banach spaces with application to constructive fixed-point theorem, J. Math. Anal. Appl, 15 (1966), 228-242.

3. On the approximative solvability of nonlinear equations, Math. Ann., 177 (1968), 156-164.

4. On the proiectional solvability and the Fredholm alternative for equations involving linear A-proper operators, Arch. Rat. Mech. Anal. 30, (1968), p. 270-284.
5. Nonlinear equations involving noncompact operators, Proc. Symp. Nonlinear Funct. Anal. Chicago 18, part I, (1968), 206-233.
6. Fixed point-theorems involving P-compact, semi-accretive and accretive operators not defined on all of a Banach space, J. Math. Anal. Appl. 23 (1968), 336-354.
7. On existence theorems for non-linear equations involving non-compact mappings, Proc. Nat. Acad. Sci., 67 (1970), 326-330.
8. Invariance domain theorem for locally A-proper mappings and its implications, J. Funct. Anal. 5 (1970), 137-159.
9. On the relationship of A-properness to mappings of monotone type with applications to elliptic equations, "Fixed point Theory and its Appl", Acad. Press, (1976), 149-174.
10. A characterization of strict convexity of Banach spaces and other uses of duality mappings, J. Funct. Anal., 6,2 (1970), 282-292.
11. Using degree theory densely defined A-proper maps in the solvability of semilinear equations with unbounded and noninvertible linear part, Nonlinear Analysis, Theory, Methods and Appl., Vol. 4, No. 2. (1980), 259-281, Pergamon Press Ltd.

Petryshyn, W.V. - Fitzpatrick, P.M.

1. A degree theory, fixed point theorems and mapping theorems for multivalued noncompact mappings, Trans. A.M.S. 194 (1974), 1-25.

Petryshyn, W.V. - Tuker T.S.

1. On the functional equations involving non-linear generalized P-compact operators, Trans. Amer. Math. Soc., 135(1969), 343-373.

Pettis, B.J.

1. A proof that every uniformly convex space is reflexive, Duke. Math J. 5 (1939) 249-253.

Phelps, R.

1. Convex functions, monotone operators and differntiability, Lecture Notes in Math., 1364, (1989).

Picard, C.

1. Opérateurs T-acretifs d' un espace de Banach reticulé, Séminaire sur les semigroupes et les opérateurs nonlinéaires, Orsay, (1970-1971).

Prüss, J.

1. A characterization of uniform convexity and applications to accretive operators, Hiroshima Math. I, 11 (1981), 229-234.

Rainwater, J.

1. Local uniform convexity of Day's norm $C_0(\Gamma)$, Proc. A. M. S., 22 (1969), 335-339.

Reich, S.

1. Product formulas, nonlinear semigroups and accretive operators, J. Funct. Anal. 36 (1980), 147-169.
2. New results concearning accretive operators and nonlinear semigroups, J. Math. Phys. Sci. 18, (1984), 91-97.

Restrepo, G.

1. Differentiable norms in Banach spaces, Bull. Amer. Math. Soc. 70 (1964).
2. Differentiable norms, Vol. Soc. Mat. Mexicana 2, 10 (1965), 47-55.

Rockafellar, R.T.

1. Extensions of Fenchel's duality theorem for convex functions, Duke Math. J., 33 (1966), 81-90.
2. Characterization of the subdifferentials of convex functions, Pacific J. Math., 33 (1966), 497-510.
3. Convex properties of non-linear maximal monotone opertators, Bull. Amer. Math Soc. 75, 1 (1969), 74-77.
4. Convex Analysis, Princeton Univ. Press, (1969).
5. Convex functions, monotone operators and variational inequalities. Proc. Venice Inst. Monotone Op., (1969).
6. Local boundedness of nonlinear monotone operators, Michigan Math. J., 16 (1969), 397-407.
7. On the maximal monotonicity of subdifferential mappings, Pacif. J. Math., 33, 1 (1970), 209-217.
8. On the virtual convexity of the domain and range of a nonlinear maximal monotone operators, Math., Anal. 185 (1970), 81-90.
9. The theory of subgradients and its applications to problems of Optimization, Heldermann, Berlin, (1981).

Sato, K.

1. On the generators of non-negative contraction semigroups in Banach lattices, J. Math. Soc. Japan 20 (1968), 423-436.

2. On dispersive operators in Banach lattices, Pacific J. of Math, 33 (1970), 429-433.

Sburlan, S.
1. Gradul topologic, Ed. Acad. Rom., Buc. (1983).

Schaefer, H. H.
1. Banach lattices and positive operators, Springer Verlag (1974).

Schoenberg. J.
1. On a theorem of Kirzbaum, Amer. Math Monthly 60, (1953), 620.

Schwartz, J.
1. Non-linear functional analysis. Gordon and Breach, (1969).

Segal, I.
1. Non-linear semigroups, Ann. of Math. 78 (1963), 330-364.

Singer, I.
1. Bases in Banach spaces, I, Springer Verlag (1970).

Smulian, V.L.
1. On some geometrical properties of the unit spere in the space of the type (B), Mat. Sbornik, 6., (1939), 77-94.
2. Sur la derivalilité de la norme dans l'espace de Banach, Dakl. Akad-Nauk, S.S.S.R., 27, (1940), 643-648.

Stampacchia, G.
1. Variational inequalities, Proc. of a N.A.T.A. Adv. Study Inst. Venice, (1968).

Sundaresan, K.
1. Smooth Banach spaces, Bull. Amer. Math. Soc., 72 (1966), 520-521.
2. Smooth Banach spaces, Math. Ann, 173, (1968), 191-199.

Trojanski, S.L.
1. Equivalent norms in unseparable B-spaces with an unconditional basis, Teor. Funkt. Functional Anal. i Priloj. Kharkov, 6 (1968), 59-65.
2. On locally uniformly convex an differentiable norms in certains non-separable Banach spaces, Studia Math. XXXVII (1971), 173-180.
3. On equivalent norms and minimal systems in nonseparable Banach spaces, Studia. Math. 43 (1972), 125-138.

Tucker, T.S.

1. Leray-Schader Theorem for P-compact operators and its consequences, J. of Math. Anal. and Appl., 23 (1968), 355-364.

Vainberg, M.M.

1. Variational methods and methods of monotone operators in the theory of nonlinear equations, Wiley (1973).

Vesely, L.

1. Some new results on accretive multivalued operators, Comment. Math. Univ. Carolinac 30, (1989), 45-55.

Watanabe, J.

1. Semigroups of non-linear operators on closed convex sets, Proc. Japan. Acad. 45, 4 (1969), 219-223.
2. On non-linear semigroups generated by cyclically dissipative sets, J. Fac. Sci. Tokyo 18 (1971), 127-137.

Webb, J.R.L.

1. On a property of duality mappings and the A-properness of accretive operators, Bull. Lonolon Math. Soc. 13, (1981), 235-238.

Webb, G.F.

1. Non-linear contraction semi-groups in weakly complete Banach spaces, Thesis Emary Univ., (1968).
2. Representation of semigroups of non-linear nonexpansive transformations in Banach spaces, J. Math. Mech., 19 (1970),159-170.
3. Non-linear evolutions and product stable operators on Banach space, Trans. Amer. Math. Soc. 148 (1970), 273-282.
4. Continuous non-linear perturbations of linear accretive operators in Banach spaces J. Funct. Anal. 10, (1972), 191-203.
5. Accretive operators and existence for nonlinear functional equations, J. Diff. Eq., 14 (1973), 57-69.

Wong, Ship Fah.

1. The topological degree for A-proper maps, Canad. J. Math. 23 (1971), 403-412.
2. A product formula for the degree of A-proper maps, J. Funct. Anal. 10 (1972), 361-371.

Zarantonello, E.H.

1. Projections on convex sets in Hilbert spaces and spectral theory, "Contributions to non-linear Analysis", Acad. Press. (1971), 237-424.

Zeidler, E.
1. Nonlinear Functional Analysis and its Applications, Part. III, Springer Verlag, (1986); Part II/A, II/B (1989).

Zizler, V.
1. On some rotundity and smoothness properties of Banach spaces, Dissertationes. Math., 87 (1971), 5-33.

INDEX

9 780792 309109